Advances in Unconventional Oil and Gas

Advances in Unconventional Oil and Gas

Editors

Shu Tao
Dengfeng Zhang
Huazhou Huang
Shuoliang Wang
Yanjun Meng

MDPI • Basel • Beijing • Wuhan • Barcelona • Belgrade • Manchester • Tokyo • Cluj • Tianjin

Editors
Shu Tao
China University of
Geosciences
China

Dengfeng Zhang
Kunming University of
Science and Technology
China

Huazhou Huang
China University of Mining
and Technology
China

Shuoliang Wang
China University of
Geosciences (Beijing)
China

Yanjun Meng
Taiyuan University of
Technology
China

Editorial Office
MDPI
St. Alban-Anlage 66
4052 Basel, Switzerland

This is a reprint of articles from the Special Issue published online in the open access journal *Energies* (ISSN 1996-1073) (available at: https://www.mdpi.com/journal/energies/special_issues/Unconventional_Oil_Gas).

For citation purposes, cite each article independently as indicated on the article page online and as indicated below:

LastName, A.A.; LastName, B.B.; LastName, C.C. Article Title. *Journal Name* **Year**, *Volume Number*, Page Range.

ISBN 978-3-0365-5283-5 (Hbk)
ISBN 978-3-0365-5284-2 (PDF)

Contents

energies

MDPI

Article

Genesis of Coalbed Methane and Its Storage and Seepage Space in Baode Block, Eastern Ordos Basin

Hao Chen [1], Wenguang Tian [1,*], Zhenhong Chen [1], Qingfeng Zhang [2] and Shu Tao [3,*]

1 PetroChina Research Institute of Petroleum Exploration and Development, Beijing 100083, China; chenhao69@petrochina.com.cn (H.C.); Chenzhenhong@petrochina.com.cn (Z.C.)
2 PetroChina Coalbed Methane Company Limited, Beijing 100028, China; zqf2012@petrochina.com.cn
3 School of Energy Resources, China University of Geosciences, Beijing 100083, China
* Correspondence: tianwg69@petrochina.com.cn (W.T.); peach888@163.com (S.T.)

Abstract: The Baode block on the eastern margin of the Ordos Basin is a key area for the development of low-rank coalbed methane (CBM) in China. In order to find out the genesis of CBM and its storage and seepage space in Baode block, the isotopic testing of gas samples was carried out to reveal the origin of CH_4 and CO_2, as well, mercury intrusion porosimetry, low temperature nitrogen adsorption, and X-ray CT tests were performed to characterize the pores and fractures in No. 4 + 5 and No. 8 + 9 coal seams. The results showed that the average volume fraction of CH_4, N_2, and CO_2 is 88.31%, 4.73%, and 6.36%, respectively. No. 4 + 5 and No. 8 + 9 coal seams both have biogenic gas and thermogenic methane. Meanwhile, No. 4 + 5 and No. 8 + 9 coal seams both contain CO_2 generated by coal pyrolysis, which belongs to organic genetic gas, while shallow CO_2 is greatly affected by the action of microorganisms and belongs to biogenic gas. The average proportion of micropores, transition pores, mesopores, and macropores is 56.61%, 28.22%, 5.10%, and 10.07%, respectively. Samples collected from No. 4 + 5 coal seams have developed more sorption pores. Meanwhile, samples collected from No. 8 + 9 coal seams exhibited a relatively low degree of hysteresis (Hg retention), suggesting good pore connectivity and relatively high seepage ability, which is conducive to gas migration. The connected porosity of coal samples varies greatly, mainly depending on the relative mineral content and the proportion of connected pores.

Keywords: genesis of coalbed methane; pore-fracture system; storage and seepage space; Baode block

Citation: Chen, H.; Tian, W.; Chen, Z.; Zhang, Q.; Tao, S. Genesis of Coalbed Methane and Its Storage and Seepage Space in Baode Block, Eastern Ordos Basin. *Energies* **2022**, *15*, 81. https://doi.org/10.3390/en15010081

Academic Editors: Nikolaos Koukouzas and Reza Rezaee

Received: 23 October 2021
Accepted: 21 December 2021
Published: 23 December 2021

1. Introduction

Following the Qinshui Basin, the Ordos Basin is another large gas field in China with proven reserves of more than 100 billion cubic meters. It is also the first demonstration area for the exploration and development of low- and medium-rank coalbed methane (CBM) in China [1,2]. The Carboniferous Permian coal seams in the Ordos Basin have undergone different degrees of subsidence, uplift, and denudation, resulting in different degrees of thermal evolution of coal seams in different regions [3–5]. The Baode block is located in the northern part of the eastern margin of the Ordos Basin. The $R_{o,max}$ of coal ranges from 0.52% to 0.89%, belonging to low- and middle-rank bituminous coal. Some CBM wells in the Baode block have obtained industrial gas flow, showing a good prospect for CBM development [6,7].

Previous studies have made preliminary discussions on the formation conditions and genetic types of CBM in the eastern margin of the Ordos Basin, and pointed out that the shallow part is a mixture of secondary biogas and thermogenic gas, while the deep part is mainly composed of thermogenic gas [8,9]. However, there is a lack of a systematic understanding of the origin of CBM in Baode block. In addition to methane, low-rank coal reservoirs often contain a certain proportion of CO_2 and N_2, which can be used as important objects for studying the genesis of CBM [10].

In the current study, based on proximate analysis, mean vitrinite reflectance (R_o) measurements and maceral analyses, the material composition (including macerals, ash, moisture, volatile) of nine coal samples collected from Baode block were characterized. At the same time, isotopic testing of 54 gas samples collected from desorption tanks in different desorption periods was carried out to reveal the gas composition characteristics and gas genesis. Finally, based on high-pressure mercury injection, low-temperature liquid nitrogen, and X-ray CT experiments, the gas storage and seepage space were finely characterized, which provides a theoretical basis for further clarifying the direction of CBM exploration and development in the Baode block.

2. Geologic Setting

The Ordos Basin, a huge intra-cratonic basin, is located in North China and contains the second largest accumulation of coal resources in China [11,12] (Figure 1a). The basin is divided into seven structural units [13,14] (Figure 1b). The eastern margin of the Ordos Basin is a N–S striking and west-inclined monocline within the following three tectonic units: Weibei uplift in the north, Yimeng uplift in the south, and Jinxi fold in the middle [15–17] (Figure 1b). The Baode block is located in the northern part of the eastern margin of the Ordos Basin (Figure 1c). The formation is gentle, sloping westward at an angle of 1° to 5°. The faults are rare and small in scale, and the faults are mainly in the northeast direction with a fault distance of 10–25 m.

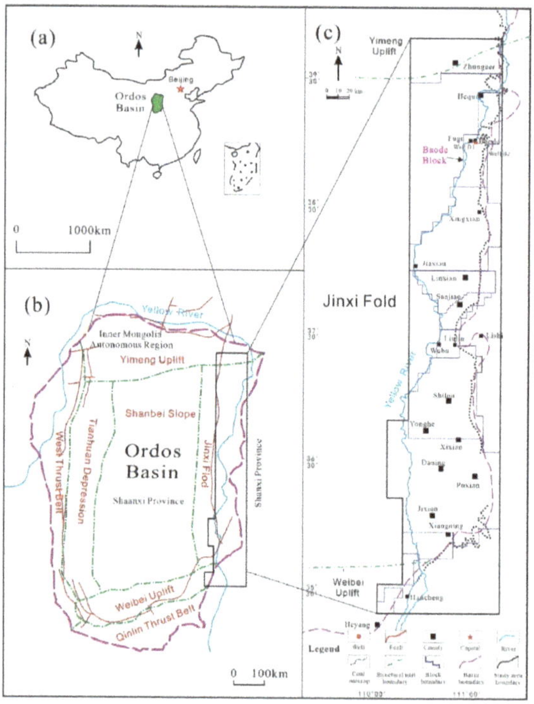

Figure 1. (**a**) Location of the Ordos Basin. (**b**) Tectonic units of the Ordos Basin. (**c**) Division of eastern margin of Ordos Basin and the location of Baode block, with the location of Wells B1 and B2.

In the Baode block, the main coal-bearing sequences occur in the Upper Carboniferous Taiyuan Formation (C_3t) and the Lower Permian Shanxi Formation (P_1s) (Figure 2). The C_3t is in integrated contact with the underlying strata, with a thickness of 50–90 m. It is mainly composed of black-gray mudstone, gray-white medium sandstone, gray coarse sandstone, and coal seam. A layer of bioclastic limestone is locally developed in the lower part, and the bottom is gray-white thick layered medium coarse sandstone and gravelly coarse sandstone. It is a set of interactive marine coal bearing deposits (Figure 2). The P_1s is 60–90 m thick. It is mainly composed of gray fine sandstone, black-gray black, sandy mudstone, and coal seam. The bottom of P_1s is gray-white coarse-grained quartz sandstone. The formation is a set of meandering river and delta plain swamp facies deposits (Figure 2). No. 4 + 5 coal seam in P_1s and No. 8 + 9 coal seam in C_3t are continuous and thick. They are the main coal seams in the Baode block and the target seam for CBM development, occurring at a depth of 400–1200 m, with a thickness ranging from 1.16 m to 20.21 m. The lithology of the roof and floor of No. 4 + 5 and No. 8 + 9 coal seams is dominated by mudstone, locally sandy mudstone. The thickness of the roof of the No. 4 + 5 coal seam is 4.3–13.9 m, and the thickness of the floor is 2.4–6.7 m, while those of the 8 + 9 coal seam are 2.8–6.8 m and 2.0–18.6 m, respectively.

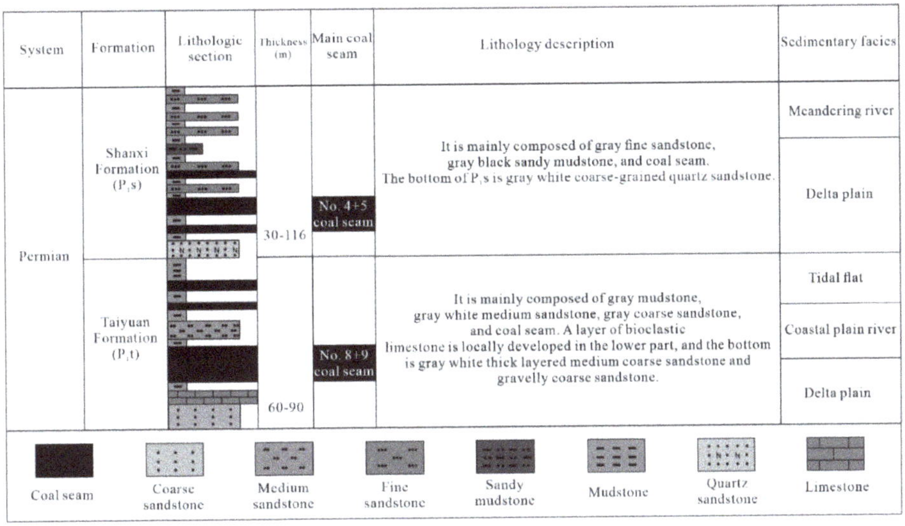

Figure 2. Lower Permian stratigraphic column in the Baode block.

3. Samples and Analytical Procedures

3.1. Coal and Gas Samples

Eleven coal samples were collected from Wells B1 and B2 (Figure 1c) in the Baode block, and the burial depths of the main coal seams in these two wells are 501.40–547.20 m and 1011.40–1068.30 m, respectively. Among them, four samples were collected form No. 4 + 5 coal seam, and seven of them were collected form No. 8 + 9 coal seam. Part of the sample was carefully packed and then immediately sent to the laboratory for experiments, and another part of the sample was immediately put into different desorption tanks to collect the desorption gas. A total of 54 gas samples were collected from eleven desorption tanks in different desorption periods for gas composition and isotopes analysis.

3.2. Material Composition

Nine of the collected coal samples were analyzed for proximate analysis on an air-dried basis following the Chinese national standard GB/T 212-2008 [18]. According to ISO 7404.3-1994 [19] and ISO 7404.5-1994 [20], mean vitrinite reflectance (R_o) measurements and maceral analyses (500 points) were performed on the same polished section of the coal samples using a Leitz MPV-3 photometer microscope. Nonlinear error: max. ±1 low significance bit. A/D conversion accuracy: ±1 low significance bit, less than 2‰.

3.3. Pore-Size Distribution

Based on the material composition analysis results, six samples were selected to analyze the characteristic pore-size distributions by using mercury intrusion porosimetry (MIP), low-temperature nitrogen adsorption (LTNA) and X-ray CT. Diameters of pores detected by LTNA range from 2 to 300 nm, and those accessed by MIP are in the diameter of 30– > 1000 nm. Thus, the adsorption pores (<10^2 nm) [21] can be determined by LTNA, while the seepage pores (>10^2 nm) [21] can be well characterized by MIP. MIP was carried out according to the national standard, SY/T 5346-2005 [22], by using Micromeritics Auto Pore IV 9500 instrument. Before the LTNA experiments, all samples were crushed and sieved to a size of 0.18–0.25 mm (60–80 mesh) (dried for 48 h), and then tested using a Micromeritics ASAP2020 instrument at 77K. The X-ray CT was carried out by using U.S. ACTIS-250/320PK/225FFI industrial CT system. The spatial resolution is close to 50 μm in the process of X-ray CT measurement, and therefore large pores and microfractures can be identified in the coal core plug. The specific experimental procedures and image processing methods were presented in detail by Tao et al. (2019) [23].

Thus, the characteristics of seepage pores (>10^2 nm) were measured by MIP, and then the characteristics of adsorption pores (<10^2 nm) were measured by LTNA. Finally, the three-dimensional models of samples' adsorption pores and percolation pores were constructed by X-ray CT experiment.

3.4. Gas Composition and Isotopes

A total of 54 gas samples were collected from eleven desorption tanks to analyze the gas composition by using an Agilent 7890B gas chromatograph, according to the Chinese national standard GB/T 13610-2014 [24]. Afterward, $\delta^{13}C$ values of CH_4 and CO_2, and δD values of CH_4 of 54 gas samples were determined on a Finnigan MAT 253 mass spectrometer. The $\delta^{13}C$ and δD values were calibrated with respect to the VPDB and VSMOW standards, and the standard deviations were ±(0.1‰~0.3‰) and ±(1‰~2‰), respectively.

4. Results and Discussion

4.1. Basic Information of Coals

As shown in Table 1, the Baode coal samples have ash yields ranging from 4.96% to 7.31%, whereas they have relatively high moisture contents (18.6–26.56%) and volatile yields (25.95–30.32%). Higher volatile content means lower coal metamorphism, with R_o of 0.62–0.76%. The vitrinite content is the highest (41.5–84.8%, mean 69.3%), followed by inertinite (8.7–54.4%, mean 22.8%) and liptinite (0.5–17.0%, mean 7.9%). Macerals in the Baode coal samples include: cutinite, microsporinite, desmocollinite, fusinite, semi-fusinite, resinite, and liptodetrinite. The cutinite is usually distributed in strips, and the microsporinite is distributed in parallel planes in a worm-like shape (Figure 3).

Table 1. Results of the proximate and maceral analyses.

Sample No.	Coal Seam No.	Depth (m)	Ro (%)	Maceral Composition (%)			Proximate Analysis (%)		
				Vitrinite	Inertinite	Liptinite	M_{ad}	A_{ad}	V_{ad}
BD-1	4 + 5	501.40–501.70	0.64	74.0	9.0	17.0	22.24	5.33	27.73
BD-2	4 + 5	502.90–503.30	0.67	45.1	54.4	0.5	26.56	4.96	26.46
BD-3	4 + 5	1011.40–1011.70	0.62	83.5	8.7	7.9	24.44	6.55	25.95
BD-4	4 + 5	1011.40–1011.70	0.62	84.8	10.0	5.2	18.60	7.31	30.32
BD-5	8 + 9	546.54–546.84	0.63	76.4	12.7	11.0	26.52	6.44	28.08
BD-6	8 + 9	546.84–547.20	0.68	82.3	14.8	2.8	22.95	7.25	30.12
BD-7	8 + 9	1061.10–1061.40	0.70	81.0	12.9	6.1	22.33	5.30	30.30
BD-8	8 + 9	1065.60–1066.00	0.76	54.7	38.8	6.5	25.73	6.51	30.02
BD-9	8 + 9	1068.00–1068.30	0.75	41.5	44.2	14.4	25.52	6.52	27.25

Notes: M_{ad} = moisture content; A_{ad} = ash yield; V_{ad} = volatile yield; ad = air dried basis.

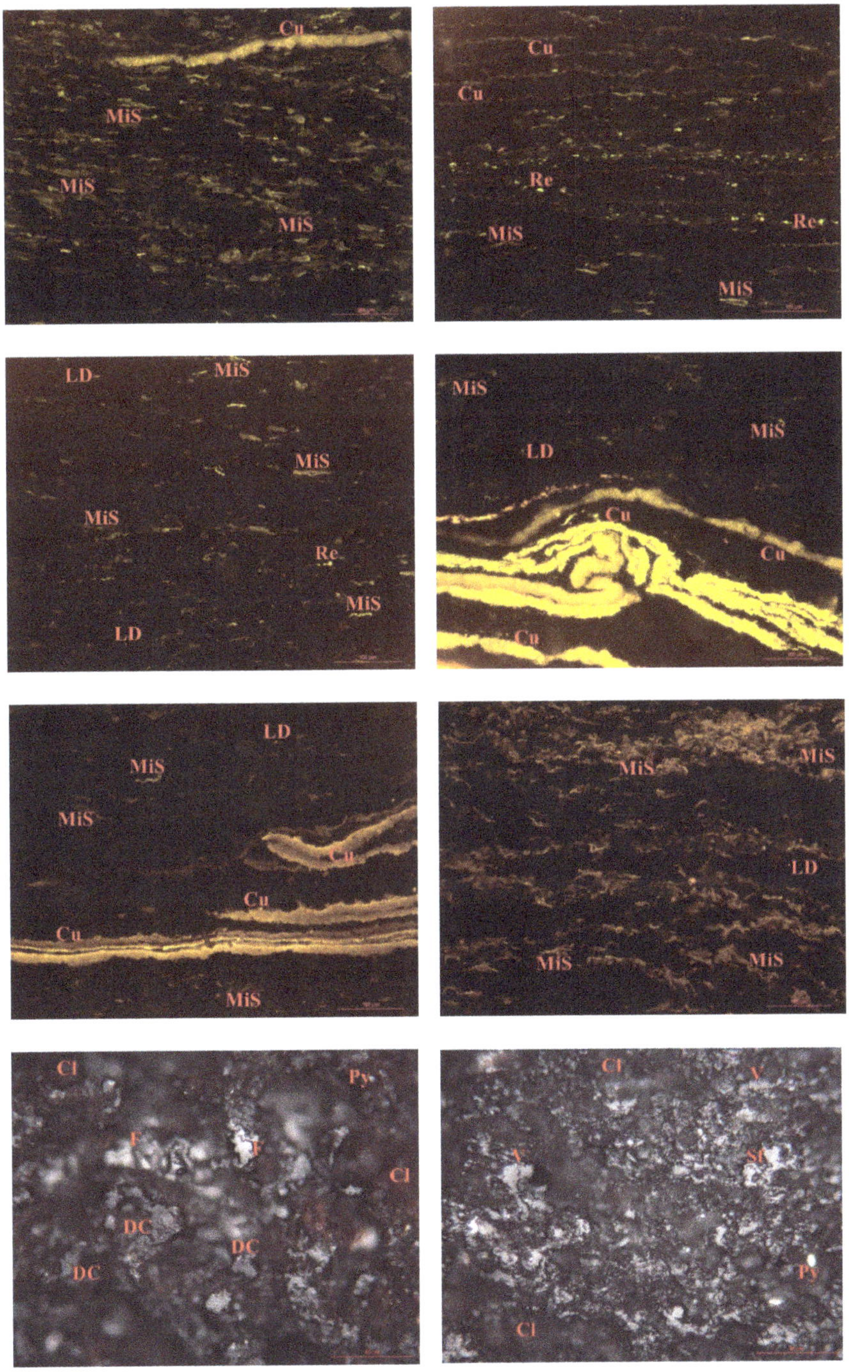

Figure 3. Characteristic macerals from Baode coals, polished section, reflected fluorescence and reflected plane-polarized light (last two pictures). Cu: cutinite; MiS: microsporinite; DC: desmocollinite; F: fusinite; Sf: Semifusinite; Py: Pyrite; V: Vitrinite; Cl: clay; Re: resinite; LD: liptodetrinite.

4.2. CBM Composition and Origin

4.2.1. Compositional Characteristics of CBM

As shown in Table 2, the gas in No. 4 + 5 and No. 8 + 9 coal seams are dominated by CH_4, with a volume fraction ranging from 80.9% to 94.7% (mean 88.32%). The volume fraction of N_2 varies between 0.6–15.9% (mean 4.95%), and the N_2 content in No. 4 + 5 coal seam is higher than that in No. 8 + 9 coal seam. Meanwhile, as the depth increases, the N_2 content decreases relatively, which shows that the gas composition of shallow CBM is affected by mixing with atmospheric air [25,26]. Figure 4 indicates that the N_2 volume fraction and CH_4 volume fraction in the two sets of coal seams are negatively correlated. The main difference between 8 + 9 coal seam (a) and 8 + 9 coal seam (b) is the different proportions of N_2 and CH_4. Different proportions can characterize the source of coal seam gas. The proportion of gas in No. 8 + 9 coal seam (a) is the same as that in No. 4 + 5 coal seam. The CO_2 content increases with increasing burial depth of No. 8 + 9 coal seam. This phenomenon is related to the action of CO_2-reducing bacteria in the shallow part which decreases with depth [27].

Table 2. CBM composition of the collected gas samples from eleven desorption tanks in different desorption periods.

Coal Seam No.	Depth/m	Sample No.		Volume Fraction of Each Component in CBM (%)			Coal Seam No.	Depth/m	Sample No.		Volume Fraction of Each Component in CBM (%)		
				CO_2	N_2	CH_4					CO_2	N_2	CH_4
4 + 5	501.40–501.70	B1-1	B1-1-1	3.0	10.2	86.7	8 + 9	546.84–547.20	B1-4	B1-4-3	3.3	10.7	86.0
			B1-1-2	3.6	8.8	87.4				B1-4-4	3.5	9.9	86.6
			B1-1-3	3.3	7.6	89.1				B1-4-5	3.6	5.6	90.8
			B1-1-4	3.4	6.0	90.6				B1-4-6	1.9	4.7	93.5
			B1-1-5	1.6	4.3	94.1		1061.10–1061.40	B2-3	B2-3-1	9.0	2.8	88.1
	502.90–503.30	B1-2	B1-2-1	3.1	15.9	80.9				B2-3-2	9.4	2.6	87.9
			B1-2-2	3.8	15.2	81.1				B2-3-3	9.0	2.5	88.3
			B1-2-3	3.2	12.7	84.1				B2-3-4	8.6	2.7	88.5
			B1-2-4	3.8	9.4	86.8				B2-3-5	8.9	2.5	88.4
			B1-2-5	2.4	5.7	91.9		1064.00–1064.35	B2-4	B2-4-1	10.6	1.9	87.3
	1011.40–1011.70	B2-1	B2-1-1	3.5	4.5	92.0				B2-4-2	10.4	1.6	87.8
			B2-1-2	4.1	3.7	92.1				B2-4-3	10.7	1.7	87.4
			B2-1-3	4.8	3.0	92.2				B2-4-4	12.0	2.1	85.7
			B2-1-4	3.8	3.0	93.2				B2-4-5	10.2	1.7	88.0
			B2-1-5	2.2	3.1	94.7		1065.60–1066.00	B2-5	B2-5-1	9.3	1.7	88.7
	1014.80–1015.20	B2-2	B2-2-1	3.0	8.2	88.6				B2-5-2	9.2	1.6	88.8
			B2-2-2	3.2	7.7	88.8				B2-5-3	10.7	1.5	87.4
			B2-2-3	2.4	6.5	90.8				B2-5-4	12.0	1.7	85.8
			B2-2-4	3.0	5.9	90.7		1066.80–1067.10	B2-6	B2-6-1	9.7	5.5	82.6
			B2-2-5	3.6	3.3	92.6				B2-6-2	13.1	2.6	82.2
8 + 9	546.54–546.84	B1-3	B1-3-1	4.3	8.5	87.1				B2-6-3	11.6	1.2	84.9
			B1-3-2	4.0	4.8	91.2				B2-6-4	11.7	1.1	85.0
			B1-3-3	4.5	2.6	92.9				B2-6-5	9.7	0.7	87.3
			B1-3-4	4.4	2.2	93.4		1068.00–1068.30	B2-7	B2-7-1	7.8	1.2	88.0
			B1-3-5	4.7	2.2	93.2				B2-7-2	9.5	0.6	87.2
	546.84–547.20	B1-4	B1-4-1	3.1	11.5	85.4				B2-7-3	9.6	2.7	84.1
			B1-4-2	2.7	14.0	83.3				B2-7-4	6.7	1.6	87.9

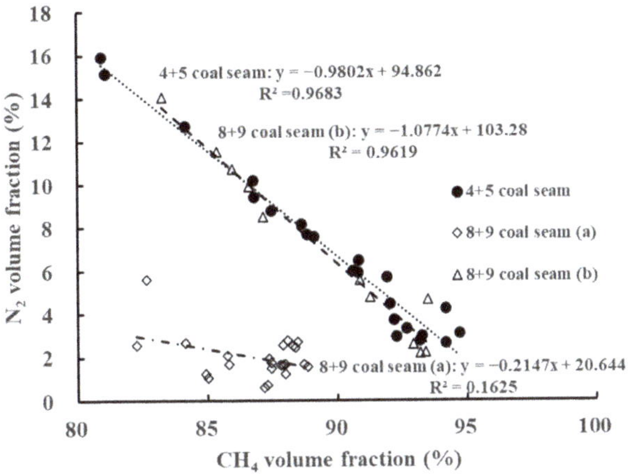

Figure 4. Relationship between N_2 volume fraction and CH_4 volume fraction.

4.2.2. Genetic Types of CH_4 and CO_2

CBM is divided into biogenic gas and thermogenic gas, and the thermogenic gas is further subdivided into early thermogenic wet gas and late thermogenic dry gas [28]. To some extent, different gas concentrations in CBM reflect the genesis of CBM which can be identified by gas composition.

As shown in Table 3, the value of $\delta^{13}C$ (CH_4) in No. 4 + 5 coal seam varies between −55.6‰ and −47.7‰, with an average of −52.5‰, and the value of that in No. 8 + 9 coal seam varies between −62.3‰ and −50.4‰, with an average of −54.5‰. The $\delta^{13}C$ (CH_4) values of the two coal seams are within the range of national $\delta^{13}C$ (CH_4) observation values of CBM (from −73.7‰ to −24.9‰) [29], belonging to light carbon isotopes. The value of $\delta^{13}D$ (CH_4) in No. 4 + 5 coal seam ranges from −256.2‰ to −241.6‰, with an average of −249.4‰, and the value of that in No. 8 + 9 coal seam ranges from −261.8‰ to −247.6‰, with an average of −252.8‰. At shallower than 550 m, the average value of $\delta^{13}C$ (CH_4) is less than −55‰, indicating that biogenic gas is the main source. With the increase in burial depth (about 1000 m in Table 3), the average value of $\delta^{13}C$ (CH_4) is around −50‰, indicating that thermogenic gas is dominant. Figure 5 also shows that No. 4 + 5 and No. 8 + 9 coal seams both have biogenic gas and thermogenic gas [30].

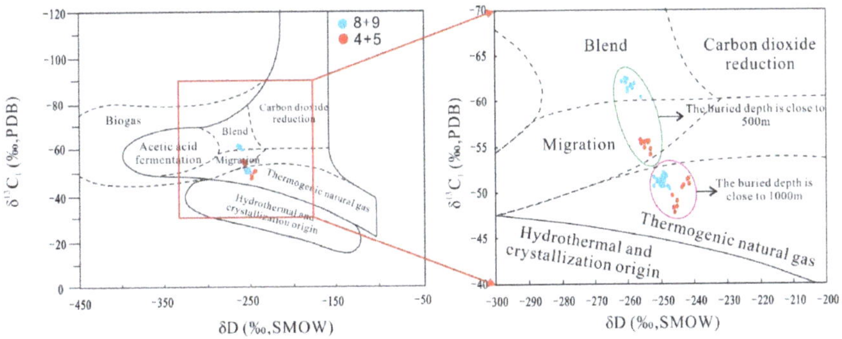

Figure 5. Genetic type discrimination map of CBM. (C_1 stands for CH_4) (modified from [31]).

Table 3. Carbon and hydrogen stable isotope compositions and CDMI values of gas samples from eleven desorption tanks in different desorption periods.

Coal Seam No.	Depth/m	Sample No.		$\delta^{13}C$ (‰) CH_4	$\delta^{13}C$ (‰) CO_2	δD (‰)	CDMI
4 + 5	501.40–501.70	B1-1	B1-1-1	−55.5	−7.2	−253.6	3.4
			B1-1-2	−54.6	−7.3	−253.2	3.9
			B1-1-3	−55.4	−6.0	−253.9	3.6
			B1-1-4	−55.5	−5.6	−254.3	3.6
			B1-1-5	−54.0	−9.8	−253.2	1.7
	average			−55	−7.2	−253.7	/
	502.90–503.30	B1-2	B1-2-1	−55.6	−10.5	−256.2	3.7
			B1-2-2	−55.4	−9.2	−255.8	4.4
			B1-2-3	−55.3	−8.0	−255.1	3.6
			B1-2-4	−55.3	−6.9	−255.9	4.2
			B1-2-5	−54.8	−9.0	−254.9	2.6
	average			−55.3	−8.7	−255.6	/
	1011.40–1011.70	B2-1	B2-1-1	−51.4	−1.8	−241.6	3.7
			B2-1-2	−51.3	−1.9	−242.1	4.3
			B2-1-3	−50.9	−0.9	−241.6	5
			B2-1-4	−50.8	−2.9	−243.1	3.9
			B2-1-5	−50.5	−2.8	−243.8	2.3
	average			−51	−2.1	−242.5	/
	1014.80–1015.20	B2-2	B2-2-1	−49.6	−6.5	−246.6	3.2
			B2-2-2	−49.0	−1.0	−245.3	3.5
			B2-2-3	−48.4	−6.6	−245.3	2.6
			B2-2-4	−48.2	−3.8	−246.1	3.2
			B2-2-5	−47.7	−0.6	−245.9	3.8
	average			−48.6	−3.7	−245.9	/
8 + 9	546.54–546.84	B1-3	B1-3-1	−61.3	−12.0	−258.4	4.7
			B1-3-2	−62.3	−10.3	−259.9	4.2
			B1-3-3	−62.0	−8.8	−261.8	4.6
			B1-3-4	−61.8	−8.8	−258.0	4.5
			B1-3-5	−61.6	−9.0	−258.9	4.8
	average			−61.8	−9.8	−259.4	/
	546.84–547.20	B1-4	B1-4-1	−62.3	−12.4	−260.7	3.5
			B1-4-2	−61.9	−11.0	−260.5	3.1
			B1-4-3	−60.9	−9.6	−259.8	3.7
8 + 9	546.84–547.20	B1-4	B1-4-4	−61.5	−9.6	−259.0	3.9
			B1-4-5	−61.3	−9.2	−259.9	3.8
			B1-4-6	−60.3	−11.3	−256.0	2.0
	average			−61.4	−10.5	−259.3	/
	1061.10–1061.40	B2-3	B2-3-1	−51.5	3.1	−252.3	9.3
			B2-3-2	−51.4	3.4	−248.9	9.6
			B2-3-3	−51.2	3.7	−250.2	9.3
			B2-3-4	−50.8	4.2	−250.6	8.9
			B2-3-5	−50.5	3.7	−252.3	9.2
	average			−51.1	3.6	−250.9	/
	1064.00–1064.35	B2-4	B2-4-1	−51.7	5.1	−248.5	10.8
			B2-4-2	−51.6	5.5	−249.4	10.6
			B2-4-3	−51.7	5.2	−249.7	10.9
			B2-4-4	−51.3	6.7	−250.3	12.3
			B2-4-5	−50.5	4.8	−249.2	10.4
	average			−51.4	5.4	−249.4	/
	1065.60–1066.00	B2-5	B2-5-1	−52.0	1.8	−249.2	9.5
			B2-5-2	−51.4	7.6	−251.2	9.4
			B2-5-3	−51.0	5.5	−251.3	10.9
			B2-5-4	−50.5	4.9	−250.0	12.3
	average			−51.2	5	−250.4	/
	1066.80–1067.10	B2-6	B2-6-1	−51.8	2.8	−248.5	10.5
			B2-6-2	−51.5	7.1	−249.2	13.7
			B2-6-3	−51.4	5.6	−249.3	12.0
			B2-6-4	−51.2	5.8	−249.1	12.1
			B2-6-5	−51.1	5.4	−250.0	10
	average			−51.4	5.3	−249.2	/
	1068.00–1068.30	B2-7	B2-7-1	−51.1	5.2	−249.7	8.1
			B2-7-2	−50.4	3.8	−247.6	9.8
			B2-7-3	−50.9	5.1	−249.2	10.2
			B2-7-4	−50.6	5.8	−248.0	7.1
	average			−50.7	5	−248.6	/

CO_2 in CBM is mainly generated in the low maturity evolution stage of organic matter, which is generated through the chemical reaction of oxygen-containing groups such as decarboxylation and carbonyl in coal molecules [31]. Previous studies have shown that the value of $\delta^{13}C$ in organic CO_2 is generally −39‰~−8‰, where the value of $\delta^{13}C$ (CO_2) produced by humic organic matter is generally −25‰~−5‰; while the value of $\delta^{13}C$ (CO_2) produced by thermal degradation of organic matter is −28‰~−10‰, and the value of $\delta^{13}C$ (CO_2) transformed by microbial reduction is more important, reaching 18‰. In the study area, the $\delta^{13}C$ (CO_2) value in No. 4 + 5 coal seam is between −10.5‰ and −0.6‰, and that in No. 8 + 9 coal seam is between −11.3‰ and −7.6‰ (Table 3), which conforms to the carbon isotope characteristics of organic (biological) genetic gas. Among them, some samples have higher $\delta^{13}C$ (CO_2) values due to microbial transformation. The burial depth also affects the $\delta^{13}C$ (CO_2) values. The burial depth of No. 4 + 5 coal seam and No. 8 + 9 coal seam in the same well is similar, but that of the same coal seam in different wells (such as No. 4 + 5 coal seam in wells B1 and B2) is quite different. Therefore, the $\delta^{13}C$ (CO_2) values of samples are separated in two groups in Figure 5.

According to the component data of CBM, the CO_2–CH_4 coefficient of each sample is calculated by:

$$\text{CDMI} = \frac{CO_2}{CO_2 + CH_4} \times 100\%$$

The CDMI value and the $\delta^{13}C$ (CO_2) are used to draw the relationship diagram reflecting the genetic types of CO_2 (Figure 6). No. 4 + 5 and No. 8 + 9 coal seams both contain CO_2 generated by coal pyrolysis, which belongs to organic genetic gas [28], while shallow CO_2 is greatly affected by the action of microorganisms and belongs to biogenic gas. In the stagnant CBM system, the water-soluble consumption of CO_2 produced by coalification (including associated gas during microbial methane production and thermal degradation gas of coal-forming material) is not complete [32], resulting in high CO_2 concentration in gas components, and increases with the increase of burial depth.

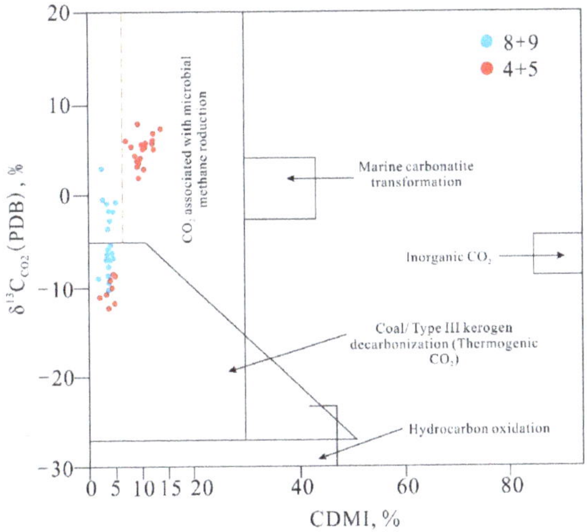

Figure 6. Relationship between $\delta^{13}C$ (CO_2) and CDMI.

4.3. Characterization of Gas Storage and Seepage Space

4.3.1. Characteristics of Seepage Pores

The experimental data of mercury injection are shown in Table 4, and the mercury injection curves are shown in Figure 7. The pores contained in the coal samples are mainly micropores and transition pores, and the development of mesopores and macropores is less. The average proportion of pores in each pore size class is 56.61%, 28.22%, 5.10%, and 10.07%, respectively. Mercury injection curves can be divided into three types. Type 1 contains samples BD-1 and BD-3. At pressures lower than 10 MPa, the amount of mercury injected increased rapidly; at pressures greater than 10 MPa, the amount of mercury injected increased slowly, indicating that micropores are more developed. The mercury intrusion curve and the extrusion curve basically overlap, indicating that the mercury removal efficiency is high and the pore connectivity is good. Type 2 contains BD-5 and BD-6. At pressures below 10 MPa, the amount of mercury injected increased slowly, and at pressures greater than 10 MPa, the amount of mercury increased rapidly, indicating that the pore structure is dominated by micropores. The large offset between the mercury injection curve and the mercury extrusion curve shows the low mercury removal efficiency and poor pore connectivity. The Type 3 mercury injection curve of BD-2 and BD-9 has a three-stage structure. At pressures less than 2 MPa, the mercury injection curve rises linearly; at pressures between 2 MPa and 10 MPa, the mercury injection rate gradually slows down; at pressures greater than 10 MPa, the mercury injection rate increases again, indicating that compared with the first two types of curves, mesopores and macropores are more developed in samples BD-2 and BD-9.

Table 4. Mercury injection data of typical samples in the Baode block.

Sample No.	Coal Seam No.	Porosity (%)	Pore Volume Percentage (%)			
			Micropore (<10 nm)	Transition Pores (10–10^2 nm)	Mesopore (10^2–10^3 nm)	Macropore (>10^3 nm)
BD-1	4 + 5	5.07	64.09	26.41	2.46	7.04
BD-2	4 + 5	5.39	61.15	22.19	3.99	12.67
BD-3	8 + 9	5.42	62.24	26.58	3.35	7.83
BD-5	8 + 9	6.89	54.89	31.38	4.65	9.08
BD-6	8 + 9	6.09	47.91	32.87	9.61	9.60
BD-9	8 + 9	5.29	49.38	29.89	6.54	14.19

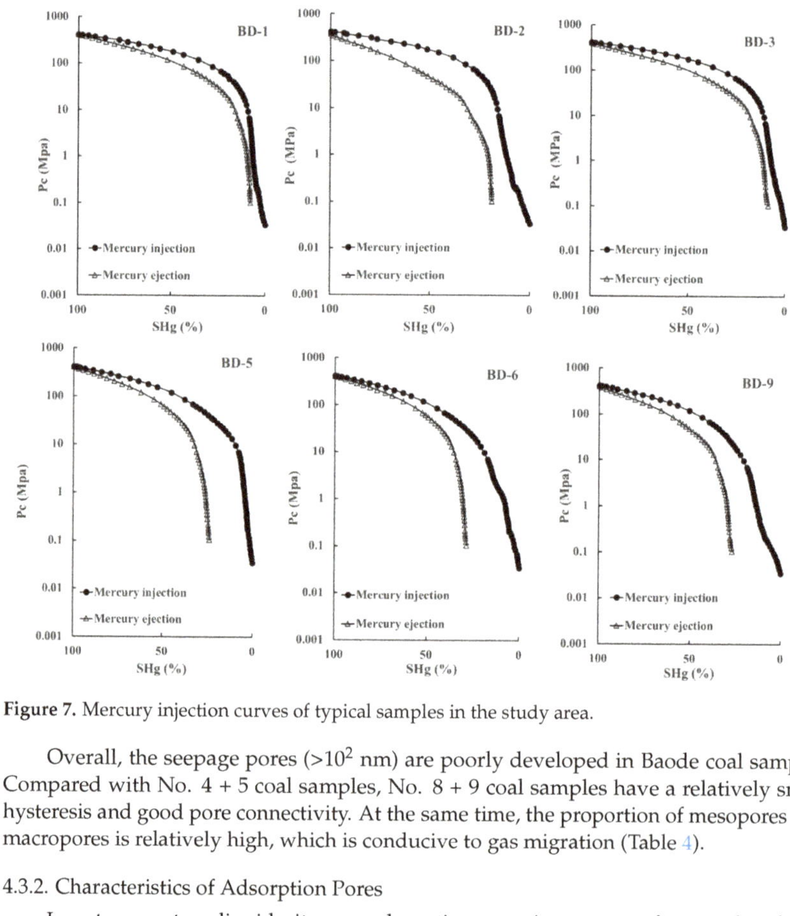

Figure 7. Mercury injection curves of typical samples in the study area.

Overall, the seepage pores (>10^2 nm) are poorly developed in Baode coal samples. Compared with No. 4 + 5 coal samples, No. 8 + 9 coal samples have a relatively small hysteresis and good pore connectivity. At the same time, the proportion of mesopores and macropores is relatively high, which is conducive to gas migration (Table 4).

4.3.2. Characteristics of Adsorption Pores

Low-temperature liquid nitrogen adsorption experiments are often used to finely characterize the adsorption pores (<10^2 nm) of coal samples [33,34]. As shown in Table 5, the BET specific surface area (SSA) is 1.41–5.14 m^2/g, and the total pore volume of BJH is 0.0063–0.0166 mL/g. The average pore diameter (APD) ranges from 11.50 nm to 19.86 nm, and the proportion of transition pores is the largest.

Table 5. Nitrogen adsorption data of typical samples in the study area.

Sample No.	Coal Seam No.	BET SSA (m^2/g)	BJH TPV (10^{-3} mL/g)	APD (nm)	Pore Volume Percentage (%)		
					<10 nm	10–100 nm	>100 nm
BD-1	4 + 5	4.04	0.0119	11.50	29.55	50.33	20.12
BD-2	4 + 5	5.14	0.0166	12.51	24.12	62.65	13.23
BD-3	8 + 9	1.81	0.0080	17.17	16.31	55.01	28.68
BD-5	8 + 9	2.93	0.0156	19.86	12.76	55.12	32.12
BD-6	8 + 9	1.76	0.0074	17.73	16.20	52.69	31.11
BD-9	8 + 9	1.41	0.0063	18.33	14.34	52.41	33.25

Notes: BJH TPV = Total pore volume; APD = Average pore diameter; BET SSA = Specific surface area.

Based on the adsorption/desorption curve of the nitrogen adsorption experiment, scholars use the hysteresis loop to classify the pore morphology in coal [35–37]. As can be seen in Figure 8, the pore morphology of the samples is divided into two types. Type 1 contains samples of BD-1 and BD-2 collected from No. 4 + 5 coal samples, which has an obvious hysteresis loop located at the relative pressure of 0.5–1, indicating that ink bottle pores are well-developed in No. 4 + 5 coal samples. Type 2 contains samples BD-3, BD-5, BD-6, and BD-9, which belong to the No. 8 + 9 coal samples. There is no hysteresis loop or an obvious hysteresis loop, which means that the inflection point of desorption curve is not obvious, thus the adsorption and desorption curves are roughly parallel, indicating that the samples mainly develop an airtight pore closed at one end.

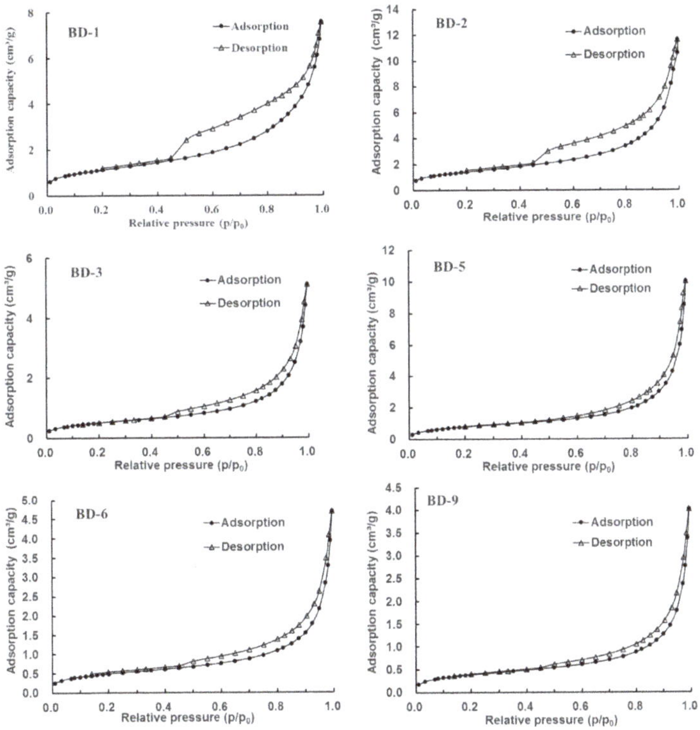

Figure 8. Typical nitrogen adsorption and desorption curve of coal samples in Baode block.

As shown in Figure 9, all samples have pores with a diameter of 1–100 nm, and micropores with a pore diameter of less than 10 nm are less developed, and there is a peak around 40–50 nm. The contribution of SSA is dominated by 1–10 nm micropores. The difference is that the 1–3 nm pores of the No. 4 + 5 coal samples have a great contribution, while the 3–10 nm pores of the No. 8 + 9 coal samples have a great contribution. Compared with No. 8 + 9 coal samples, No. 4 + 5 coal samples have developed more adsorbed pores such as ink bottle pores, so the SSA is relatively large, which is conducive to the adsorption of CBM but not conducive to desorption.

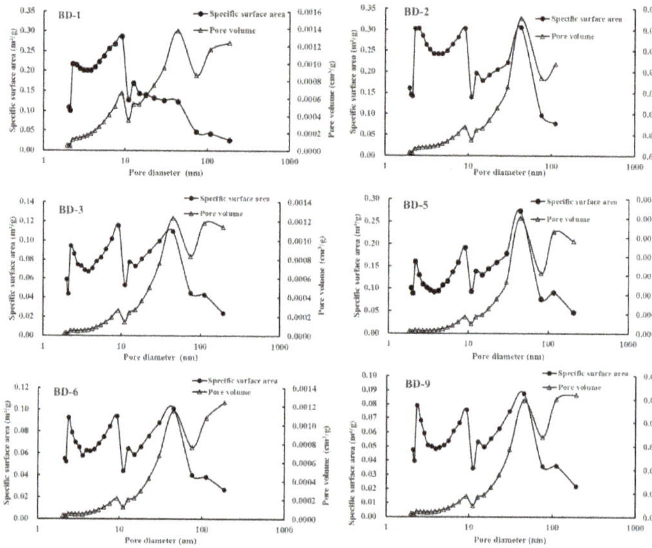

Figure 9. Relationship between pore diameter, SSA, and pore volume of typical coal samples in Baode block.

The BET model, FHH model, and thermodynamic model are often used to calculate the fractal dimension of micropores of coal [38,39]. In this paper, the FHH model is used to calculate the fractal dimension based on nitrogen adsorption data;

$$\ln(V/V_0) = A \ln(\ln P/P_0) + C$$

where V is the volume of gas molecules adsorbed at equilibrium pressure; V_0 is the volume of gas adsorbed by the monolayer; P_0 is the saturated vapor pressure of gas adsorption; A is the slope of the double logarithm curve of $\ln V$ and $\ln(\ln(P_0/P))$; C is a constant.

In the process of liquid nitrogen adsorption, micropores are filled first, and then monolayer adsorption occurs, followed by multi-molecular layer adsorption. When the relationship between the relative pressure and pore diameter conforms to the Kelvin equation, capillary condensation occurs [40]. Therefore, taking the relative pressure as the boundary of 0.5 [40], the fractal dimensions D_1 and D_2 can be calculated respectively (Table 6). The fractal dimension D should meet $2 \leq D \leq 3$, in which the larger the fractal dimension, the rougher the coal surface and the stronger the adsorption capacity [41,42].

Table 6. Fractal dimension of typical samples in the study area.

Sample No.	Relative Pressure (P/P_0): 0~0.5			Relative Pressure (P/P_0): 0.5~1		
	A_1	$D_1 = 3 + A_1$	R_1^2	A_2	$D_2 = 3 + A_2$	R_2^2
BD-1	−0.45	2.55	0.9997	−0.41	2.59	0.9864
BD-2	−0.46	2.54	0.9996	−0.45	2.55	0.9996
BD-3	−0.47	2.53	0.9975	−0.49	2.51	0.9996
BD-5	−0.51	2.49	0.9908	−0.54	2.46	0.9991
BD-6	−0.41	2.59	0.9962	−0.47	2.53	0.9999
BD-9	−0.44	2.56	0.9947	−0.48	2.52	0.9987

As shown in Figure 10, the fractal dimension D_1 has no obvious correlation with the SSA and APD. The fractal dimension D_1 reflects the porosity of the sample with a relative pressure of 0~0.5, which cannot characterize all the pore characteristics of the sample. The fractal dimension D_2 is positively correlated with the SSA and negatively correlated with the APD, which indicates that the larger the SSA, the smaller the average pore size, and the higher the overall micropore proportion of the sample. It indicates that the more developed the micropore, the stronger the adsorption capacity of coal. As the value of D_2 increases, the pore structure becomes more complex, and the adsorption capacity of coal becomes stronger.

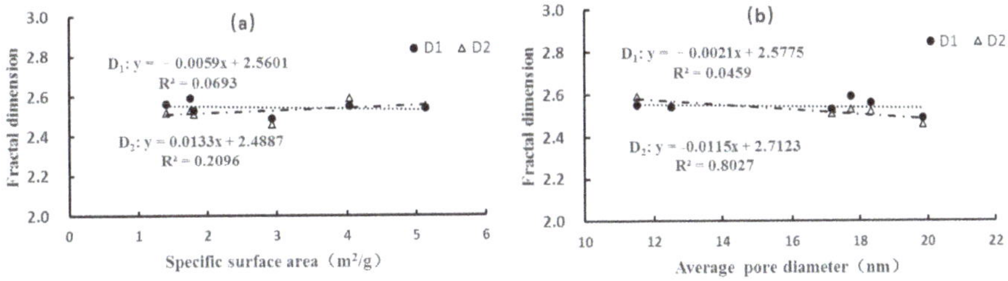

Figure 10. (**a**) Relationship between fractal dimension and SSA; (**b**) Relationship between fractal dimension and APD.

4.3.3. Three-Dimensional Model of Storage and Seepage Space

In order to intuitively obtain the distribution characteristics of pores and fractures, the samples are selected for X-ray CT imaging tests (Table 7). The results show that the porosity of the selected coal samples varies from 0.76% to 4.39%, with an average of 2.38%. However, the proportion of connected pores is extremely low (0–35.04%), resulting in an extremely low connected porosity of 0–1.21%. There is a negative correlation between the porosity and the mineral content. Sample BD-3 has the lowest porosity, because most of the fractures inside the sample are filled with minerals (Figure 11).

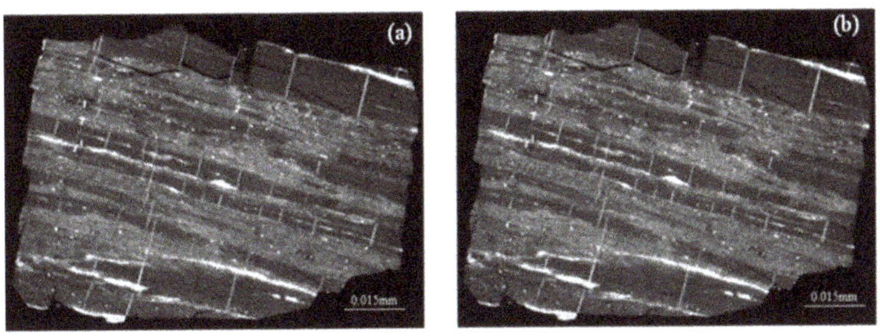

Figure 11. Two-dimensional slices of mineral filling fractures in sample BD-3 ((**a**) is the first slice and (**b**) is the 45th slice; white means the fractures are filled with minerals).

Table 7. X-ray CT imaging test results of typical samples.

Sample No.	Porosity (%)	Connected Porosity (%)	Proportion of Connected Pores (%)	Mineral Content (%)
BD-1	1.97	0	0	3.31
BD-2	3.92	0.79	20.15	0.27
BD-3	0.76	0.07	9.21	4.5
BD-5	4.39	1.21	27.56	0.26
BD-6	1.17	0.41	35.04	3.52
BD-9	3.47	0.17	4.85	2.17

After three-dimensional reconstruction, the distribution of the coal matrix, pores, fractures, and minerals in coal samples can be more easily displayed (Figure 12). The distribution direction of minerals is similar to the direction of densely developed pores, especially in samples BD-3 and BD-6, and the distribution of pores and fractures is extremely uneven. Meanwhile, although some samples have relatively high porosity, such as BD-9, due to the extremely low proportion of connected pores, the connected porosity is still very low. On the contrary, if the mineral content in coal is small and distributed in a dispersed state, the proportion of connected pores will be high, and the connected porosity will be high, which is conducive to gas flow, such as sample BD-5. Therefore, it is not the measured porosity that plays a key role in the development of CBM, but it depends on the connectivity of the pore and fracture system.

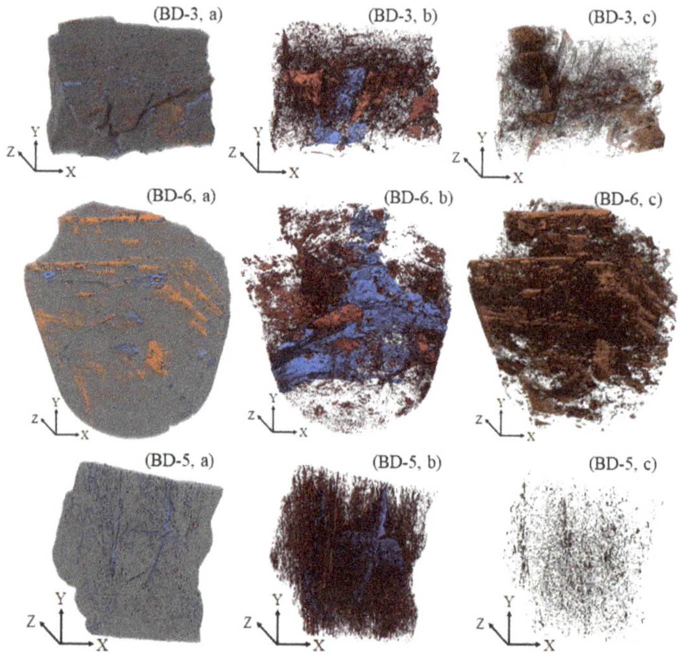

Figure 12. (**a**) Three-dimensional reconstruction image of coal samples (gray: matrix, blue: pores and fractures, orange: mineral). (**b**) Three-dimensional image of connected and isolated pores and fractures (blue: connected pores and fractures, red: isolated pores and fractures). (**c**) Three-dimensional image of minerals.

5. Conclusions

(1) Gases disrobed from No. 4 + 5 and No. 8 + 9 coal seams in Baode block are dominated by CH_4, followed by N_2 and CO_2. Under the influence of air mixing, the N_2 content decreases with the increase of burial depth. Meanwhile, the CO_2 content increases with increasing burial depth of No. 8 + 9 coal seam, which related to the action of CO_2-reducing bacteria in the shallow part.

(2) The $\delta^{13}C$ (CO_2) value in No. 4 + 5 coal seam is between −10.5‰ and −0.6‰, and that in No. 8 + 9 coal seam is between −11.3‰ and −7.6‰. No. 4 + 5 and No. 8 + 9 coal seams both have biogenic gas and thermogenic methane. Meanwhile, No. 4 + 5 and No. 8 + 9 coal seams both contain CO_2 generated by thermal maturation of coal (thermogenic gas), while shallow CO_2 is likely to result from the action of microorganisms (microbial gas). With the increase of burial depth, the content of CO_2 increases.

(3) The seepage pores (>10^2 nm) are poorly developed in the Baode coal samples. Samples collected from No. 4 + 5 coal seams have developed more sorption pores such as ink bottle pores, so the SSA is relatively large, which is conducive to the adsorption of CBM. The mercury intrusion and extrusion curves of samples from No. 8 + 9 coal seams exhibit a relatively low degree of hysteresis (Hg retention), indicating good pore connectivity. At the same time, the proportion of mesopores and macropores is relatively high, which is conducive to gas migration.

(4) The porosity of the coal samples is inversely related to the mineral content and the occurrence of the mineral. If the mineral content in coal is small and distributed in a dispersed state, the proportion of connected pores will be high, and the connected porosity will be high, which is conducive to gas flow.

Author Contributions: Conceptualization, H.C. and W.T.; methodology, H.C.; software, Z.C.; validation, H.C., W.T. and Q.Z.; formal analysis, S.T.; investigation, S.T.; resources, W.T.; data curation, W.T.; writing—original draft preparation, H.C.; writing—review and editing, W.T.; visualization, Z.C.; supervision, Q.Z.; project administration, H.C.; funding acquisition, W.T. All authors have read and agreed to the published version of the manuscript.

Funding: This research was funded by [the National Natural Science Foundation Project] grant number [No. 41772132, 41502157, 41530314], [the Key Project of the National Science & Technology] grant number [No. 2016ZX05043-001], [the Fundamental Research Funds for the Central Universities] grant number [No. 2652019095] and The APC was funded by [the National Natural Science Foundation Project]. Key Technologies R&D Programme of PetroChina Company Limited (2019E-25).

Institutional Review Board Statement: Not applicable.

Informed Consent Statement: Not applicable.

Data Availability Statement: Not applicable.

Conflicts of Interest: The authors declare no conflict of interest.

References

1. Tao, S.; Chen, S.D.; Pan, Z.J. Current status, challenges, and policy suggestions for coalbed methane industry development in China: A review. *Energy Sci. Eng.* **2019**, *7*, 1059–1074. [CrossRef]
2. Tao, S.; Pan, Z.J.; Tang, S.L.; Chen, S.D. Current status and geological conditions for the applicability of CBM drilling technologies in China: A review. *Int. J. Coal Geol.* **2019**, *202*, 95–108. [CrossRef]
3. Qin, Y.; Moore, T.A.; Shen, J.; Yang, Z.B.; Shen, Y.L.; Wang, G. Resources and geology of coalbed methane in China: A review. *Int. Geol. Rev.* **2017**, *1*, 777–812.
4. Shao, L.Y.; Wang, X.T.; Wang, D.D.; Li, M.P.; Wang, S.; Li, Y.J.; Shao, K.; Zhang, C.; Gao, C.X.; Dong, D.X.; et al. Sequence stratigraphy, paleogeography, and coal accumulation regularity of major coal-accumulating periods in China. *Int. J. Coal Sci. Technol.* **2020**, *7*, 240–262. [CrossRef]
5. Chen, Y.; Ma, D.M.; Xia, Y.C.; Guo, C.; Yang, F.; Shao, K. Characteristics of the mud shale reservoirs in coal-bearing strata and resources evaluation in the eastern margin of the Ordos Basin, China. *Energy Explor. Exploit.* **2020**, *38*, 372–405. [CrossRef]
6. Tao, S.; Tang, D.Z.; Xu, H.; Li, S.; Geng, Y.G.; Zhao, J.L.; Wu, S.; Meng, Q.; Kou, X.; Yang, S.Y.; et al. Fluid velocity sensitivity of coal reservoir and its effect on coalbed methane well productivity: A case of Baode Block, northeastern Ordos Basin, China. *J. Petrol. Sci. Eng.* **2017**, *152*, 229–237. [CrossRef]

7. Yang, X.C.; Mao, J.S.; Lin, W.J.; Hao, S.; Zhao, L.M.; Wang, Y.; Li, L. Exploration history and enlightenment of coalbed methane in baode block. *Xinjiang Petrol. Geol.* **2021**, *42*, 381–388.
8. Tian, W.G.; Tang, D.Z.; Wang, Z.L.; Sun, B. Origin of coalbed methane in Baode, Northeastern Ordos Basin. *Geol. J. China Univ.* **2012**, *18*, 479–484.
9. Guo, H.G.; Yu, Z.S.; Zhang, H.X. Phylogenetic diversity of microbial communities associated with coalbed methane gas from Eastern Ordos Basin, China. *Int. J. Coal Geol.* **2015**, *150*, 120–126. [CrossRef]
10. Liu, S.M.; Tan, F.R.; Huo, T.; Tang, S.H.; Zhao, W.X.; Chao, H.D. Origin of the hydrate bound gases in the Juhugeng Sag, Muli Basin, Tibetan Plateau. *Int. J. Coal Sci. Technol.* **2020**, *7*, 43–57. [CrossRef]
11. Xu, H.; Tang, D.; Liu, D.; Tang, S.; Yang, F.; Chen, X.; Deng, C. Study on coalbed methane accumulation characteristics and favorable areas in the Binchang area, southwestern Ordos Basin, China. *Int. J. Coal Geol.* **2012**, *95*, 1–11. [CrossRef]
12. Yang, C.; Zhang, J.; Tang, X. Microscopic pore types and its impact on the storage and permeability of continental shale gas, Ordos Basin. *Earth Sci. Front.* **2013**, *20*, 240–250.
13. Xue, G.; Liu, H.; Li, W. Deformed coal types and pore characteristics in Hancheng coalmines in Eastern Weibei coalfields. *Int. J. Min. Sci. Technol.* **2012**, *22*, 681–686. [CrossRef]
14. Tang, X.; Zhang, J.; Shan, Y.; Xiong, J. Upper Paleozoic coal measures and unconventional natural gas systems of the Ordos Basin, China. *Geosci. Front.* **2012**, *3*, 863–873. [CrossRef]
15. Wang, L.; Jiang, B.; Qu, Z. Structural control on gas content distribution in eastern margin of Ordos basin. *Coal Geol. Explor.* **2013**, *41*, 14–19.
16. Jiang, B.; Xu, J.; Zhu, K.; Wang, J.; Wang, J.; Qu, Z. Structural and hydrogeological controls of coalbed methane preservation in the eastern Ordos Basin. *Geol. J. China Univ.* **2012**, *3*, 438–446.
17. Cao, D.Y.; Wang, A.M.; Ning, S.Z.; Li, H.T.; Guo, A.J.; Chen, L.M.; Liu, K.; Tan, J.Q.; Zheng, Z.H. Coalfield structure and structural controls on coal in China. *Int. J. Coal Sci. Technol.* **2020**, *7*, 220–239. [CrossRef]
18. GB/T 212-2008. *Chinese National Standard: Proximate Analysis of Coal*; General Administration of Quality Supervision, Inspection and Quarantine: Beijing, China, 2008. (In Chinese)
19. ISO 7404.3-1994. *Methods for the Petrographic Analysis of Bituminous Coal and Anthracite—Part 3: Method Of Determining Maceral Group Composition*; International Organization for Standardization: Geneva, Switzerland, 1994.
20. ISO 7404.5-1994. *Method for the Petrographic Analysis of bituminous Coal and Anthracite—Part 5: Method of Determining Microscopically the Reflectance of Vitrinite*; International Organization for Standardization: Geneva, Switzerland, 1994.
21. Tao, S.; Zhao, X.; Tang, D.Z.; Deng, C.M.; Meng, Q.; Cui, Y. A model for characterizing the continuous distribution of gas storing space in low-rank coals. *Fuel* **2018**, *233*, 552–557. [CrossRef]
22. SY/T 5346-2005. *Chinese Petroleum and Natural Gas Industry Standard: Rock Capillary Pressure Measurement*; National Energy Administration: Beijing, China, 2005. (In Chinese)
23. Tao, S.; Pan, Z.J.; Chen, S.D.; Tang, S.L. Coal seam porosity and fracture heterogeneity of marcolithotypes in the Fanzhuang Block, southern Qinshui Basin, China. *J. Nat. Gas Sci. Eng.* **2019**, *66*, 148–158. [CrossRef]
24. GB/T 13610-2014. *Composition Analysis of Natural Gas by Gas Chromatography*; Standardization Administration of the People's Republic of China: Beijing, China, 2014. (In Chinese)
25. Yang, Z.; Grace, J.R.; Lim, C.J.; Zhang, L. Combustion of low-concentration coal bed methane in a fluidized bed. *Energy Fuels* **2011**, *25*, 975–980. [CrossRef]
26. Wang, Q.; Dong, S.; Wang, H.; Yang, J.; Huang, H.; Dong, X.; Yu, B. Hydrogeochemical processes and groundwater quality assessment for different aquifers in the Caojiatan coal mine of Ordos Basin, northwestern China. *Environ. Earth Sci.* **2020**, *79*, 199. [CrossRef]
27. Li, Z.W.; Tang, D.Z.; Tang, S.L.; Pu, Y.F.; Zhang, A.B. Study on formation mechanism of CO_2-enriched CBM reservoirs in low-rank coal seams from southern Zhunggar Basin. *Coal Sci. Technol.* **2021**, *49*, 174–180.
28. Scott, A.R.; Kaiser, W.R.; Walter, B.; Ayers, J. Thermogenic and secondary biogenic gases, San Juan Basin, Colorado and New Mexico; implications for coalbed gas producibility. *AAPG Bull.* **1994**, *78*, 1186–1209.
29. Qi, Y.; Jiao, S.; Tang, Y.; Ye, J. Characteristics and origins of stable carbon isotope in coalbed methane of China. *J. China Univ. Min. Technol.* **2000**, *29*, 10–17.
30. Whiticar, M.J. Carbon and hydrogen isotope systematics of bacterial formation and oxidation of methane. *Chem. Geol.* **1999**, *161*, 291–314. [CrossRef]
31. Xu, T.; Xie, Q.; Kang, Y. Heat effect of the oxygen-containing functional groups in coal during spontaneous combustion processes. *Adv. Powder Technol.* **2017**, *28*, 1841–1848. [CrossRef]
32. Gao, C.; Liu, D.; Li, Z.; Cai, Y.; Fang, Y. Fluid performance in coal reservoirs: A comprehensive review. *Geofluids* **2021**, *2021*, 6611075. [CrossRef]
33. Tao, S.; Chen, S.D.; Tang, D.Z.; Zhao, X.; Xu, H.; Li, S. Material composition, pore structure and adsorption capacity of low-rank coals around the first coalification jump: A case of eastern Junggar Basin, China. *Fuel* **2018**, *211*, 804–815. [CrossRef]
34. Men, X.Y.; Tao, S.; Liu, Z.X.; Tian, W.G.; Chen, S.D. Experimental study on gas mass transfer process in a heterogeneous coal reservoir. *Fuel Process. Technol.* **2021**, *216*, 106779. [CrossRef]

35. Chen, S.D.; Tang, D.Z.; Tao, S.; Chen, Z.L.; Xu, H.; Li, S. Coal reservoir heterogeneity in multi-coal seams of the Panguan syncline, Western Guizhou, China: Implication for the development of superposed CBM-bearing systems. *Energy Fuels* **2018**, *32*, 8241–8253. [CrossRef]
36. Chen, S.D.; Tang, D.Z.; Tao, S.; Xu, H.; Li, S.; Zhao, J.L.; Jiang, Q.; Yang, H.X. Pore structure characterization of different rank coals using N_2 and CO_2 adsorption and its effect on CH_4 adsorption capacity: A case in Panguan syncline, western Guizhou, China. *Energy Fuels* **2017**, *31*, 6034–6044. [CrossRef]
37. Wang, C.; Hao, S.; Sun, W.; Chu, W. Fractal dimension of coal particles and their CH_4 adsorption. *Int. J. Coal Sci. Technol.* **2012**, *22*, 855–858.
38. Kim, D.; Seo, Y.; Kim, J.; Han, J.; Lee, Y. Experimental and simulation studies on adsorption and diffusion characteristics of coalbed methane. *Energies* **2019**, *12*, 3445. [CrossRef]
39. Friesen, W.I.; Mikula, R.J. Fractal dimensions of coal particles. *J. Colloid Interface Sci.* **1987**, *120*, 263–271. [CrossRef]
40. Zelenka, T. Adsorption and desorption of nitrogen at 77 K on micro- and mesoporous materials: Study of transport kinetics. *Micropor. Mesopor. Mater.* **2016**, *227*, 202–209. [CrossRef]
41. Chen, S.D.; Tang, D.Z.; Tao, S.; Ji, X.Y.; Xu, H. Fractal analysis of the dynamic variation in pore-fracture systems under the action of stress using a low-field NMR relaxation method: An experimental study of coals from western Guizhou in China. *J. Petrol. Sci. Eng.* **2019**, *173*, 617–629. [CrossRef]
42. Chen, Y.; Tang, D.Z.; Xu, H.; Tao, S.; Li, S.; Yang, G.H.; Yu, J.J. Pore and fracture characteristics of different rank coals in the eastern margin of the Ordos Basin, China. *J. Nat. Gas Sci. Eng.* **2015**, *26*, 1264–1277. [CrossRef]

energies

MDPI

Article

Investigation of Flowback Behaviours in Hydraulically Fractured Shale Gas Well Based on Physical Driven Method

Wei Guo, Xiaowei Zhang *, Lixia Kang, Jinliang Gao and Yuyang Liu

PetroChina Research Institute of Petroleum Exploration & Development, Beijing 100083, China; guowei69@petrochina.com.cn (W.G.); kanglixia@petrochina.com.cn (L.K.); gaojinliang@petrochina.com.cn (J.G.); yuyangliu@petrochina.com.cn (Y.L.)
* Correspondence: zhangxw69@petrochina.com.cn; Tel.: +86-010-83599457

Abstract: Due to the complex microscope pore structure of shale, large-scale hydraulic fracturing is required to achieve effective development, resulting in a very complicated fracturing fluid flowback characteristics. The flowback volume is time-dependent, whereas other relevant parameters, such as the permeability, porosity, and fracture half-length, are static. Thus, it is very difficult to build an end-to-end model to predict the time-dependent flowback curves using static parameters from a machine learning perspective. In order to simplify the time-dependent flowback curve into simple parameters and serve as the target parameter of big data analysis and flowback influencing factor analysis, this paper abstracted the flowback curve into two characteristic parameters, the daily flowback volume coefficient and the flowback decreasing coefficient, based on the analytical solution of the seepage equation of multistage fractured horizontal Wells. Taking the dynamic flowback data of 214 shale gas horizontal wells in Weiyuan shale gas block as a study case, the characteristic parameters of the flowback curves were obtained by exponential curve fittings. The analysis results showed that there is a positive correlation between the characteristic parameters which present the characteristics of right-skewed distribution. The calculation formula of the characteristic flowback coefficient representing the flowback potential was established. The correlations between characteristic flowback coefficient and geological and engineering parameters of 214 horizontal wells were studied by spearman correlation coefficient analysis method. The results showed that the characteristic flowback coefficient has a negative correlation with the thickness × drilling length of the high-quality reservoir, the fracturing stage interval, the number of fracturing stages, and the brittle minerals content. Through the method established in this paper, the shale gas flowback curve containing complex flow mechanism can be abstracted into simple characteristic parameters and characteristic coefficients, and the relationship between static data and dynamic data is established, which can help to establish a machine learning method for predicting the flowback curve of shale gas horizontal wells.

Keywords: shale gas; flowback; big-data analysis; horizontal well; fracturing fluids

Citation: Guo, W.; Zhang, X.; Kang, L.; Gao, J.; Liu, Y. Investigation of Flowback Behaviours in Hydraulically Fractured Shale Gas Well Based on Physical Driven Method. *Energies* **2022**, *15*, 325. https://doi.org/10.3390/en15010325

Academic Editor: Maxim Tyulenev

Received: 10 November 2021
Accepted: 23 December 2021
Published: 4 January 2022

1. Introduction

Multistage fracturing of horizontal wells is widely used in the exploration and development of shale gas. It typically takes tens of thousands of square meters of fracturing fluids and thousands of tons of sand to fracture a shale gas horizontal well, and the flowback of the fracturing fluids directly affects shale gas production. Hence, the flowback curve of fracturing fluids is a key issue in shale gas exploration [1]. The pores and fractures in shale gas reservoirs have different scales [2], and new multiscale pore-split systems are created by multistage fracturing. Thus, the shale gas flow mechanism is very complex, including pressure sensitivity, adsorption, diffusion, slip, imbibition, and seepage [3–6]. Currently, no consensus has been reached on this flow mechanism. Moreover, there is gas–water two-phase flow in the formation after large-scale fracturing. At present, there is no commercial software to realize the numerical simulation of flowback curve. It may take several years to develop such software because, when developing such calculation programs, the solutions of the flow equations, unstructured meshing, equation discretization, and large-scale irregular sparse matrix are very complex. The gas flow equation for shale gas involves many flow mechanisms. The solution of this equations requires many known parameters which are difficult to obtain directly or measure experimentally. For example, shale gas is a multicomponent mixture containing hydrocarbons and nonhydrocarbon gases (CO_2 and N_2) [7], and its molecular free path at a high temperature and pressure cannot be obtained for unknown component proportions. Even if there is a perfect numerical simulation program, inaccurate parameters will affect the reliability of the results.

In recent years, with the development of big-data-related technology, researchers began to introduce machine learning into the research of oil and gas development and have mainly used it as a prediction tool. Kohli et al. [8] took well-log data as input parameters and trained the multilayer forward neural network by using the least square Levenberg–Marquardt optimization algorithm to predict the formation permeability; the predicted permeability is consistent with field data. Jia et al. [9] studied the water channel problem resulting from long-term water injection. A density peak clustering algorithm based on streamline clustering was used to quantify the flow area for water flooding in the oil reservoir and thereby effectively identify the invalid water injection circulation channels between the injection and production wells, as well as areas with development potential. Adibifard et al. [10] carried out a Chebyshev polynomial interpolation of pressure derivative data and input the results to an artificial neural network to estimate reservoir parameters. Ghaffarian et al. [11] processed the pseudo pressure of gas wells and used the pseudo pressure derivative data as the input of single and coupled multilayer perceptron network to identify the condensate gas reservoir model. Tian et al. [12] used recurrent neural network learning to train the data collected by permanent downhole pressure gauge (PDG) for the inversion of reservoir permeability and other parameters and production prediction. Hung et al. [13] introduced the application of Gaussian process regression (GPR), support vector machine (SVM), and random forest (RF) to predict CO_2 trapping efficiency in saline formations.

As the fracturing fluids flowback is influenced by many factors, big data technology has attracted increasing attention. Bai et al. [14] used exponential and harmonic functions to fit the water production in the flowback stage and produced the water stages of 32 wells and developed a prediction tool for accumulated water production based on the fitting results. Zhou et al. [15] used various binary and multivariate methods to analyse the data from 187 wells in the Marcellus block in the USA. The flowback rate during the first three weeks of data collection was found to increase with the thermal maturity and decrease with increasing strata thickness. Lin et al. [16] used a back propagation neural network to model the static data and flowback rate for 74 wells in Sichuan Province, China. The flowback rate and gas production in the first month were estimated using six parameters with relatively high weights in the model. Liu et al. [17] developed two neural networks with different structures to predict the flowback rate at specific time points and compared the predictions. In summary, current big data-based methods used to determine the flowback pattern focused on static or dynamic data of many wells for a given block, and a combined analysis of static and dynamic data has not been conducted. Some studies have been performed by using a fixed data point or a periodic average of dynamic data, which is a static data methodology in essence. All these lead to incomplete data utilization and limited application scope of the method.

In order to simplify the time-dependent flowback curve into simple parameters, as the target parameters of big data analysis and flowback influencing factor analysis, this paper combined the flowback dynamic data and static data of shale gas fracturing horizontal wells to study the flowback curve. The approaches and processes used in this study are shown in Figure 1. The first part of this paper briefly introduces the geological background of the research area and the relevant data collected. In the second part, the multi-stage fracturing physical model of shale gas horizontal wells was established. The flow equation in Laplace space was obtained from the convolution formula, and the time-dependent flowback rate was obtained. In the third part, the flowback data of 214 horizontal wells in the Weiyuan block were fitted to obtain the characteristic parameters. The calculation formula of characteristic coefficient including characteristic parameters was established. Then, the correlation analysis was carried out by using a flowback characteristic coefficient, and the influencing factors of the flowback characteristic coefficient were comprehensively studied. Through the method established in this paper, the shale gas flowback curve containing complex flow mechanism can be abstracted into simple characteristic parameters and characteristic coefficients, and the relationship between static data and dynamic data is established, which can help to establish a machine learning method for predicting the flowback curve of shale gas horizontal wells.

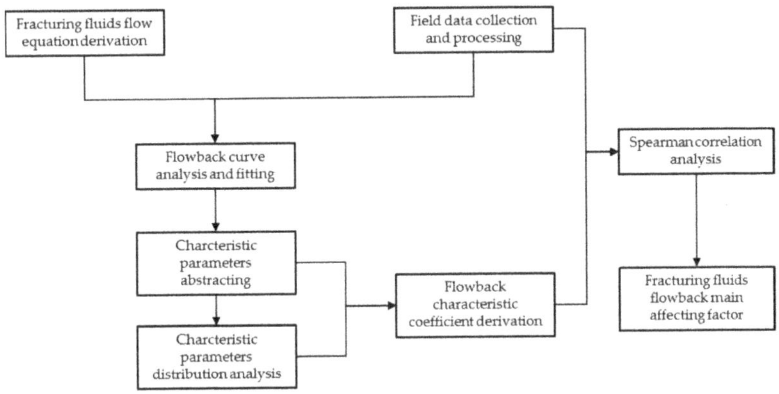

Figure 1. Schematic diagram of the approach used in the study 2. Research area background and analysis data.

Weiyuan shale gas block is located in the southwest of Sichuan Basin of China. The whole Weiyuan block has a monoclinic structure inclined to the southeast. The Longmaxi Formation (LF) is the main exploration target layer. The LF can be divided into two layers: L1 and L2. The first layer of L1 can be subdivided into two sublayers, namely, $L1_1$ and $L1_2$. The $L1_1$ sublayer can be further subdivided into four layers, namely, $L1_1{}^1$, $L1_1{}^2$, $L1_1{}^3$ and $L1_1{}^4$, from bottom to top. The middle and lower parts of $L1_1{}^1$ have the best reservoir quality.

Table 1 shows the data from the study area. A total of 20 different types of data were collected from 282 wells in Weiyuan shale gas block, including the fracturing lateral length, the number of fracturing stages, first-year average daily production rate, vertical depth, TOC, porosity, high-quality reservoir thickness, gas saturation, pressure coefficient, brittle mineral content, average fracturing stage interval, fracturing fluid intensity, proppant intensity, average hydraulic fracturing fluid displacement, drilling length in high-quality reservoir, EUR (Estimated Ultimate Recovery), 30 day flowback rate, 90 day flowback rate, 180 day flowback rate, 360 day flowback rate, and peak gas production flowback rate. The data volume ranges from 214 to 282. The minimum data volume is 214, and the corresponding data of these 214 wells is used in subsequent analysis. The 30 day flowback rate, 90 day flowback rate, 180 day flowback rate, 360 day flowback rate, peak gas production flowback rate, and first-year average daily production rate are taken from the daily production reports of the horizontal wells. The fracturing lateral length, number of fracturing stages, average fracturing stage interval, fracturing fluid intensity, proppant intensity, and average hydraulic fracturing fluid displacement are taken from the drilling and completion reports of horizontal wells. The vertical depth, TOC, porosity, high-quality reservoir thickness, gas saturation, pressure coefficient, brittle mineral content and drilling length in high-quality reservoirs are taken from well logging interpretation.

Table 1. Statistical tables of data for the study area. Copyright permission: The copyright of Table 1 belongs to the publisher and authors.

Data set	Data Volume	Minimum Value	Maximum Value	Average Value	Standard Deviation
30 day flowback rate	260	5%	77%	24%	0.12
90 day flowback rate	240	8%	103%	35%	0.16
180 day flowback rate	231	10%	116%	42%	0.18
360 day flowback rate	214	13%	129%	48%	0.20
Peak gas production flowback rate	243	3%	83%	23%	0.15
Fracturing lateral length (m)	281	502.00	2577.00	1515.87	308.47
Number of fracturing stages	281	3.00	36.00	22.63	5.56
First-year average daily production rate ($10^4m^3/d$)	281	0.98	34.02	9.47	5.28
Vertical depth (m)	279	2200.00	3800.00	3008.24	352.49
TOC content (%)	282	3.30	7.80	5.39	0.76
Porosity (%)	282	5.00	8.90	7.11	0.96
High-quality reservoir thickness (m)	282	2.30	7.50	5.17	1.36
Gas saturation (%)	279	60.00	83.00	75.70	3.14
Pressure coefficient	282	1.35	2.05	1.74	0.19
Brittle mineral content (%)	282	62.50	96.00	78.67	7.79
Average fracturing stage interval (m)	281	43.10	479.33	69.94	27.81
Fracturing fluid Intensity (m^3/m)	281	7.12	47.94	27.25	4.51
Proppant intensity (t/m)	281	0.29	3.00	1.63	0.38
Average hydraulic fracturing fluid displacement (m^3/min)	281	6.19	15.00	11.80	1.33
Drilling length in high-quality reservoir (m)	206	40.60	2380.50	1011.98	452.00

2. Basic Theory

The time-dependent daily flowback volume curve for the Weiyuan shale gas block in southern Sichuan Province, China, shows that the daily flowback volume is large during the early stages of continuous production and decreases to a stable value during the late stage of production. The daily flowback volume during early-stage production is two to three orders of magnitude higher than that during the later stage, and the relationship between daily flowback and time is generally exponential. The flowback of a fracturing fluid after large-scale fracturing is essentially the seepage of fluid from the stimulated reservoir volume (SRV) area to the wellbore. The corresponding law can also be obtained from the seepage equation of fracturing fluid.

2.1. Bottomhole Flow Equation

The formula derivation in this paper considers single-phase liquid seepage, assumes constant flowing bottom hole pressure (FBHP) and continuous production, and is based on the multi-stage fracturing seepage equation of horizontal well [18]. The following dimensionless parameters are defined:

Dimensionless pressure:

$$P_D = \frac{2\pi kh[p_i - p(r,t)]}{qB\mu} \tag{1}$$

Dimensionless time:

$$t_D = \frac{kt}{\phi\mu C_t L^2} \tag{2}$$

Dimensionless distance:

$$x_D = \frac{x}{L} \tag{3}$$

$$y_D = \frac{y}{L} \tag{4}$$

where K denotes the permeability, m^2; ϕ denotes the porosity; μ denotes the viscosity, Pa.s; C_t denotes the comprehensive compressibility coefficient, 1/Pa; x_{fi} denotes the half-length of the ith fracture, m; and $L = \sum_{i=1}^{n} x_{fi}$ denotes the sum over the half-lengths of n fractures, m.

For multistage fractured horizontal wells, the total flowback volume is the sum of the flowback volumes of all the individual fractures. The flowback volume can be related to the FBHP in Laplace space as follows:

$$\sum_{j=1}^{n} \overline{q}_{Dj} = \frac{1}{s} \tag{5}$$

$$\overline{P}_{WDi}(f(u)) = \sum_{j=1}^{n} s\overline{q}_{Dj}\overline{P}_{Dij}(f(u)) \tag{6}$$

where:

$$\overline{P}_{Dij}(f(u)) = \pi \int_0^\infty P_D(t_D)e^{-f(u)t_D}dt_D \tag{7}$$

$$P_D(t_D) = \int_0^{t_D} G_{xD}(\tau_D)G_{yD}(\tau_D)d\tau_D \tag{8}$$

The expressions for G_{xD} and G_{yD} vary with the external boundary conditions. For enclosed strata with rectangular boundaries, G_{xD} and G_{yD} are defined as follows [19]:

$$G_{xD}(\tau_D) = \frac{1}{x_{eD}}\left[1 + 2\sum_{n=1}^{\infty} exp\left(\frac{-n^2\pi^2(\tau_D)}{2x_{eD}^2}\right)cos\frac{n\pi x_{wD}}{x_{eD}}cos\frac{n\pi x_D}{x_{eD}}\right] \tag{9}$$

$$G_{yD}(\tau_D) = \frac{2}{y_{eD}}\left[1 + \frac{2y_{eD}}{\pi}\sum_{n=1}^{\infty}\frac{1}{n}exp\left(\frac{-n^2\pi^2(\tau_D)}{2y_{eD}^2}\right)sin\frac{n\pi}{y_{eD}}cos\frac{n\pi y_{wD}}{y_{eD}}cos\frac{n\pi y_D}{y_{eD}}\right] \tag{10}$$

Equations (5) and (6) can be used to obtain the following linear equation:

$$\begin{bmatrix} f(u)\overline{P}_{D1,1}\ f(u)\overline{P}_{D1,2}\ f(s)\overline{P}_{D1,3}\cdots f(u)\overline{P}_{D1,n} \\ f(u)\overline{P}_{D2,1}\ f(u)\overline{P}_{D2,2}\ f(s)\overline{P}_{D2,3}\cdots f(u)\overline{P}_{D2,n} \\ \cdots\cdots \\ f(u)\overline{P}_{Dk,1}\ f(u)\overline{P}_{Dk,2}\ f(u)\overline{P}_{Dk,3}\cdots f(u)\overline{P}_{Dk,n} \\ \cdots\cdots \\ f(u)\overline{P}_{Dn,1}\ f(u)\overline{P}_{Dn,2}\ f(u)\overline{P}_{Dn,3}\cdots f(u)\overline{P}_{D,n} \\ u\ \ u\ \ u\ \cdots\cdots u \end{bmatrix}\begin{bmatrix} \overline{q}_{1D} \\ \overline{q}_{2D} \\ \cdots\cdots \\ \overline{q}_{kD} \\ \cdots\cdots \\ \overline{q}_{nD} \\ \overline{P}_{WD} \end{bmatrix} = \begin{bmatrix} 0 \\ 0 \\ \cdots \\ 0 \\ \cdots \\ 0 \\ 0 \\ 1 \end{bmatrix} \tag{11}$$

The solution of Equation (11) without considering the wellbore storage and the skin factor yields the FBHP as the dimensionless pressure $\overline{P}_{WD}$. The dimensionless pressure $\overline{P}_{CWD}$ is obtained considering the skin factor and wellbore storage for multistage fracturing of horizontal wells in low permeability reservoirs as:

$$\overline{P}_{CWD} = \frac{f(u)\overline{P}_{WD}(f(u)) + S}{u\{1 + C_D f(u)S\overline{P}_{WD}(f(u)) + S\}} \tag{12}$$

where C is the wellbore storage constant, Pa/m^3; $C_D = \frac{C}{2\pi\phi C_t hL^2}$ denotes the dimensionless wellbore storage constant; S is the total skin factor of the fracture system; u is the Laplace transform variable; and $f(u)$ is a function that characterizes the properties of the strata. For a homogeneous stratum, $f(u) = u$.

For production wells with a variable flowback volume, the dimensionless pressure can be expressed in the following convolution form:

$$P_D(x_D, y_D, z_D, t_D) = \int_0^{t_D} Q_D(t_D - \tau)\frac{\partial P_u(x_D, y_D, z_D, \tau)}{\partial \tau}d\tau \tag{13}$$

where $P_u(x_D, y_D, z_D, t)$ denotes the dimensionless pressure of multistage fractured horizontal wells per unit flowback volume; $Q_D(t_D) = \frac{q(t)}{q_I}$ denotes the dimensionless flowback volume; and q_I represents the unit flowback volume.

Using the properties of Laplace transform for convolutions yields the relationship between the pressure and flowback volume in Laplace space:

$$\overline{P}_D(x_D, y_D, z_D, u) = \overline{Q}_D(u)u\overline{P}_u(x_D, y_D, z_D, u) \tag{14}$$

Equation (14) can be used to obtain the flowback volume in Laplace space:

$$\overline{Q}_D(u) = \frac{\overline{P}_D(x_D, y_D, z_D, u)}{u\overline{P}_u(x_D, y_D, z_D, u)} \tag{15}$$

Equation (15) can also be expressed in terms of the FBHP as follows:

$$\overline{Q}_D(u) = \frac{\overline{P}_{wfD}(u)}{u\overline{P}_{wfu}(u)} \tag{16}$$

where $\overline{P}_{wfu}(u)$ is the $\overline{P}_{WD}$ obtained by solving Equation (11) for a unit flowback using G_{xD} and G_{yD}, which are both exponential functions of time. Therefore, the flowback can be simplified to $Q_l(t) = \alpha exp(-\beta t)$. Thus, the time-dependent flowback curve is defined by two characteristic parameters, α and β. Figure 2 shows the flowback volume versus time for horizontal segments of a 1000-m horizontal well with 20 stages, an effective fracture half-length of 40 m, and an effective fracture permeability of 0.25 mD. Fitting the curve yields the following equation for the flowback volume:

$$q(t) = 109.23exp(-0.007t) \tag{17}$$

where t denotes the flowback time, (d); and $q(t)$ is the daily flowback volume, (m^3/d).

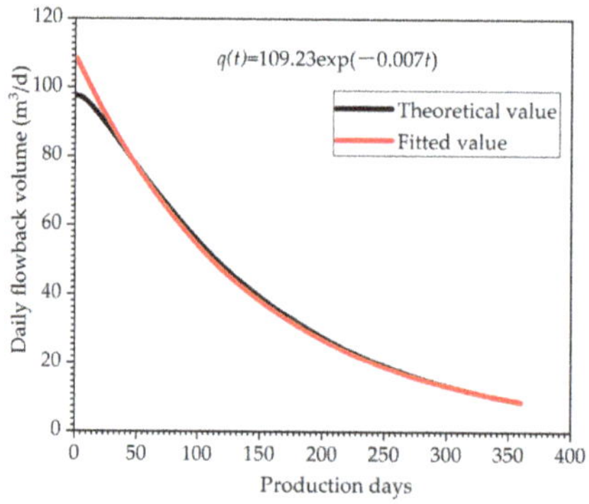

Figure 2. Theoretical flowback curve for multistage fracturing of horizontal wells.

Equation (9) accurately describes the characteristics of the flowback curve of horizontal wells, which indicates that α and β can be effectively used to parametrize the dynamic flowback curves. In addition, the flowback pattern is consistent with field measurements

(Figure 3). The field data in Figure 3 are from one of the 282 Wells in Table 1. The abscissa is production days, the ordinate is daily fracturing fluid flowback volume, and the blue dot represents the daily flowback volume of a specific day.

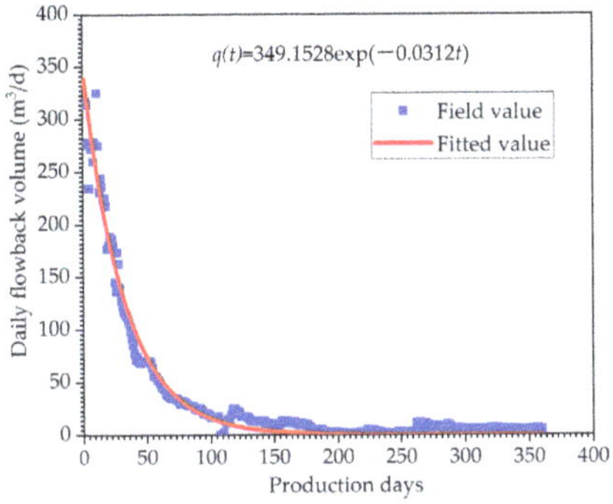

Figure 3. Field flowback curve for multistage fracturing of horizontal wells.

2.2. Fitting

The exponential expression for the flowback volume can be used to obtain the daily flowback rate R_l for a single well as follows:

$$R_l(t) = \frac{q}{Q_{total}} = \frac{\alpha}{Q_{total}} exp(-\beta t) = \gamma exp(-\beta t) \tag{18}$$

where q denotes the daily flowback volume, m^3/d; Q_{total} denotes the total quantity of injected fracturing fluid, m^3; α is a daily flowback volume coefficient, m^3/d; β is a decreasing flowback coefficient, 1/d, which is a fitting parameter; $\gamma = \alpha/Q_{total}l$ is the daily flowback rate coefficient, 1/d, which is also a fitting parameter; and t denotes the production time (d).

The curves for the daily flowback rate and daily flowback volume have the same shape but different magnitudes. For a total daily flowback volume that is recorded only once a day, the range of t in Equation (18) is $\{t \geq 1, t \in Z\}$. The cumulative flowback rate $\hat{S_N}$ for the previous N days can be then calculated as:

$$\hat{S_N} = \sum_{t=1}^{N} \gamma exp(-\beta t) \frac{\gamma}{e^{\beta} - 1}\left(1 - \frac{1}{e^{N\beta}}\right) \tag{19}$$

The cumulative flowback rates on the 30th, 90th, 180th and 360th day of production for a shale gas well are used as the characteristic flowback rates, and the estimated values of α and β are obtained by solving Equation (20):

$$\begin{cases} S_{30} = \hat{S_{30}} \\ S_{90} = \hat{S_{90}} \\ S_{180} = \hat{S_{180}} \\ S_{360} = \hat{S_{360}} \end{cases} \tag{20}$$

The values on the left and right sides of Equation (20) correspond to measured data and the calculation results from Equation (19), respectively. As there are fewer unknowns than equations in Equation (20), exact solutions for γ and β cannot be obtained. Therefore, γ and β are estimated by minimizing the sum of squares of the differences between the two sides of Equation (20):

$$(\hat{\gamma}, \hat{\beta}) = \underset{(\gamma,\beta)}{\operatorname{argmin}} \sum_{N} \left(\hat{S_N} - S_N\right)^2 \tag{21}$$

where the values of N are 30, 90, 180, and 360.

2.3. Big Data Analysis

A large quantity of data is generated during the exploration and development of shale gas, i.e., from drilling to fracturing and from well closing in to flowback. Various factors affect the flowback, which cannot be solved using currently available mathematical equations. Big data can be used to solve this problem. In this study, the Spearman correlation coefficient and distribution estimation were used to perform a big data analysis.

2.3.1. Correlation Analysis

The Spearman correlation coefficient has two advantages. First, sortable variables are assumed, and a normal sample distribution of the samples is not required, which is appropriate for the dataset used in this study. Second, the Spearman correlation corresponds to a monotonous, rather than linear, correlation between two random variables X and Y and is therefore likely to reveal a nonlinear relationship between the variables. The Spearman correlation coefficient is defined as follows [18]:

$$\rho_{XY} = 1 - \frac{6\sum_{i=1}^{N}[rank(x_i) - rank(y_i)]^2}{N(N^2-1)} \tag{22}$$

where N denotes the number of samples, and rank denotes the rank number of an observed value of X or Y.

The range of ρ_{XY} is $[-1, 1]$. A positive or negative ρ_{XY} indicates a positive or negative correlation between X and Y, respectively. The higher the absolute value of ρ_{XY} is, the stronger the correlation is. The significance of the Spearman correlation coefficient can be tested by the following hypotheses [20]:

$$H_0 : \rho_{XY} = 0 \tag{23}$$

$$H_1 : \rho_{XY} \neq 0 \tag{24}$$

If H_0 is satisfied, then:

$$\rho_{XY} \dot{\sim} \text{ Normal}\left(0, \frac{1}{n-1}\right) \tag{25}$$

Generally, if $p \leq 0.05$, H_0 is rejected, and ρ_{XY} is considered significant, whereas $p > 0.05$ indicates there is insufficient evidence to reject H_0, and ρ_{XY} may be false.

2.3.2. Estimation of Distribution of Fitted Parameters

In order to test whether the study sample is statistically significant, it is necessary to analyse the probability distribution of parameter fitting results. Consider a candidate distribution for the set of fitted parameters with a probability density function $f(x/p)$ and a log-likelihood function $l(p/x)$. The optimal parameter $\hat{p}$ of this candidate distribution can be obtained using the following equation:

$$\hat{p} = \underset{p}{\arg max} l(p|x) \tag{26}$$

Let $\hat{f}(x|\hat{p})$ denote the optimal estimate of the probability density function $f(x|p)$. The quality of the estimate can be tested using the following equation:

$$SSE = \sum_{i=1}^{N_H}\left(\hat{f}(x_i|\hat{p}) - h(x_i)\right)^2 \tag{27}$$

where SSE denotes the sum of the squares of the errors; $h(x)$ denotes the probability density obtained from a histogram; and N_H denotes the number of bins of the histogram.

The optimal estimate minimizes SSE and is denoted as $\hat{\mathcal{P}}$. The correctness of the estimation can be evaluated by the Kolmogorov–Smirnov test. For a dataset $\mathcal{D}$ with a sample size n and a reference distribution $\mathcal{P}$, the Kolmogorov–Smirnov test is [20]: $H_0 : \mathcal{D}$ and $\mathcal{P}$ have the same distribution and $H_1 : \mathcal{D}$ and $\mathcal{P}$ have different distributions.

The tested statistics are as follows:

$$D_n = \sup_x |F_n(x) - F(x)| \tag{28}$$

where $F_n(x)$ is an empirical distribution function for the data, and $F(x)$ is the cumulative distribution function of the reference distribution $\mathcal{P}$. Let the fitted parameters corresponding to the dataset $\mathcal{D}$ and $\hat{\mathcal{P}}$ denote the optimal reference distribution for the reference distribution $\mathcal{P}$. When $p \leq 0.05$ in Equation (28), H_0 is rejected, that is, the estimated results may not conform to the distribution of the fitted parameters. For $p > 0.05$, H_0 cannot be rejected, that is, the estimated results may conform to the distribution of the fitted parameters.

3. Results and Discussion

The Weiyuan shale gas block was considered as a case study. Flowback, production, fracturing, drilling, and geological data were collected for 214 horizontal wells as shown in Table 1. The flowback data of each well was fitted to yield the following daily flowback rate:

$$R_l(t) = 0.0121exp(-0.0293t) \tag{29}$$

where 0.0121 and 0.0293 are the mean values of $\hat{\gamma}$ and $\hat{\beta}$. The mean R^2 for all the fitting results is 0.9212, and the mean value of the mean absolute error is 0.0157, indicating a good fit. The distributions of the fitted parameters, $\hat{\gamma}$ and $\hat{\beta}$, are asymmetrical. Therefore, it is inappropriate to use normal distributions for $\hat{\gamma}$ and $\hat{\beta}$. The theoretical domain of $\hat{\gamma}$ and $\hat{\beta}$ is (0,+∞). To obtain the distributions of $\hat{\gamma}$ and $\hat{\beta}$, seven candidate distributions with support sets of (0,+∞) or (0,+∞) were selected, and the optimal parameters of each distribution were calculated by considering scaling but not translation, as shown in Table 2.

Table 2. Estimated probability densities for $\hat{\gamma}$ and $\hat{\beta}$.

Distribution		SEE, $\hat{\gamma}$	SEE, $\hat{\beta}$
Contrast	Normal	200,347.01	86,414.97
Candidate	Chi-square	109,125.65	52,896.92
	Exponential	147,857.30	73,137.37
	F	96,919.53	52,771.35
	Gamma	109,125.86	52,896.91
	Rayleigh	126,688.09	54,432.71
	Rice	126,703.31	55,320.32
	Inverse Gaussian	127,636.96	64,920.44

Table 2 shows that the F distribution is the optimal distribution for $\hat{\gamma}$ and $\hat{\beta}$, and the estimated results and optimal parameters are shown in Figure 4.

The Kolmogorov–Smirnov test was conducted on the empirical distributions of $\hat{\gamma}$ and $\hat{\beta}$, and the optimal distributions had p values of 0.8209 and 0.4793, respectively, which were both above 0.05. Therefore, the distributions in Figure 3 can be used as empirical distributions of $\hat{\gamma}$ and $\hat{\beta}$. As can be seen from Figure 3, the fitting results of 214 wells are statistically significant and can be further used for big data analysis. For the Weiyuan shale gas block, this empirical distribution can be used to predict the most likely distribution range of γ and β of wells that are about to be put into production and further predict the flowback curve.

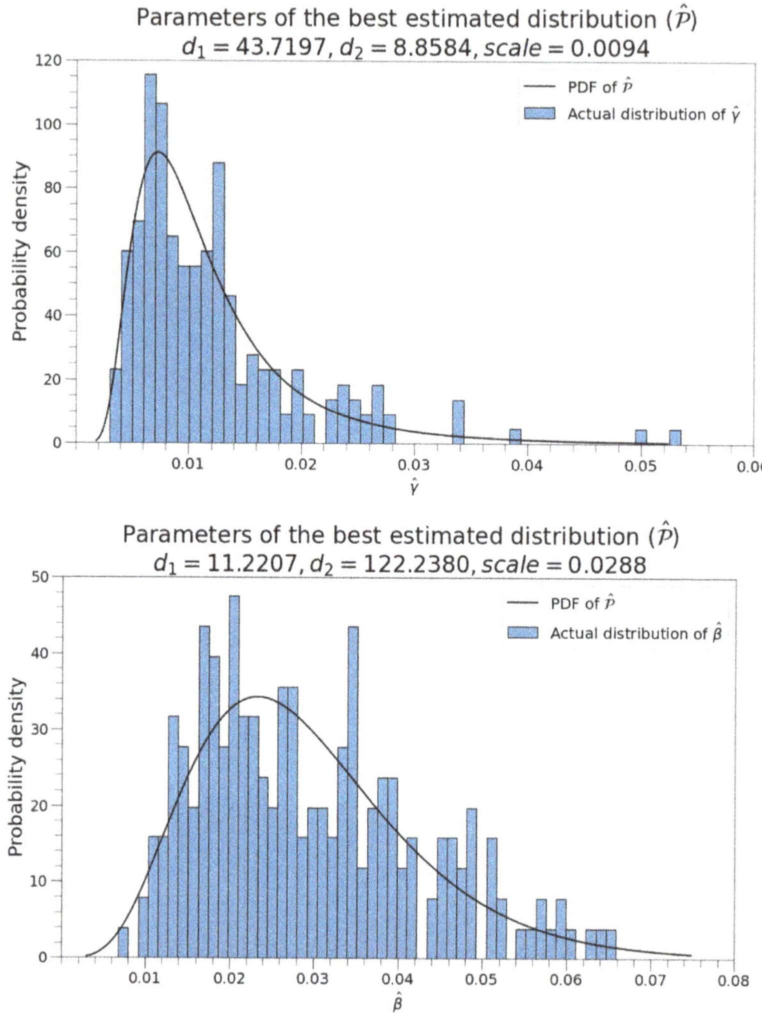

Figure 4. Best estimated distribution of $\hat{\gamma}$ and $\hat{\beta}$.

A physical interpretation of the results is that $\gamma e^{(-\beta)}$ corresponds to the flowback rate on the first day, which is also the peak daily flowback rate, and β corresponds to the speed at which the theoretical daily flowback decreases. Thus, the larger γ is, the higher the starting point of the flowback curve is; the larger β is, the lower the starting point of the flowback curve is, and the more rapidly the flowback decreases. Figure 5 shows the relationship between $\hat{\gamma}$ and $\hat{\beta}$.

Figure 5 shows a strong and significant correlation between $\hat{\gamma}$ and $\hat{\beta}$. Both parameters are related to the shape of the theoretical flowback curve and are therefore very likely to depend on each other. As the curve shape is jointly determined by two factors, the height of the starting point and the speed of decrease, neither factor can fully reflect the flowback volume. Hence, a variable composed of γ and β was used in this study to directly describe the flowback volume or flowback rate and is used in conjunction with static variables to perform a binary analysis. The fitting results are as follows: the minimum and maximum values of $\hat{\beta}$ are 0.0068 and 0.0658, respectively, and the 0.01th and 99.99th percentiles of the estimated distribution are 0.0052 and 0.0686, respectively.

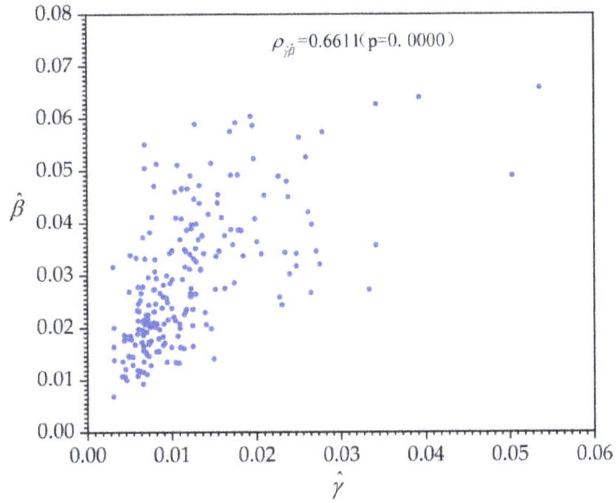

Figure 5. Correlation between $\hat{\gamma}$ and $\hat{\beta}$.

Figure 6 shows that the higher the value of β is for the theoretical curve, the more rapidly $f(N;\beta) = 1 - \frac{1}{e^{N\beta}}$ converges to 1. For β = 0.0052, $f(N;\beta)$ reaches 0.9945 at N = 1000, and for β = 0.0686, $f(N;\beta)$ reaches 0.9990 at N = 100, indicating that there is a minimal fracturing fluid volume after 1000 days of flowback for most shale gas wells. In Equation (19), $\gamma/(e^{\beta} - 1)$ corresponds to the approximate long-term cumulative flowback rate; thus, a flowback characteristic coefficient is defined, $\eta = \gamma/(e^{\beta} - 1)$, that reflects the flowback potential of a horizontal well.

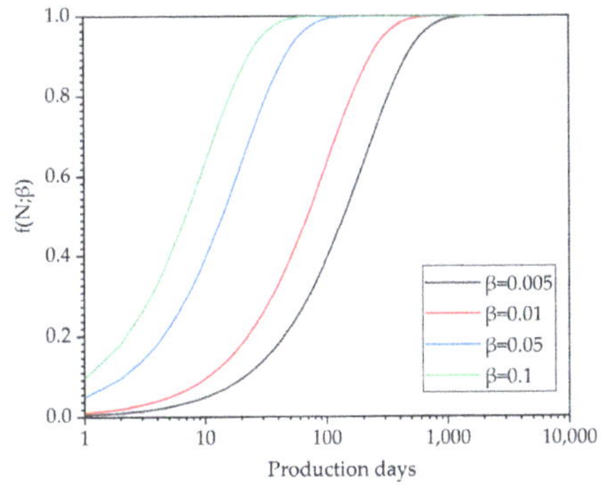

Figure 6. Sensitivity analysis of $f(N;\beta)$.

The flowback curve can be simplified by flowback characteristic coefficient. With η as the analysis object, the Spearman coefficient can be used to analyse the correlation between flowback rates and geological and engineering parameters and find the main affecting factors.

The correlation between η and the 14 static parameters in Table 1 was analysed using the Spearman correlation coefficient method. For comparison, the correlation between the first-year average daily production rate and 14 static parameters was also studied. Table 3 shows the correlations between these two variables and the static parameters, where the p value is given in parentheses. Static parameters with nonsignificant correlations with the two variables were excluded.

Table 3. Correlations between η, first-year average daily production rate, and static parameters.

Variable	η	First-Year Average Daily Production Rate
Fracturing lateral length	−0.1767 (0.0096)	0.4435 (0.000)
Number of fracturing stages	−0.2102 (0.0020)	0.4068 (0.000)
Fracturing stage intervals	0.1991 (0.0034)	−0.1036 (0.1308)
Gas saturation	−0.1061 (0.1216)	0.2536 (0.0002)
brittle mineral content	−0.1746 (0.0105)	0.3302 (0.0000)
Pressure gradient × vertical depth	0.1923 (0.0047)	−0.0335 (0.6265)
Thickness × drilling length of high-quality reservoir	−0.2928 (0.0002)	0.6400 (0.000)

Table 2 shows that the first-year average daily production rate is highly correlated with the thickness × reservoir drilling length of high-quality reservoir, and the correlation coefficient exceeds 0.5, indicating that the larger the thickness × drilling length of the high-quality reservoir, the higher the first-year average daily production rate, which is consistent with the results of Ma et al. [21]. The correlation between the static parameters and the first-year daily production rate can be sorted by the correlation coefficient as follows: thickness × drilling length of high-quality reservoir > fracturing lateral length > number of fracturing stages > brittle mineral content > gas saturation > fracturing stage interval > pressure coefficient × vertical depth. The drilling length of the high-quality reservoir, the fracturing lateral length, the number of fracturing stages, and the fracturing stage interval are engineering parameters. The better the drilling effect, the better fracturing effect and the higher the first-year average daily production rate. High-quality reservoir thickness, brittle mineral content, gas saturation and pressure coefficient × vertical depth are geological parameters. The better physical properties of shale gas reservoir, the higher first-year average daily production rate. This is why there is a good correlation between the first-year average daily production rate and various parameters.

Although the correlation coefficients between η and the static parameters are less than 0.5, relatively speaking, η and the thickness × drilling length of high-quality reservoir are best correlated, and the correlation symbol is opposite to that of the first-year daily production rate, which means that the higher the first-year average daily production rate, the lower the flowback rate. The correlation between static parameters and η is sorted by coefficient, which is basically consistent with that of first-year average production rate, but the positions of individual parameters are interchanged. The order according to the absolute value of correlation is thickness × drilling length of high-quality reservoir > number of fracturing stages > pressure coefficient× vertical depth > fracturing stage interval > fracturing lateral length > brittle mineral content > gas saturation. η is positively correlated with the fracturing stage interval, which means that the larger the fracturing stage interval, the worse the fracturing effect. The hydraulic fracturing process does not create complex fracture networks, and the fracturing fluids mainly concentrate near the main fracture and wellbore, resulting in high flowback rate. η is positively correlated with the pressure coefficient × vertical depth. The pressure coefficient × vertical depth represents the formation pressure. The higher the formation pressure, the stronger the liquid carrying capacity of the shale gas well and the higher the flowback rate. η is negatively correlated with the fracturing lateral length, the number of fracturing stages, and the content of brittle minerals. The higher the brittle minerals, the easier the reservoir is to be fractured; the longer the fracturing lateral length is and the more the number of fracturing stages is, the more complex the fracture network will be formed after hydraulic fracturing and the larger the SRV will be. The fracturing fluids are bound in complex micro fractures and cannot be discharged, resulting in a low flowback rate.

Table 4 shows the correlations between η and the whole static parameters in Table 1. Figure 6 shows the hot map of correlations between η and static parameters. From Table 1 and Figure 7, it can be seen that η is well correlated with the 30 day, 90 day, 180 day, 360 day, and peak gas production flowback rates. The longer the flowback time, the better the correlation between η and the flowback rate. The correlation between η and the 30 day, 90, 180, and 360 day flowback rate confirmed that it is reasonable and scientific to use η to characterize the flowback rate. η is positively correlated with vertical depth, TOC content, porosity, pressure coefficient, and average fracturing intervals, which means the higher these parameters are, the higher the flowback rates are. η is negatively correlated with high-quality reservoir thickness, gas saturation, brittle mineral content, fracturing fluid intensity, proppant intensity, average hydraulic fracturing fluid displacement, and drilling length in high-quality reservoirs, which means the higher these parameters are, the lower flowback rates are. η is negatively correlated with the first-year average daily production rate and EUR, but the correlation coefficients are low. It means that flowback rate is not merely correlated with gas production.

Table 4. Correlations between η and static parameters.

Variable	η
30 day flowback rate	0.7871 (0.0000)
90 day flowback rate	0.9294 (0.0000)
180 day flowback rate	0.9759 (0.0000)
360 day flowback rate	0.9950 (0.0000)
Peak gas production flowback rate	0.7352 (0.0000)
Fracturing lateral length (m)	−0.1767 (0.0096)
Number of fracturing stages	−0.2102 (0.0020)
First-year average daily production rate ($10^4 m^3/d$)	−0.0544 (0.4282)
Vertical depth (m)	0.1954 (0.0041)
TOC content (%)	0.1154 (0.0922)
Porosity (%)	0.0613 (0.3722)
High-quality reservoir thickness (m)	−0.1478 (0.0037)
Gas saturation (%)	−0.1061 (0.1216)
Pressure coefficient	0.1849 (0.0067)
Brittle mineral content (%)	−0.1746 (0.0105)
Average fracturing interval (m)	0.1991 (0.0034)
Fracturing fluid intensity (m3/m)	−0.0914 (0.1827)
Proppant intensity (t/m)	−0.1021 (0.1366)
Average hydraulic fracturing fluid displacement (m^3/min)	−0.1020 (0.1370)
Drilling length in high-quality reservoir (m)	−0.2640 (0.0009)
EUR ($10^8 m^3$)	−0.0918 (0.1811)

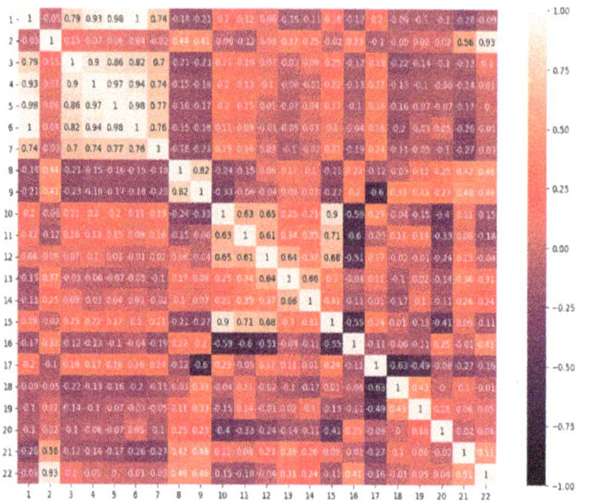

Figure 7. Hot map of correlations between η and static parameters.

From the correlation between various static parameters and the first-year average daily production rate and η, it is reasonable to simplify and characterize the flowback characteristics of shale gas wells by the flowback characteristic coefficient η. In our previous study [17], based on the actual data in the Weiyuan shale gas block, combined with a deep learning algorithm, two different depth feedforward neural networks were designed, which are the single output neural network and the multi-output neural network, to predict the shale gas well flowback rate. The 30 day, 90, 180, 360 day, and peak gas production flowback rates of shale gas wells were predicted. This method can predict the flowback rate of a specific day, but the main factors affecting the flowback rate at any time cannot be obtained. Combining the deep learning algorithm and the backflow characteristic coefficients proposed in this article, the flowback rate at any time in the future can be predicted. The method proposed in this paper also has some limitations. For gas wells with intermittent shut-in or long-term shut-in or for long-term flooded gas wells, there will be large errors in the prediction of the flowback rate. In addition, the research does not consider the impact of gas–liquid two-phase flow on flowback.

4. Summary and Conclusions

Based on the seepage theory of fracturing fluid in multi-stage fractured horizontal wells, the study uses convolution and Laplace transform methods to abstract the flowback curve into two characteristic parameters, the daily flowback rate coefficient γ and the flowback decline coefficient β. Taking the Weiyuan shale gas block as a study case, the flowback data of 214 wells were fitted, and the distribution characteristics of the fitted characteristic parameters were studied. The flowback characteristic coefficients to characterize the flowback potential of shale gas wells were established. The Spearman correlation coefficient method was used to study the correlation between the geological and engineering static parameters of 214 wells, the characteristic flowback coefficients, and the first-year average daily production rate. There are several conclusions obtained from this research:

(1) The fitting results of the flowback curve for 214 production wells show that the average daily flowback rate coefficient and the flowback decline coefficient of all wells are 0.0121 and 0.0293, and the average value of R^2 of all fitting results is 0.9212, the mean value of the mean absolute error is 0.0157, and the fitting effect is better. Both the daily flowback rate coefficient γ and the flowback decline coefficient β are right-skewed distributions.

(2) The comparative study on the correlation between the flowback characteristic coefficient and the first-year average daily production rate and static parameters shows that the thickness $\times$ the drilling length of a high-quality reservoir is best correlated with the characteristic flowback coefficient and the first-year average daily production rate, and their correlation symbol is opposite, which means that the larger the thickness $\times$ the drilling

length of the high-quality reservoir, the higher the average first-year daily production rate and the lower the flowback rate of the shale gas well. The factors affecting the flowback rate mainly include geological factors and engineering factors. The order of correlation coefficients is as follows: thickness × drilling length of high-quality reservoir, number of fracturing stages, pressure coefficient × vertical depth, fracturing stage interval, fracturing lateral length, brittle mineral content, and gas saturation.

(3) Through the method established in this paper, the shale gas flowback curve containing complex flow mechanism can be abstracted into simple characteristic parameters and characteristic coefficients. The method proposed in this paper can provide a novel way for machine learning and other big data analysis methods to study the flowback characteristics of shale gas horizontal wells. In future studies, the time-dependent gas output will be related to the flowback rate. Combining with machine learning, the flowback rate and gas output at any time can be predicted in the future.

Author Contributions: Conceptualization, W.G.; methodology, X.Z.; formal analysis, L.K.; investigation, Y.L.; resources, J.G.; data curation, W.G.; writing—original draft preparation, W.G.; writing—review and editing, X.Z. All authors have read and agreed to the published version of the manuscript.

Funding: This research was financially supported by the R&D Department of Petrochina (No. 2021DJ2005).

Institutional Review Board Statement: Not applicable.

Informed Consent Statement: Not applicable.

Data Availability Statement: Not applicable.

Conflicts of Interest: The authors declare no conflict of interest.

Nomenclature

K	permeability, m^2
Φ	porosity, dimensionless
t	time, day
μ	viscosity, pa.s
C_t	comprehensive compressibility coefficient, dimensionless
x_{fi}	half-length of the ith fracture, m
P	pressure, pa
P_D	dimensionless pressure, dimensionless
t_D	dimensionless time, dimensionless
x_D	dimensionless distance in x direction, dimensionless
y_D	dimensionless distance in y direction, dimensionless
C	wellbore storage constant, Pa/m^3
C_D	dimensionless wellbore storage constant, dimensionless
S	total skin factor of the fracture system, dimensionless
u	Laplace transform variable
α	daily flowback volume coefficient, m^3/d
β	decreasing flowback coefficient, 1/d
q	daily flowback volume, m^3/d
Q_{total}	total quantity of injected fracturing fluids, m^3
γ	daily flowback rate coefficient, 1/d
ρ	Spearman correlation coefficient, dimensionless
R_l	daily flowback rate, dimensionless
η	flowback characteristic coefficient, dimensionless

References

1. Williams-Kovacs, J.D.; Clarkson, C.R. A modified approach for modeling two-phase flowback from multi-fractured horizontal shale gas wells. *J. Nat. Gas Sci. Eng.* **2016**, *30*, 127–147. [CrossRef]
2. Liu, J.; Li, P.; Sun, Z.; Lu, Z.; Du, Z.; Liang, H.; Lu, D. A new method for analysis of dual pore size distributions in shale using nitrogen adsorption measurements. *Fuel* **2017**, *210*, 446–454. [CrossRef]
3. Rongze, Y.; Xiaowei, Z.; Yanan, B.; Yang, L.; Mingxiang, H. Flow mechanism of shale gas reservoirs and influential factors of their productivity. *Nat. Gas. Ind.* **2012**, *32*, 10–15.
4. Zoback, M.D.; Kohli, A.; Das, I.; McClure, M. The importance of slow slip on faults during hydraulic fracturing stimulation of shale gas reservoirs. In Proceedings of the SPE Americas Unconventional Resources Conference, Pittsburgh, PA, USA, 5–7 June 2012.
5. Du, W.Y.; Fu, J.; Sun, Z.; Qiao, N. Shale gas seepage mechanism and transient pressure analysis. *Chin. J. Comput. Phys.* **2015**, *32*, 51–57.
6. Jia, B.; Chen, Z.; Xian, C. Investigations of CO_2 storage capacity and flow behavior in shale formation. *J. Pet. Sci. Eng.* **2022**, *208*, 109659. [CrossRef]
7. Dai, J.; Zou, C.; Liao, S.; Dong, D.; Ni, Y.; Huang, J.; Wu, W.; Gong, D.; Huang, S.; Hu, G. Geochemistry of the extremely high thermal maturity Longmaxi shale gas, southern Sichuan Basin. *Org. Geochem.* **2014**, *74*, 3–12. [CrossRef]
8. Kohlia, A.; Arora, P. Application of artificial neural net-works for well logs. In Proceedings of the International Petroleum Technology Conference, Doha, Qatar, 19 January 2014.
9. Hu, J.; Lihui, D. Oil reservoir water flooding flowing area identification based on the method of streamline clustering artificial intelligence. *Pet. Explor. Dev.* **2018**, *45*, 312–319.
10. Adibifard, M.; Tabatabaei-Nejad, S.; Khodapanah, E. Artificial Neural Network (ANN) to estimate reservoir parameters in Naturally Fractured Reservoirs using well test data. *J. Pet. Sci. Eng.* **2014**, *122*, 585–594. [CrossRef]
11. Ghaffarian, N.; Eslamloueyan, R.; Vaferi, B. Model identification for gas condensate reservoirs by using ANN method based on well test data. *J. Pet. Sci. Eng.* **2014**, *123*, 20–29. [CrossRef]
12. Tian, C.; Horne, R.N. Recurrent neural networks for permanent downhole gauge data analysis. In Proceedings of the SPE Annual Technical Conference and Exhibition, San Antonio, TX, USA, 9–11 October 2017.
13. Thanh, H.V.; Lee, K.K. Application of machine learning to predict CO_2 trapping performance in deep saline aquifers. *Energy* **2021**, *239*, 122457. [CrossRef]
14. Bai, B.; Goodwin, S.; Carlson, K. Modeling of frac flowback and produced water volume from Wattenberg oil and gas field. *J. Pet. Sci. Eng.* **2013**, *108*, 383–392. [CrossRef]
15. Zhou, Q.; Dilmore, R.; Kleit, A.; Wang, J.Y. Evaluating fracture-fluid flowback in Marcellus using data-mining technologies. *SPE Prod. Oper.* **2016**, *31*, 133–146. [CrossRef]
16. Lin, B.; Guo, J.; Liu, X.; Xiang, J.; Zhong, H. Prediction of flowback ratio and production in Sichuan shale gas reservoirs and their relationships with stimulated reservoir volume. *J. Pet. Sci. Eng.* **2020**, *184*, 106529. [CrossRef]
17. Yuyang, L.; Xinhua, M.; Xiaowei, Z.; Wei, G.; Lixia, K.; Rongze, Y.; Yuping, S. Shale gas well flowback rate prediction for Weiyuan field based on a deep learning algorithm. *J. Pet. Sci. Eng.* **2021**, *203*, 108637. [CrossRef]
18. Li, Q.; Li, P.; Pang, W.; Li, D.; Liang, H.; Lu, D. A new method for production data analysis in shale gas reservoirs. *J. Nat. Gas Sci. Eng.* **2018**, *56*, 368–383. [CrossRef]
19. Lu, D.T. *Modern Well Testing Theory and Application*; Petroleum Industry Press: Beijing, China, 2009.
20. Zwillinger, D.; Kokoska, S. *Standard Probability and Statistics Tables and Formulae*; Chapman & Hall/CRC: Boca Raton, FL, USA, 2000; pp. 346, 366–367.
21. Xinhua, M.A.; Xizhe, L.I.; Liang, F.; Yujin, W.A.N.; Qiang, S.H.I.; Yonghui, W.A.N.G.; Zhang, X.; Mingguang, C.H.E.; Wei, G.U.O. Dominating factors on well productivity and development strategies optimization in Weiyuan shale gas play, Sichuan Basin, SW China. *Pet. Explor. Dev.* **2020**, *47*, 594–602.

energies

MDPI

Article

Performance Evaluation of Multistage Fractured Horizontal Wells in Tight Gas Reservoirs at Block M, Ordos Basin

Li Wu [1], Jiqun Zhang [1], Deli Jia [1], Shuoliang Wang [2,*] and Yiqun Yan [1]

1 PetroChina Research Institute of Petroleum Exploration and Development, Beijing 100083, China; wuli47@petrochina.com.cn (L.W.); zhangjiq@petrochina.com.cn (J.Z.); jiadeli422@petrochina.com.cn (D.J.); yanyiqun@petrochina.com.cn (Y.Y.)
2 School of Energy, China University of Geosciences (Beijing), Beijing 100083, China
* Correspondence: wangshuoliang@cugb.edu.cn

Abstract: Block M of the Ordos Basin is a typical low-permeability tight sandstone gas accumulation. To develop these reservoirs, various horizontal well fracturing technologies, such as hydra-jet fracturing, open-hole packer multistage fracturing, and perf-and-plug multistage fracturing, have been implemented in practice, showing greatly varying performance. In this paper, six fracturing technologies adopted in Block M are reviewed in terms of principle, applicability, advantages, and disadvantages, and their field application effects are compared from the technical and economic perspectives. Furthermore, the main factors affecting the productivity of fractured horizontal wells are determined using the entropy method, the causes for the difference in application effects of the fracturing technologies are analyzed, and a comprehensive productivity impact index (CPII) in good correlation with the single-well production of fractured horizontal wells is constructed. This article provides a simple and applicable method for predicting the performance of multi-frac horizontal wells that takes multiple factors into account. The results can be used to select completion methods and optimize fracturing parameters in similar reservoirs.

Keywords: Ordos Basin; low-permeability tight sandstone gas reservoir; horizontal well; multistage fracturing; comprehensive productivity impact index

Citation: Wu, L.; Zhang, J.; Jia, D.; Wang, S.; Yan, Y. Performance Evaluation of Multistage Fractured Horizontal Wells in Tight Gas Reservoirs at Block M, Ordos Basin. *Energies* **2022**, *15*, 613. https://doi.org/10.3390/en15020613

Academic Editor: Junqian Li

Received: 4 December 2021
Accepted: 4 January 2022
Published: 16 January 2022

1. Introduction

Low-permeability tight sandstone gas reservoirs have high potential due to their wide distribution and large reserves. However, they usually have no or low natural flow rate that meets economic boundaries due to poor physical properties. Horizontal drilling and hydraulic fracturing are two technologies widely used to improve the ultimate recovery of these reservoirs by maximizing reservoir contact [1,2]. In the past few decades, with the widespread application of horizontal wells in unconventional reserves, especially shales and other tight rock formations, the level of multi-frac technology has also been continuously improved. A series of horizontal completion methods have been formed on the basis of hydra-jet, open-hole multistage system (OHMS), and plug and perf (P-n-P), including coiled tubing (CT) hydra-jet, fixed-string hydra-jet, open-hole (OH) packer and ball-activated sliding sleeve, OH packer and infinite sliding sleeve, cementing sliding sleeve, and quick-drill or dissolvable bridge plug [3,4]. Field tests verified that multistage fracturing in these wells has been proven to be a key technique for the efficient development of these resources [5–7]. However, what are the advantages and disadvantages of each process? Which method yields better gas production? What factors affect the performance of multistage fractured horizontal wells?

Multistage fracturing has been used to increase the production rate of shale gas since about 2000 in the United States. Scholars have worked extensively to compare different completion methods to determine which approach achieves a higher output, mainly from two aspects: comparisons derived from field studies and comparisons based on analytical models.

Many methods such as historical production data, sonic anisotropy and radioactive tracer logs, or resistivity and acoustic imaging can be used to estimate the fracturing effect in field research. For example, East et al. [8] presented a comparison of different completion technologies through microseismic fracture mapping. Evidence from mapping of two wells in the Barnett Shale illustrates that horizontal completions were favorably stimulated by a hydra-jet/water-frac procedure with light sand and large volume. There is no doubt that microseismic monitoring and logging could be effective means to assess the effectiveness of fracturing, but they are costly and require extra on-site work. In contrast, comparisons based on historical data are a much more efficient approach. Publications have compared multiple indicators including initial and long-term production rate, ultimate recovery, and ROI.

Some scholars focused on the application of a certain process in oil and gas reservoirs. Mcdaniel et al. [9] reviewed applications of hydra-jet perforating in the years 2003–2009, particularly their practice on horizontal and highly deviated wells, and the research showed that hydra-jet perforating is widely used in horizontal wells, and that this technique is extremely superior in hard and very hard formations if high nozzle pressures are attained. Xiude et al. [10] proposed a hydra-jet fracturing technology with bottom packer on coiled tubing, and they verified its successful application in low-permeability gas reservoirs in the Sichuan Basin, China. Xue [11] and Jiang et al. [12] studied and demonstrated the advantages of hydra-jet fracturing technology, and they described its good performance in low-permeability hydrocarbon fields in China. Li et al. [13] reviewed the experimental, field, and numerical simulation studies of hydro-frac of unconventional reservoirs. Qin et al. [14] adopted a new OH horizontal well sliding sleeve multistage fracturing tool according to the geological conditions, fracture treatment, well trajectory, and diameter changes of the open-hole horizontal wells with long horizontal sections in the Daniudi gas field and Honghe oilfield, which was then satisfactorily applied in more than 20 horizontal wells in the Ordos Basin. Parshin et al. [15] demonstrated the successful application of the coiled tubing multistage hydraulic fracturing technology in the AS3 reservoir of Vinogradova Oilfield, which significantly improved the production performance of wells.

On the other hand, some scholars were more interested in a comparison between completion methods and fracturing processes. The most common topic was a performance comparison between OHMS and P-n-P, the two most common and widely used technologies; the former is often regarded as a representative of an open-hole approach, and the latter often represents cemented completion. Samuelson et al. [2] compared the applications of the two methods in the Cleveland tight gas sand formation of the Texas panhandle founded on 3 months of cumulative gas production. Lohoefer et al. [16] conducted a long-term comparison of production (cumulative gas production in 3 years) between the two completions in Barnett shale of Newark Field. Edwards et al. [17] compared the average 1 year cumulative gas production of 30 wells; six of the wells were completed with OHMS and the others with P-n-P, in the center of the Granite Wash tight sand reservoir located in western Oklahoma and the northern Texas Panhandle. The results of these three studies were consistent, all showing that wells completed with OHMS performed better. Wilson et al. [18] compared the daily gas production rates of 15 wells of two tight gas districts in the Lower Montney formation situated in southeast of Dawson Creek, British Columbia. The observation showed that the average performance of OHMS in these two areas was better than P-n-P completions; however, when it came to a single well, the results were not in the same direction. Furthermore, Augustine, Theppornprapakorn, and Vasudevan compared the output of OHMS and P-n-P analytical methods. Augustine [19] analyzed the difference using a 2D reservoir model. The study concluded that the biggest difference in production would occur when the reservoir permeability is in the range of millidarcy to microdarcy, while the production penalty in the range of microdarcy and nanodarcy is negligible. This range of permeability means that cementing has a greater impact on the production of fractured horizontal wells in tight gas reservoirs than in shales. Theppornprapakorn [20] constructed a 3D gas reservoir model with a single transverse fracture based on the assumption of steady-state production and no formation damage. Results of the CFD simulation were consistent with Augustine's conclusion, i.e., OHMS outproduced P-n-P, but the difference between the two was much smaller than in previous studies. Vasudevan [21] assessed the gas yield of horizontal wells with both transverse and longitudinal fracture in a relatively high-permeability reservoir using CFD simulations. This study obtained the same results as earlier discussions.

Another hot topic is the comparison of various fracturing processes and their applications. Li et al. [22] compared the field plays of open-hole preset external string packer multistage fracturing, casing cementing multistage fracturing, and OH packer and CT with BOT packer multistage fracturing in the Daniudi gas field, showing that the last method was the best performer. Thomson et al. [23] analyzed four different horizontal completion systems for the Montney tight gas formation in NE British Columbia, namely, packer isolation and frac sleeves, P-n-P, hydrajet perforating on CT with sand plug isolation, and CT deployed bridge plugs and tubing conveyed perforating; they inferred that fracture stimulation is a powerful approach to raise the productivity of horizontal wells in this area. Gas rate per interval suggested that wells with different completion procedures yielded almost similar rates. Mcdaniel et al. [24] studied the cost, completion risk, and ROI of six hydraulic frac multistage isolation methods in horizontal wells, including hydra-jet, ball-activated sliding sleeve, and bridge plug, and they observed a strong relationship between higher initial rates and larger volumes of proppant being placed, longer laterals, and more stages in Haynesville shale completions. Yet, the reservoir quality varies from well to well in many low- to ultralow-permeability formations, making it very hard to evaluate if a certain completion process practically brings about more hydrocarbons. Stanojcic et al. [25] reviewed the advantages and disadvantages of the CT, jointed-pipe, sleeve, and perf-and-plug techniques, as well as their applications, suggesting that nearly 20 pinpoint fracturing technologies were available, and they recommended fracturing treatments depending on well types and formation conditions. Xing et al. [26] studied the cased-hole mechanical packer, OH hydra-jet, and OH packer completion technologies, and they developed the key technology of horizontal well multistage fracturing for the Daniudi gas field. Kennedy et al. [27] found little difference in the initial production rate, regardless of the completion process used, according to field studies carried out by some operators in tight sand and shale gas. Research conducted by Burton [28] confirmed three completion techniques as the most effective and efficient in these types of formations, namely, P-n-P, ball-activated completions, and coiled tubing-activated completions. Each completion method has its advantages and limitations, and there is no single solution for every application in unconventional reservoirs. Salah et al. [29] comparatively analyzed three fracturing technologies, i.e., cemented P-n-P, cemented shiftable sliding sleeves with dissolvable isolation drop ball, and abrasive jetting perforation and annular pumping (AJPAP) with sand plug diversion, in terms of principle, advantages and disadvantages, field application, and effectiveness, and they concluded that there is no fit-for-all fracturing treatment; instead, the fracturing process should be designed comprehensively from economic and technical aspects. Subsequently, Prudskiy et al. [30] demonstrated the application effect of horizontal well multistage hydraulic fracturing technology in the Sakhalin Shelf oilfield in contrast to traditional horizontal well completion methods, while Li et al. [31] compared the characteristics and application of P-n-P with multi-cluster perforation with early applied fracturing technology and OH hydra-jet multistage fracturing with fixed string in the Sulige gas field. Both studies coincidentally indicated that the latter techniques significantly improved the productivity of horizontal wells. Some other publications focused on factors influencing the output of multi-fractured horizontal wells. Since late 2008, multi-fractured horizontal wells have been distinctly successful in the Cardiun Formation, Western Canadian Sedimentary Basin. Omatsone et al., [32] studied the performance of 120 multistage fractured horizontals; the study was quite thorough, as it looked at detailed reservoir parameters, fracture parameters, and well performance. Results concluded that the early performance of the wells was related to net pay thickness and horizontal well length. Gilbert and Baree noted that a complicating factor for determining the drainage area of multi-fractured horizontals is the rate at which interference occurs between adjacent fracture stages. Their study suggested that more fracture stages do not always result in an improvement in ultimate recovery. There were over 2500 horizontal wells drilled in Bakken/Three Forks oil play, Montana, North Dakota, and Saskatchewan from 2000 to 2009. Rankin et al. [33] reviewed the development philosophy and described the evolution of the completion strategies of these oil plays, recommending that ultimate recovery increments can be ascribed to an improvement in the completion design because of the similar reservoir quality in the study area. Taylor et al. [34] suggested that initial production rates might be a quick and simple guide to relative long-term well performance, but making critical economic decisions developed on initial production rates alone can be misleading; it is, therefore, essential to take reservoir characterization and well completion design into account to optimize ultimate recovery. The study also suggested that the fracture size and stage space had an impact on the rates. Hamm [35] studied and compared the results of horizontal multistage development, in various Western Canada plays. Type curves indicated that initial rates did not necessarily mean more ultimate recoveries; production rate differences between Manitoba Bakken and North Dakota were thought to be as a result of differences in permeability and reservoir pressure, while the differences between newer and older wells were greatly affected by well interference. Al-Ghazal [36] analyzed the practical application data of the multistage fracturing of horizontal wells in a Saudi Arabia tight gas field and revealed that borehole trajectory, reservoir parameters, and completion design are important factors affecting single-well productivity. Alekseev [37] divided the factors affecting the productivity of horizontal wells into three categories, reservoir quality, drilling quality, and completion quality, pointing out that the fracturing technologies differ significantly in parameters and processes; thus, their stimulation effects vary greatly from well to well under the combined influence of reservoir and drilling factors. Chodzicki's [38] results from over 120 wells in the Spearfish low-perm sandstone/siltstone formation, Southwest Manitoba, showed great production results from the application of horizontal well and multistage fracturing techniques, but individual fractured horizontal wells also yield variable results, making it difficult to correlate any strong trends that related initial production rates and estimated ultimate reserves to fracture size, number of stages, or total proppant.

These studies remarkably promoted the innovation of completion methods and processes, as well as enabled the horizontal well fracturing to be more efficient, intelligent, and infinite. However, although many scholars have adopted different indicators to compare the effect of different fracturing process horizontal wells based on field production data, there is no agreement on either indicators or performance. Some researchers concluded that one completion practice is better than another, while others held that there is no notable disparity with different completion processes in the production period. The results of comparisons that utilized analytical methods appeared to reach a consensus, i.e., open-hole completions exhibit better performance than cemented completions. However, the conclusion was obtained by assuming equal fracture geometry and absences of natural fracture. This clearly shows that the model has many limitations when applied in practice.

At the same time, various studies on multi-stage fracturing horizontal wells indicated that some wells performed better for various reasons. Several authors tried to analyze the interactions between hydrocarbon output and various parameters, such as net pay thickness, horizontal well length, reservoir pressure permeability, and completion design, but no obvious relationships were revealed. Some suggested these were related to rock and reservoir properties, while others suggested they were related to drilling and completion quality. It appears that there was no one quantitative evaluation method that could be applied to predict the well performance because of the multiple effect of geological heterogeneities, reservoir quality variations, and engineering complexity.

Therefore, it is necessary to carefully review various horizontal well fracturing technologies, compare their application effects in the same block, and deeply analyze the reasons resulting in difference well performance to identify quantitative description methods for fractured horizontal well production.

This paper takes 24 fractured horizontal wells in the S interval, Block M, the Ordos Basin, as examples. Block M, which is adjacent to the Sulige, Zizhou, Changbei, and Daniudi gas fields, is a typical gas accumulation with low porosity, low permeability, and low abundance. The main reservoirs were developed by the horizontal well plus multistage fracturing in the whole horizontal section. This block is highly representative for reservoir conditions or development strategy. Because of the short time in commercial development, the multistage fracturing technology of horizontal wells is mainly selected with reference to the practices in similar adjacent gas fields. So far, six horizontal well fracturing techniques have been attempted, including hydra-jet multistage fracturing, OH packer fracturing, and perf-and-plug multistage fracturing. It is found that wells with different completion methods vary significantly in gas production. In this paper, the six technologies are reviewed in terms of principle, applicability, advantages, and disadvantages, and their application effects are compared. Furthermore, the causes of yield difference of these technologies are analyzed, the main factors affecting the average gas rate of fractured horizontal wells are quantitatively studied, and a comprehensive productivity impact index (CPII) in good correlation with the 1 year average gas rate is constructed. This article provides a simple and applicable method for predicting the performance of multi-frac horizontal wells that takes multiple factors into account. The results can be used to select completion methods and optimize fracturing parameters in similar reservoirs.

2. Well Completion Methods

2.1. Common Multistage Fracturing Technologies

2.1.1. CT with BOT Packer Hydra-Jet Multistage Fracturing

Hydra-jet multistage fracturing is a basic technology of horizontal well stimulation. Generally, there is no need to adopt a mechanical seal method; instead, one can rely on the highly concentrated and strong focus of dynamic fluid force [9]. It is a method that combines hydraulic sandblasting perforation, fracturing, and interval isolation in one technique according to the Bernoulli principle. In the fracturing process, small flow channels are formed in the reservoir by hydraulic sandblasting perforation [39–41], and then the pressure in the annulus is increased while the flow channels are pressurized by the injection of high-pressure water jets. Once the pressure in the flow channels exceeds the formation fracture pressure, the formation is fractured instantly to realize hydra-jet fracturing. In hydra-jet fracturing, continuous fluid injection into the annulus is needed to help realize the propagation of fractures, and its pressure control is the key to realize effective perforation and dynamic isolation [42], as shown in Figure 1.

This technology adopts hydra-jet directional perforation, which can accurately lower the jetting tool to the designed position and exactly create fractures, without using mechanical packing tools. It also has the advantages of lower operation risk, shorter working cycle, and lower cost. It is appropriate for a wide range of completion processes, including open-hole, casing perforation, and screen pipe. The wide-ranging and successful use of this procedure in the field reveals its popularity and successful application [11].

However, for formations with high in situ stress, high fracture pressure, and high rock confining pressure, this technology may cause problems such as insufficient sandblasting penetration depth, difficult fracture initiation, and poor isolation effect between intervals; hence, the application of this technology in deep formations is limited.

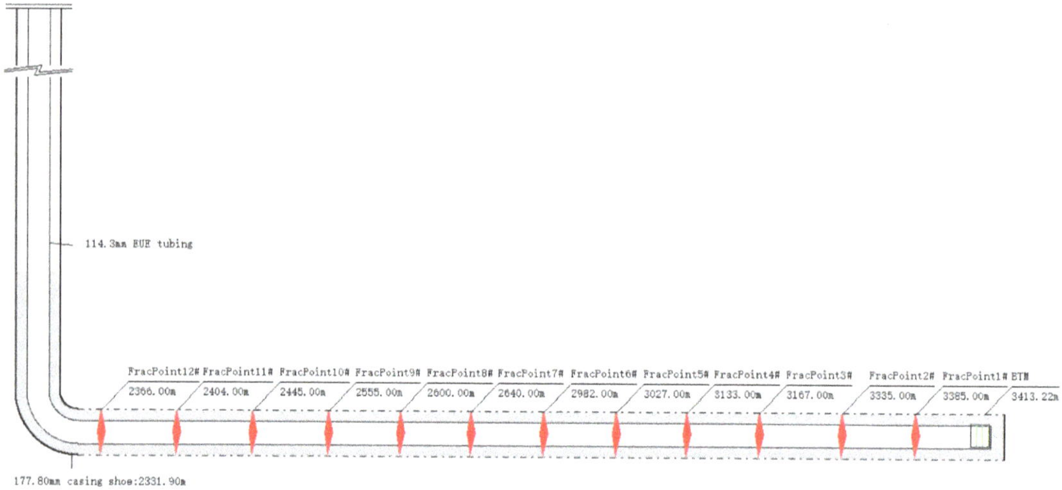

Figure 1. Schematic of coiled tubing with bottom packer hydra-jet multistage fracturing.

2.1.2. Fixed-String Hydra-Jet Multistage Fracturing

The fixed-string hydra-jet multistage fracturing technology introduces the ball-activated sliding sleeves used in packer multilayer fracturing into conventional hydra-jet fracturing treatment. It adopts multiple sets of jet gun groups, together with supporting sliding sleeves, and it can realize sandblasting perforation and fracturing through dropping balls, without moving the jet string. The sliding sleeve-type hydra-jet tool is run in advance, and the sliding sleeves at corresponding intervals are opened step by step during fracturing. Following the principle of the multistage opening of the hydra-jetting tool, it can be ensured that only the predetermined intervals are fractured. After the fracturing stimulation of the predetermined interval is completed, the plugging balls can be cast in hole to block the flow channels of the interval and activate the upper sliding sleeve to realize the stimulation of the upper interval. By repeating this process, the multistage fracturing stimulation of the whole horizontal well can be completed [43], as shown in Figure 2.

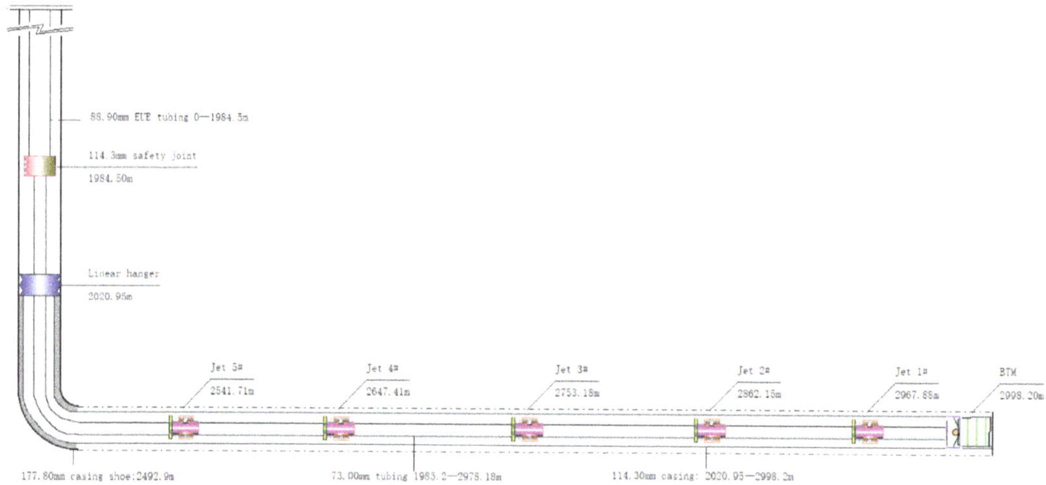

Figure 2. Schematic of fixed-string hydra-jet multistage fracturing.

This technology combines the advantages of hydra-jet technology and sliding sleeve multilayer fracturing, i.e., under the condition of not moving the pipe string, multistage fracturing can be continuously carried out in one trip without using packers, suggesting a high operation efficiency. In addition, by using the fracturing pipe string, the positive circulation and backwashing channels can be established. This technology has the advantages of simple process, low cost, fast production, and easy retrieval of the pipe string, and it subsequently allows repeated fracturing stimulation. It overcomes the disadvantages of conventional hydra-jetting operation, such as the need for snubbing units and moving the pipe string, long operation period, the reservoir damage caused by well killing, and the low displacement limited by coiled tubing. Therefore, this technology is suitable for stimulation in deep high-pressure reservoirs. However, its applicability is poor under the condition of complex borehole trajectory, and the number of fracturing stages needs to be improved [44].

2.1.3. OH Packer and Ball-Activated Sliding Sleeve Multistage Fracturing

Multistage fracturing with OH packer in horizontal wells is a technology to realize selective isolation and staged fracturing stimulation of horizontal wells in only one trip. Essentially, the open-hole packer is used to separate the horizontal section into multiple stages, and then balls with small size difference are dropped stage by stage to open the sliding sleeves at each stage to realize multistage fracturing [45–47], as shown in Figure 3.

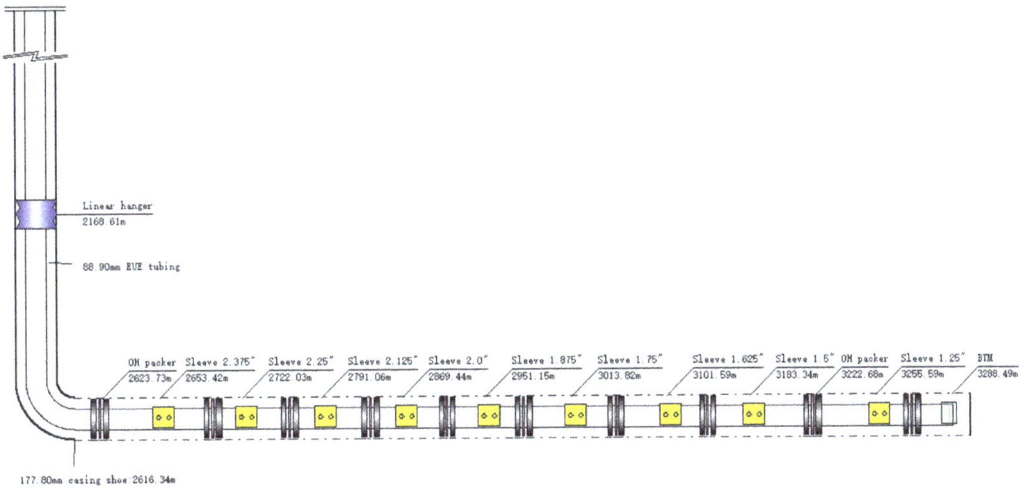

Figure 3. Schematic of open-hole packer multistage fracturing completion string.

Multistage fracturing with OH packer in horizontal wells can realize multistage fixed-point fracturing stimulation in one trip, saving significant completion time and money. It is advantageous for its fewer downhole tools, simple process, and high safety and efficiency, without the necessity of cementing and perforation in the horizontal section, which can reduce time and cost and increase the return on investment. However, it requires packers with high performance, and the completion string below the suspension packer cannot be retrieved; furthermore, due to the graduated ball seat sizes for each additional zone, each ball seat generates its own backpressure in the system, thereby limiting the number of fracturing stages [48]. During the process, the balls must be dropped in sequence from small to large and cannot be operated flexibly, making the operation difficult and risky for a completion string and ball seat of sliding sleeves that are small in inner diameter. Therefore, the technology is not suitable for post-fracturing production tests or other oil and gas wells requiring repeated fracturing [49].

2.1.4. OH Packer and Infinite Sliding Sleeve Multistage Fracturing

In infinite sliding sleeve multistage fracturing, stages are essentially isolated with OH packer or cement sheath, and then the infinite sliding sleeves installed on the casing or tubing string at pay zones are activated by the bottomhole assembly (BHA) conveyed by CT to establish the flow channel. The sliding sleeves, as a part of the casing, are lowered by the drilling rig and landed at the predetermined fracturing depth, and the position of the sliding sleeves corresponds to the position of the fracturing intervals.

OH packer and infinite sliding sleeve multistage fracturing is achieved by installing open-hole packers and infinite sliding sleeves on the tubing string, and the intervals are separated by activating and setting open-hole packers, as shown in Figure 4.

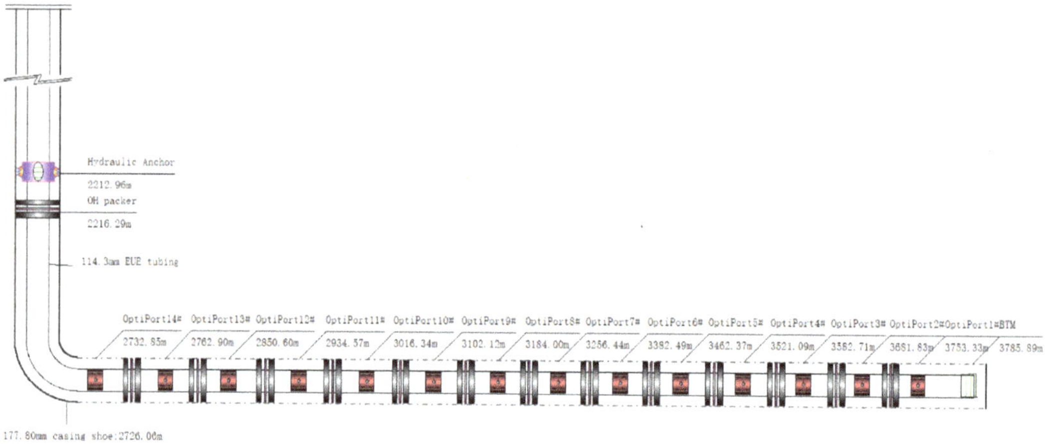

Figure 4. Schematic of open-hole packer and infinite sliding sleeve multistage fracturing completion string.

The number of stages in infinite sliding sleeve multistage fracturing is not limited by the process [48,50] by adopting same size balls and ball-seats for all zones. This technique can considerably lessen frictional forces and facilitates a more effective operation, in addition to realizing an inside diameter (ID) extremely close to the host tubular string; therefore, a much lower surface fracturing pressure can be used [48]. Due to the use of CT, it is convenient and easy to flow back after fracturing. Moreover, CT can be used to monitor the bottomhole pressure in real time, which is conducive to the timely detection of the risk of sand plugging. A sandblasting perforator is also integrated in the switch tool of the sliding sleeves, which can be used as a preventive measure; that is, once the formation cannot be fractured, the sandblasting perforation can be used as a remedial measure, and the displacement fluid at the interval is the preflush in the next interval, which reduces the use of fracturing fluid. However, this technology requires high isolation performance of the packers in the string, and it may suffer a risk that the production casing cannot be run to the design position, resulting in misalignment of the fracturing sliding sleeves with the corresponding fracturing intervals.

2.1.5. Infinite Cementing Sliding Sleeve Multistage Fracturing

Infinite cementing sliding sleeve multistage fracturing is another form of infinite sliding sleeve multi-frac technology, and its principle, process, and parameters are basically the same as OH packer and infinite sliding sleeve multistage fracturing [51,52]. For this technology, infinite sliding sleeves are installed and run with the casing string to complete cementing operation. It differs from the OH packer and infinite sliding sleeve multistage fracturing in that the intervals are isolated by consolidated cement before multistage fracturing is carried out, as shown in Figure 5.

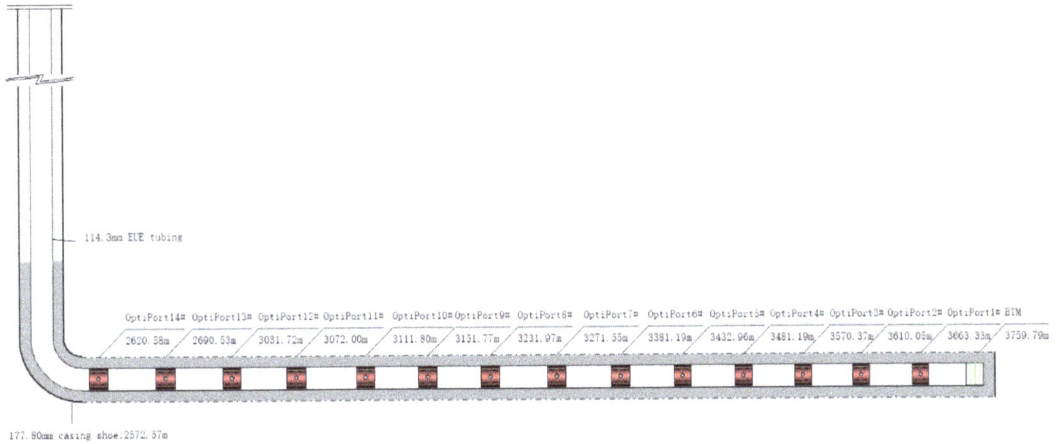

Figure 5. Schematic of infinite sliding sleeve multistage fracturing cementing.

2.1.6. Perf-and-Plug Multistage Fracturing

Perf-and-Plug is the most widely used strategy for multistage fracturing in unconventional reservoirs. It is regarded as a fully developed approach and is usually employed in cemented casing or liner completion horizontal wells [20]. CT conveyed perforation is performed, and then fracturing is realized by pumping fluid through the casing. After the fracturing operation in the interval, the bridge plug tool string with a perforation gun is pumped to the designated isolation position of the horizontal section by means of liquid injection, and the perf-and-plug operation is realized through cables. Then, the fracturing operation in the next interval can be initiated. The tool assembly is run in the wellbore stage by stage, and the fracturing operation is implemented accordingly. After the stimulation of the entire horizontal section is completed, the well is put into production from all pay zones together [53], as shown in Figure 6.

Due to the different isolation methods and pipe string structure, perf-and-plug multistage fracturing is advantageous to some extent. First, it is not necessary to run the tool in advance along with the casing, and bridge plugs are used to reliably isolate the intervals. Second, fixed-point fracturing is realized through perforation, with accurate fracture placement. Third, the bridge plugs are generally drillable or soluble, leaving a full borehole for subsequent operation and production. Compared with other technologies, this technology can realize multi-cluster fracturing in one stage, achieving large displacement, large-scale fracturing, and high operating efficiency [54–56]. Despite the simplicity of plug-and-perf completion operations, producers are faced with a few challenges, including unproductive clusters and poor perf cluster efficiency, excess fluid volumes and over-displacement, time-consuming operations, operational risk, and high costs [57].

2.2. *Comparison of Technologies*

All the abovementioned multistage fracturing technologies for horizontal wells can achieve the purpose of reservoir stimulation and production enhancement, but each technology has its applicability, advantages, and disadvantages. Table 1 compares these technologies, in which infinite cementing sliding sleeve multistage fracturing and OH packer and infinite sliding sleeve multistage fracturing are combined into infinite sliding sleeve multistage fracturing technology, since they are basically identical in principle, process, and parameters.

Table 1. Applicability, advantages, and disadvantages of horizontal well multistage fracturing technologies.

Technology	CT with BOT Packer Hydra-Jet	Fixed-String Hydra-Jet	OH Packer and Ball-Activated Sliding Sleeve	Infinite Sliding Sleeve (Infinite Cementing Sleeve and OH Packer and Infinite Sliding Sleeve)	Perf-and-Plug
Applicability	Various horizontal wells, such as open holes and cased holes.	Various horizontal wells, such as open holes and cased holes. The drift diameter of the fracturing string is small, which can effectively meet the production of gas wells with a certain quantity of water produced.	Open-hole horizontal wells with relatively regular boreholes. The drift diameter of the fracturing string is small, which can effectively meet the production of gas wells with a certain quantity of water produced.	Cemented or open-hole horizontal wells. After fracturing, full bore tubing or casing is used for production, which is more suitable for gas wells with high production.	Cased horizontal wells. After fracturing, full bore tubing or casing is used for production, which is more suitable for gas wells with high production.
Advantages	1. Isolation is realized automatically, without mechanical isolation tools, suggesting low tool risk. 2. It is not limited by completion technique. 3. Circulation can be established, and backwashing channels are available. 4. Directional perforation is realized by hydra-jetting, and fractures are created accurately.	1. Isolation is realized automatically, without mechanical isolation tools, suggesting low tool risk. 2. Multistage fracturing can be carried out in one trip, which can shorten the operation time and mitigate reservoir damage. 3. Circulation can be established, and backwashing channels are available. 4. The pipe string is easy to retrieve, and repeated fracturing can be carried out at a later stage. 5. The fracturing string has a small drift diameter and a strong liquid-carrying capacity after fracturing.	1. Intervals are isolated by open-hole packers, the operation is simple and convenient, and fracturing completion is done safely and efficiently in one trip. 2. No cementing or perforation is conducted in the horizontal section, thus saving time and cost. 3. Multistage fracturing completion is realized. 4. The fracturing string has a small drift diameter and a strong liquid-carrying capacity after fracturing, which can effectively meet the production of gas wells with a certain quantity of water produced.	1. Fracturing completion is done safely and efficiently in one trip. 2. The number of stages is not limited by the process, and multistage fracturing completion can be achieved satisfactorily. 3. No perforation is required, saving time and cost. 4. Sand plugging can be detected in advance, and large-scale fracturing can be realized. 5. Coiled tubing is used, making flowback easier. 6. The fracturing and production string is adopted with large diameter, which is convenient for subsequent treatment. 7. Any sand plugging can be removed immediately by circulation.	1. Intervals are isolated reliably. 2. The number of stages is not limited by the process, and multistage fracturing completion can be achieved satisfactorily. 3. Coiled tubing is used, making flowback easier. 4. The string has a large diameter, which is convenient for subsequent treatment. 5. Fixed-point fracture initiation is realized by perforation, and fracture placement and fracturing position are accurate. 6. Large displacement, large fluid volume, and multi-cluster perforating volume fracturing can be realized. 7. Any sand plugging can be removed immediately by circulation.

Table 1. *Cont.*

Technology	CT with BOT Packer Hydra-Jet	Fixed-String Hydra-Jet	OH Packer and Ball-Activated Sliding Sleeve	Infinite Sliding Sleeve (Infinite Cementing Sleeve and OH Packer and Infinite Sliding Sleeve)	Perf-and-Plug
Disadvantages	1. The string will be moved, operation time is long, and damage to the reservoir may be caused by well killing. 2. A wellhead with snubbing units is required, or the wellhead working pressure needs to be high. 3. Simple fractures are created, and the scale of fracturing is limited. 4. Application in deep formations is limited.	1. High requirements for borehole trajectory, etc. 2. Simple fractures are created, and the scale of fracturing is limited. It cannot create complex fractures.	1. High requirements for open-hole packer and well trajectory. 2. To prevent sand plugging, the fracturing scale is small, and there are few emergency plans for sand plugging. 3. The number of fracturing stages is limited by borehole size. 4. The displacement is limited by the opening differential pressure of the ball seat.	1. There is a risk that the production casing cannot be run to the design position, and the fracturing sliding sleeves cannot be aligned with the corresponding fracturing section. 2. Coiled tubing is used, and there are high requirements for the performance of the tool string and packers.	1. It is not suitable for open holes. 2. Multiple trips of coiled tubing lead to high operation cost. 3. Unproductive clusters and poor perf cluster efficiency. 4. Excess fluid volumes and over-displacement. 5. Time-consuming. 6. Operational risk.

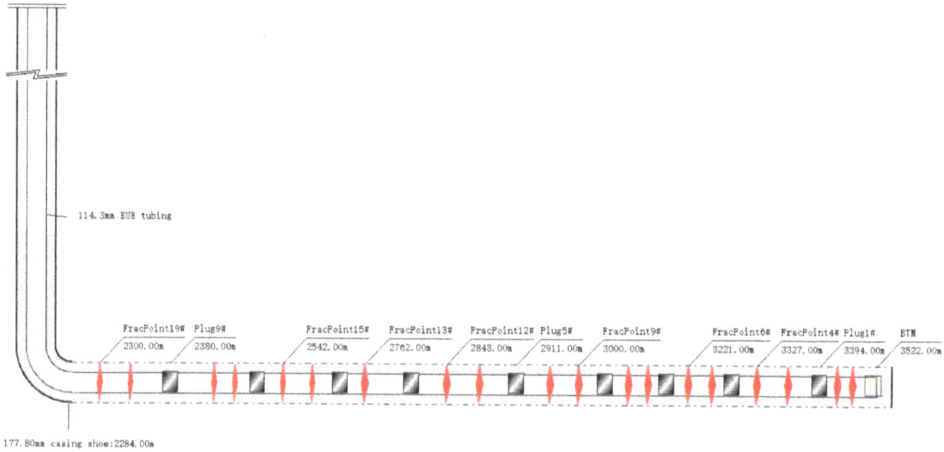

Figure 6. Schematic of perf-and-plug multistage fracturing completion string.

3. Application

3.1. Block Profile

Block M, located in the transitional zone between the Yi-Shaan slope and Western Shanxi flexural fold belt in the Ordos Basin, is a wide and gentle regional west-dipping monocline, with simple structure and undeveloped faults. It is a constant-volume elastic-drive tight gas accumulation controlled by lithology and physical properties under the local tectonic setting. The block covers an area of 1524.34 km^2, with proven reserves of 127.57 billion cubic meters. The main pay zones are S2, S1, and H8, with a superimposed gas-bearing area of 929 km^2 and an average reserve abundance of 137 million m^3/km^2. The horizontal wells in this block have no natural productivity and require fracturing to obtain commercial gas flow. The block is a typical low-permeability tight sandstone gas reservoir.

Block M was put into production in March 2014, predominantly by horizontal wells (contributing more than 90% of the total output). Trial production has been carried out in H8, S1, and BX, while horizontal wells are mainly used to develop S2. The S2 reservoir has a good continuity, connectivity, and stable distribution. The buried depth is 2330 m, the average formation pressure is 20.53 MPa, the reservoir temperature is 67.6 °C, the reservoir thickness is 5–25 m, and the average effective thickness is 9.9 m. The porosity is 3.0–12.4%, with an average of 6.8% and a median range of 4–10%. The permeability ranges from 0.1 to 1.2 $\times$ 10^{-3} μm^2, with an average of 0.47 $\times$ 10^{-3} μm^2. The clay is mainly composed of kaolinite/illite (K/I), with a high content of kaolinite. The microseismic results show that the fractures mainly strike between 75° and 90°. The interpretation results of the dipole array acoustic logging infer that the direction of the maximum principal stress of the formation is nearly NE–SW, which is basically consistent with the microseismic results. According to logging calculation, the Young's modulus is 2.0–2.8 $\times$ 10^4 MPa, the Poisson's ratio is 0.15–0.22, and the fracture pressure gradient and closure pressure gradient are 0.018–0.022 MPa/m.

Block M is adjacent to the Sulige and Zizhou gas fields in the Ordos Basin. It is highly representative for reservoir conditions or development mode. Table 2 compares the parameters between Block M and adjacent blocks.

Table 2. Comparison of parameters between Block M and adjacent blocks in Ordos Basin.

Item	Block M	Yulin Gas Field	Zizhou Gas Field	Sulige Gas Field	Shenmu Gas Field
Trap type	Lithologic/ stratigraphic	Lithologic/ stratigraphic	Lithologic/ stratigraphic	Lithologic/ stratigraphic	Lithologic/ stratigraphic
Structural location	Transition zone between the Western Shanxi flexural fold belt and the Yi-Shaan slope	Yi-Shaan slope	Yi-Shaan slope	Yi-Shaan slope	Yi-Shaan slope
Structural characteristics	West-dipping monocline	West-dipping monocline	West-dipping monocline	West-dipping monocline	West-dipping monocline
Stratum name	Shanxi Formation	Shanxi Formation	Shanxi Formation	Shanxi Formation, Shihezi Formation	Shanxi Formation, Taiyuan Formation
Sedimentary environment	Delta facies	Fluvial facies	Delta facies	Fluvial/delta facies	Fluvial/delta facies
Rock type	Quartz and lithic quartz sandstone	Quartz and lithic quartz sandstone	Quartz and lithic quartz sandstone	Quartz and lithic quartz sandstone	Quartz and lithic quartz sandstone
Effective thickness (m)	9.9	12.3	7.2	7.5	22.4
Porosity (%)	6.8	6.2	6.7	6.9	6.5
Permeability (10^{-3} μm^2)	0.47	4.8	1.1	0.52	0.51
Gas saturation (%)	68.5	78.1	70.1	53.2	54.5
Mid-reservoir depth (m)	2330	2950	2700	3000	2900
Formation static pressure (MPa)	20.53	27.5	23.7	30.1	22.1

3.2. Application Effect

Six multistage fracturing technologies were attempted, namely, CT with bottom packer hydra-jet, fixed-string hydra-jet, OH packer and ball-activated sliding sleeve, OH packer and infinite sliding sleeve, infinite cementing sliding sleeve, and perf-and-plug.

The fractured horizontal wells studied in this paper were all located in the enrichment zone of S2 in Block M, as shown in Figure 7, with similar physical properties and fluid properties. Table 3 shows the parameters of the 24 wells that have been producing for more than 1 year without engineering failures.

As seen from Table 3, the net fracturing fluid volume injected was 1896.0–8698.0 m^3, the displacement was 2.5–7.5 m^3/min, the number of fracturing stages was 5–16, and the total sand addition was 188.5–960.3 m^3, suggesting greatly variable ranges for different horizontal wells. It can be inferred that the parameters were similar for the same fracturing technology, but different among fracturing processes. As shown in Table 4, the scales of perf-and-plug multistage fracturing and infinite sliding sleeve multistage fracturing (infinite cementing sleeve and OH packer and infinite sliding sleeve) were large, while the displacement, total sand addition, and total fracturing fluid volume injected of fixed-string hydra-jet multistage fracturing were significantly lower than those of other technologies.

On the basis of the historical production data and the single-well completion fracturing cost, the production performances of fractured horizontal wells were preliminarily compared to explore whether fracturing technologies significantly enhanced the well production.

3.2.1. Production Comparison

The annual average daily gas production (annual cumulative gas production divided by production days) was used to measure the production of each fractured horizontal well, as shown in Figure 8.

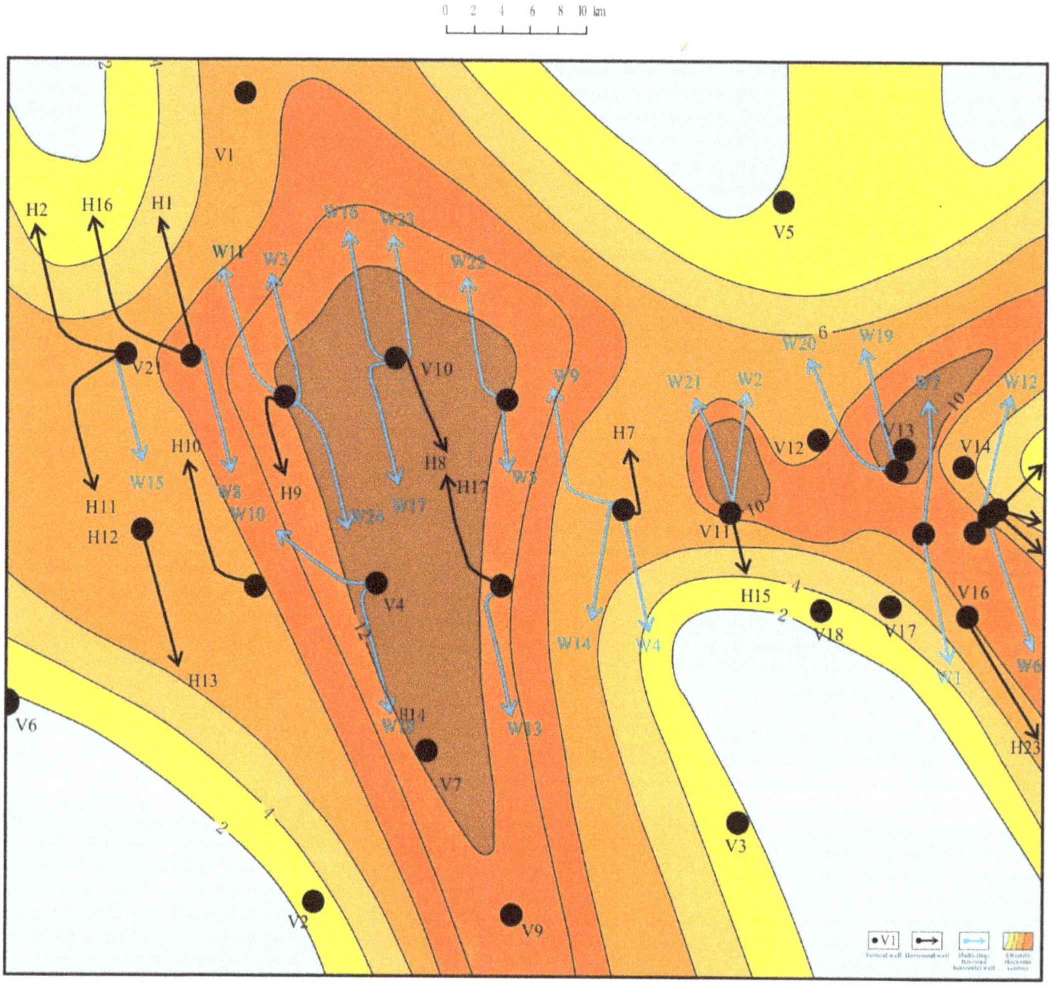

Figure 7. Effective thickness contour and well location map of the enrichment area, S2 formation, Block M.

Table 3. Parameters of multistage fractured horizontal wells.

No.	Well	Multistage Fracturing Technology	OD of Production Casing (mm)	Number of Fracturing Stages	Net Fracturing Fluid Volume Injected (m^3)	Displacement (m^3/min)	Total Sand Addition (m^3)	Average Sand Ratio (%)	Cumulative Liquid Production (m^3)	Flowback Rate (%)
1	W1	CT with BOT packer Hydra-jet	114.30	13	5839.5	3.5–4.2	619.8	16.6	1753.0	30.0
2	W2	CT with BOT packer Hydra-jet	114.30	11	6435.0	3.3–3.9	587.1	18.4	1818.0	30.0
3	W3	CT with BOT packer Hydra-jet	114.30	12	3768.4	2.5–3.9	508.5	21.1	887.5	23.0
4	W4	Fixed-string Hydra-jet	88.90	7	2932.6	2.4–2.6	360.6	16.3	1021.5	34.8
5	W5	Fixed-string Hydra-jet	88.90	5	1896.0	2.4–2.6	188.5	16.0	454.0	23.9
6	W6	OH packer and ball-activated sliding sleeve	88.90	6	3539.5	3.4–5.0	387.7	15.2	2020.0	56.0

Table 3. *Cont.*

No.	Well	Multistage Fracturing Technology	OD of Production Casing (mm)	Number of Fracturing Stages	Net Fracturing Fluid Volume Injected (m^3)	Displacement (m^3/min)	Total Sand Addition (m^3)	Average Sand Ratio (%)	Cumulative Liquid Production (m^3)	Flowback Rate (%)
7	W7	OH packer and ball-activated sliding sleeve	88.90	9	3810.7	3.5–5.5	478.6	16.5	2090.0	53.0
8	W8	OH packer and ball-activated sliding sleeve	88.90	11	4725.1	4.6–6.0	543.0	19.7	918.0	19.4
9	W9	OH packer and ball-activated sliding sleeve	88.90	10	3257.7	4.6–5.7	436.2	18.2	460.6	14.0
10	W10	OH packer and ball-activated sliding sleeve	88.90	9	3249.0	5.0–5.7	407.7	17.5	1401.3	43.0
11	W11	OH packer and ball-activated sliding sleeve	88.90	12	5129.5	5.0–5.8	603.3	18.6	1480.0	29.0
12	W12	OH packer and ball-activated sliding sleeve	88.90	8	4775.9	4.5–5.6	578.4	17.9	1127.0	23.6
13	W13	OH packer and ball-activated sliding sleeve	88.90	10	3872.4	4.5–5.5	324.4	15.4	310.0	8.0
14	W14	OH packer and infinite sliding sleeve	114.30	14	5997.1	3.4–3.9	713.6	20.5	2040.0	33.0
15	W15	OH packer and infinite sliding sleeve	114.30	13	5825.8	3.2–4.0	680.2	18.5	1661.0	28.0
16	W16	OH packer and infinite sliding sleeve	114.30	16	7261.8	3.0–3.9	960.3	22.1	995.0	13.0
17	W17	OH packer and infinite sliding sleeve	114.30	13	5872.6	2.7–3.8	830.0	21.4	1518.0	26.0
18	W18	OH packer and infinite sliding sleeve	114.30	12	5139.3	3.4–3.9	766.0	22.6	1130.0	22.0
19	W19	Infinite cementing sliding sleeve	114.30	15	5897.5	3.4–3.9	434.6	18.6	1264.0	20.0
20	W20	Infinite cementing sliding sleeve	114.30	15	5209.5	3.9–4.2	510.3	19.4	1127.0	21.0
21	W21	Infinite cementing sliding sleeve	114.30	16	5409.4	3.0–5.0	383.2	18.4	1818.0	33.0
22	W22	Perf-and-plug	114.30	10	7154.0	4.0–7.0	765.5	18.8	1458.0	21.0
23	W23	Perf-and-plug	114.30	10	6377.9	3.0–7.0	798.8	19.0	1732.0	27.0
24	W24	Perf-and-plug	114.30	12	8698.0	3.5–7.5	730.2	19.5	800.0	9.2

Table 4. Parameters of horizontal well multistage fracturing technologies.

Multistage Fracturing Technology	Number of Producing Wells	OD of Production Casing (mm)	Number of Fracturing Stages	Net Fracturing Fluid Volume Injected (m^3)	Displacement (m^3/min)	Total Sand Addition (m^3)	Average Sand Ratio (%)	Liquid Production (m^3)
CT with BOT packer hydra-jet	3	114.30	12	5347.6	3.6	571.8	18.7	1486.2
Fixed-string hydra-jet	2	88.90	6	2414.3	2.5	274.6	16.2	737.8
OH packer and ball-activated sliding sleeve	8	88.90	9	4045.0	5.0	469.9	17.4	1225.9
OH packer and infinite sliding sleeve	5	114.30	14	6019.3	3.5	790.0	21.0	1468.8
Infinite cementing sliding sleeve	3	114.30	15	5505.5	3.9	442.7	18.8	1403.0
Perf-and-plug	3	114.30	11	7410.0	5.3	764.8	19.1	1330.0

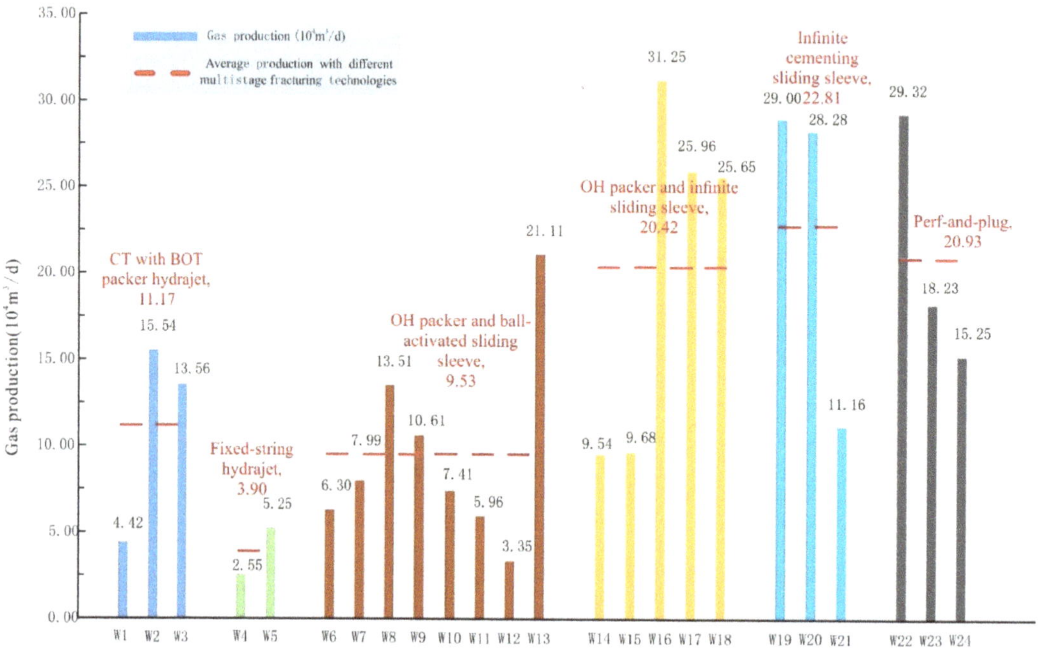

Figure 8. Production of horizontal wells with different multistage fracturing technologies.

It can be seen intuitively from Figure 8 that the annual average daily gas production was 2.55–31.25 $\times$ 10^4 m^3, and it was significantly different from well to well. When neglecting the influence of geological reservoir parameters of horizontal wells, infinite cementing sliding sleeve multistage fracturing exhibited the best performance, followed by perf-and-plug multistage fracturing and OH packer and infinite sliding sleeve multistage fracturing, while CT with BOT packer hydra-jet multistage fracturing and fixed-string hydra-jet multistage fracturing demonstrated relatively poor effects.

3.2.2. Economic Benefits

To evaluate the economics of each fracturing technology, the average payback periods for fracturing technologies were calculated according to the actual completion fracturing costs and cumulative gas productions of wells. The results are shown in Figure 9.

It can be seen from Figure 9 that the payback periods of perf-and-plug, infinite cementing sliding sleeve, and OH packer and infinite sliding sleeve multistage fracturing were similar, about 30–35 days, while the payback periods of CT with BOT packer hydra-jet and fixed-string Hydra-jet multistage fracturing were longer, about 74–97 days.

In summary, the multistage fracturing technologies of horizontal wells significantly affected the production and benefits of wells. When neglecting the influence of geological reservoir parameters of horizontal wells, from the aspects of production increase and return on investment, it can be concluded that the infinite sliding sleeve multistage fracturing (cemented and OH) and perf-and-plug had good application effects, while fixed-string hydra-jet multistage fracturing had the worst performance.

Preliminary analysis of the production data revealed that, as stated in most previous studies, there were indeed large differences in the production of horizontal wells completed with different methods. However, the impact of open hole or cementing completion on the performance in this area was much smaller than expected, suggesting that production may be influenced by factors other than completion method.

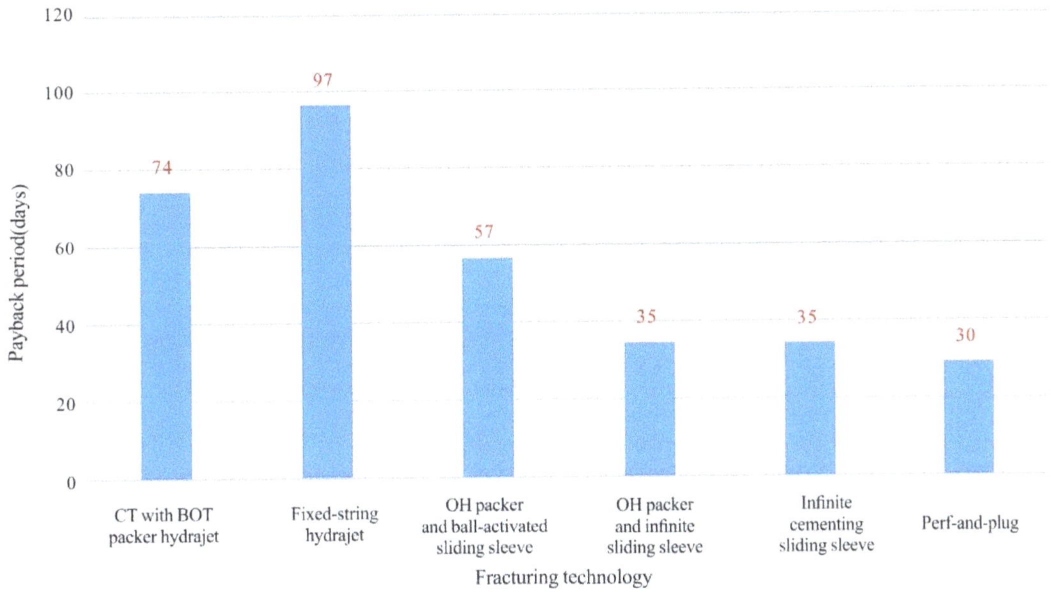

Figure 9. Payback periods for horizontal well multistage fracturing technologies.

4. Analysis of Reasons for Performance Difference

The reliability of the above conclusions needed to be further demonstrated, since they were derived from comparative analysis without consideration of the geological reservoir conditions of wells. It can be seen from Figure 8 that the same fracturing technology exhibited significantly different performance from well to well. Taking W19, W20, and W21 as examples, given the same fracturing technology and similar parameters, the production performance of W21 was much worse than that of W19 and W20. This indicates that the well production is comprehensively influenced by factors other than fracturing technology.

4.1. Main Factors Controlling the Productivity of Fractured Horizontal Wells

The gas flow regime of fractured horizontal wells is complex–turbulent flow occurring near the wellbore, which is also known as a non-Darcy flow. Due to the presence of fractures, the coupling effects between fractures and the wellbore result in a very complicated seepage. Therefore, the productivity of fractured horizontal wells is always difficult to calculate. In this paper, the most typical horizontal well productivity calculation formula, Joshi's formula, was implemented [58]. Firstly, the steady-state productivity formula of horizontal wells in homogeneous and isotropic reservoirs was derived according to the potential energy theory. Then, to consider the influence of the eccentricity of actual horizontal wells and the anisotropy of reservoirs, the modified Joshi's formula was established as Equation (1):

$$J_h = \frac{0.543K_h h\Delta P/(B_0\mu)}{ln[\frac{a+\sqrt{a^2-(L/2)^2}}{L/2}] + \frac{\beta h}{L}ln[\frac{(\frac{\beta h}{2})^2+(\beta\delta)^2}{\frac{\beta h r_w}{2}}]} \quad (1)$$

where K_h is the permeability of horizontal section (10^{-3} μm^2), h is the oil zone thickness (m), ΔP is the production pressure difference (MPa), Bo is the oil volume factor (m^3/m^3), u is the oil viscosity (mPa·s), L is the horizontal section length (m), a is the half length of the elliptical major axis, defined as $a = 0.5L[0.5 + \sqrt{0.25 + (\frac{2r_{eh}}{L})^4}]^{0.5}$ (m), r_{eh} is the quasi-circular driving radius (m), β is the reservoir anisotropy coefficient, defined as $\beta = \sqrt{K_h/K_v}$, δ is the eccentricity of the horizontal wellbore (m), and r_w is the borehole radius (m).

According to Equation (1), the productivity of horizontal wells is jointly controlled by many factors, such as the length of the horizontal section, oil zone thickness, drainage radius, production pressure difference, oil viscosity, and permeability, but there is often no good correlation between a single factor and productivity. For horizontal wells in low-permeability and tight sandstone gas reservoirs, which are generally treated by large-scale fracturing, the gas drainage radius and fluidity are significantly affected by completion technology and fracturing scale. Therefore, it is proposed to construct a comprehensive productivity impact index (CPII) to explore the relationship between the production of horizontal wells and the geological, engineering, and development factors [59–61].

This study dealt with 24 horizontal wells in S2 of Block M, which are basically similar in terms of original permeability, oil viscosity, and volume factor. To construct the CPII, a total of 11 indicative parameters (Table 5) were considered to meet Joshi's formula. Specifically, horizontal section length, encountering rate of net pay zone, reservoir thickness, nozzle size, bottomhole static pressure, and tubing pressure represent the geological and development factors, while the OD of production casing represents the borehole radius. Moreover, parameters such as the number of fracturing stages, displacement, and sand addition are used to characterize the effects of fracturing stimulation on drainage radius, induced fractures, and permeability.

The comprehensive productivity impact index (CPII) was constructed in three steps: (1) standardize the original data; (2) use the entropy method to calculate the weight of each indicator; (3) calculate the CPII from the comprehensive evaluation formula.

4.1.1. Data Processing

The range method was used for data standardization to ensure the uniformity of data. The standardization formula see Equations (2) and (3).

For positive indicators, use Equation (2).

$$Y_i(j) = \frac{X_i(j) - \min X_i(j)}{\max X_i(j) - \min X_i(j)} \tag{2}$$

For negative indicators, use Equation (3).

$$Y_i(j) = \frac{\max X_i(j) - X_i(j)}{\max X_i(j) - \min X_i(j)} \tag{3}$$

where j represents the well number, i represents the indicator layer, $X_i(j)$ represents the original value of indicator i of well j, and $Y_i(j)$ represents the standardized value of indicator i of well j. Since the interval of the standardized data is [0, 1], the standardized data are possibly equal to 0. Considering that the logarithm is introduced into the subsequent calculation using the entropy method, the standardized data were shifted by 0.5 unit.

4.1.2. Determination of Weights

Weights can be determined using many methods. Gray correlation, principal component analysis (PCA), analytic hierarchy process (AHP), and expert scoring were mainly used in reservoir engineering studies, which, however, are too subjective, with the optimal values of some indicators difficult to define and the results nonobjective enough.

Table 5. Indicative parameters of CPII for horizontal wells.

Well	Geological Factors			Engineering Factors				Development Factors			
	Horizontal Section Length (m)	Encountering Rate of Net Pay Zone (%)	Reservoir Thickness (m)	OD. of Production Casing (mm)	Number of Fracturing Stages	Net Fracturing Fluid Volume Injected (m^3)	Displacement (m^3/min)	Total Sand Addition (m^3)	Nozzle Size (mm)	Bottom Hole Static Pressure (MPa)	Initial Tubing Pressure (MPa)
W1	1173.0	65.0	4.5	114.3	13	5839.5	3.9	619.8	16	15.6	3.9
W2	1156.3	61.5	7.5	114.3	11	6435.0	3.6	587.1	10	20.0	11.1
W3	1175.1	35.3	9.5	114.3	12	3768.4	3.2	508.5	10	22.5	16.3
W4	933.0	73.7	4.0	88.90	7	2932.6	2.5	360.6	14	15.1	9.6
W5	652.0	73.2	10.0	88.90	5	1896.0	2.5	188.5	14	21.0	12.1
W6	1181.7	42.1	8.0	88.90	6	3539.5	4.2	387.7	10	18.5	5.9
W7	1188.9	42.8	8.0	88.90	9	3810.7	4.5	478.6	10	16.6	8.1
W8	1399.5	34.9	8.0	88.90	11	4725.1	5.3	543.0	10	16.9	10.5
W9	1349.2	60.0	8.0	88.90	10	3257.7	5.2	436.2	12	17.4	8.6
W10	1329.9	61.4	8.0	88.90	9	3249.0	5.4	407.7	10	17.5	6.8
W11	1336.5	58.2	8.0	88.90	12	5129.5	5.4	603.3	12	18.0	7.3
W12	1068.0	75.0	5.0	88.90	8	4775.9	5.1	578.4	12	16.2	2.3
W13	1200.0	90.0	12.0	88.90	10	3872.4	5.0	324.4	14	22.9	16.9
W14	1081.0	61.3	6.5	114.3	14	5997.1	3.7	713.6	14	19.1	9.5
W15	969.0	39.0	6.5	114.3	13	5825.8	3.6	680.2	14	21.1	8.9
W16	1245.6	80.0	11.0	114.3	16	7261.8	3.5	960.3	10	22.9	17.0
W17	1211.9	51.3	12.0	114.3	13	5872.6	3.3	830.0	16	23.2	16.8
W18	897.5	61.4	9.5	114.3	12	5139.3	3.7	766.0	16	22.7	17.3
W19	1192.4	64.6	8.0	114.3	15	5897.5	3.7	434.6	16	19.6	13.0
W20	1193.2	72.4	8.0	114.3	15	5209.5	4.1	510.3	10	20.9	13.1
W21	1222.5	50.2	8.0	114.3	16	5409.4	4.0	383.2	12	20.3	9.2
W22	1229.7	61.5	9.5	114.3	10	7154.0	5.5	765.5	16	22.5	17.8
W23	1200.7	58.9	10.0	114.3	10	6377.9	5.0	798.8	16	22.3	16.3
W24	834.0	67.9	11.0	114.3	12	8698.0	5.5	730.2	14	20.0	7.4

In this paper, the entropy method was adopted to determine the weight of each indicator. Entropy, a concept derived in thermodynamics, is used to measure the uncertainty of the system state. Information is a measure of the order degree of a system, while entropy is a measure of the disorder degree of a system, in information theory. The two are equal in absolute value but opposite in sign. Entropy is a measure of uncertainty. A smaller entropy denotes a lower disorder degree of information, less uncertainty, a larger utility value of information, and a larger weight of the indicator. The entropy method is an objective weighting approach because it only considers the discreteness of the data itself.

The entropy method is critical to determine the weights of indicators. The weight of each indicator can be calculated to provide a basis for comprehensive evaluation of indicators. The calculation process and formula are as follows:

(1) On the basis of the above standardized data, calculate the proportion of indicator i using Equation (4).

$$Z_{\mathrm{i}}(j) = \frac{Y_{\mathrm{i}}(j)}{\sum_{i=1}^{n} Y_{\mathrm{i}}(j)} \tag{4}$$

(2) According to Equation (5), calculate the information entropy of each indicator.

$$E_{\mathrm{i}} = -\frac{\sum_{i=1}^{n}\sum_{j=1}^{m} Z_{\mathrm{i}}(j)\ln Z_{\mathrm{i}}(j)}{\ln nm} \tag{5}$$

(3) Determine the weight of indicator i of CPII using Equation (6).

$$W_{\mathrm{i}} = \frac{1 - E_{\mathrm{i}}}{\sum_{j=1}^{m}(1 - E_{\mathrm{i}})} \tag{6}$$

The original data of the 24 fractured horizontal wells shown in Table 4 were processed using the entropy method, and the weights of the indicators were determined, as shown in Table 6 and Figure 10.

Table 6. Weights of CPII indicators for fractured horizontal wells.

Level-1 Indicator	Level-2 Indicator	Level-3 Indicator	Weight of Level-3 Indicator
Comprehensive productivity impact index (CPII)	Geological factors	Horizontal section length, m	0.21
		Encountering rate of net pay zone, %	0.03
		Reservoir thickness, m	0.10
	Engineering factors	OD of production casing, mm	0.07
		Number of fracturing stages	0.12
		Net fracturing fluid volume injected, m^3	0.04
		Average displacement, m^3	0.12
		Total sand addition, m^3	0.06
	Development factors	Nozzle size, mm	0.01
		Bottomhole static pressure, MPa	0.11
		Initial tubing pressure, MPa	0.12

From Table 6 and Figure 10, it can be seen that the productivity of fractured horizontal wells is comprehensively controlled by multiple indicators of geological, development, and engineering factors. In terms of Level-3 indicators, horizontal section length, initial tubing pressure, displacement, number of fracturing stages, bottomhole static pressure, and reservoir thickness exhibited the largest weights.

This result is basically consistent with previous studies, which means it is reasonable to select impact indicators using a theoretical formula (Joshi's formula, See Equation (1)). Unlike the previous study, which was only a qualitative analysis of a few indicators, this study used entropy quantification to obtain the weights of each impact indicator.

Meanwhile, the method to calculate indicator weights is objective, making it easy to apply to other reservoirs

4.2. Cause Analysis for Performance Difference

In terms of engineering factors, displacement and the number of fracturing stages had the greatest weights. Previous studies showed that large displacements and fracturing stages can only be achieved by perf-and-plug and the new open-hole multistage hydraulic fracturing system [50]. The actual fracturing parameters of the six fracturing technologies of 24 wells in Block M indicated that displacement and number of fracturing stages were largely dependent on fracturing technology. Moreover, it can be seen from Figure 11 that the difference in well production was consistent with the difference in parameters of horizontal wells. This suggests that the fracturing technology largely determined the ranges of parameters, which in turn affected the fracturing performance and well production. Therefore, the key step for successful horizontal well fracturing is the selection of fracturing technology.

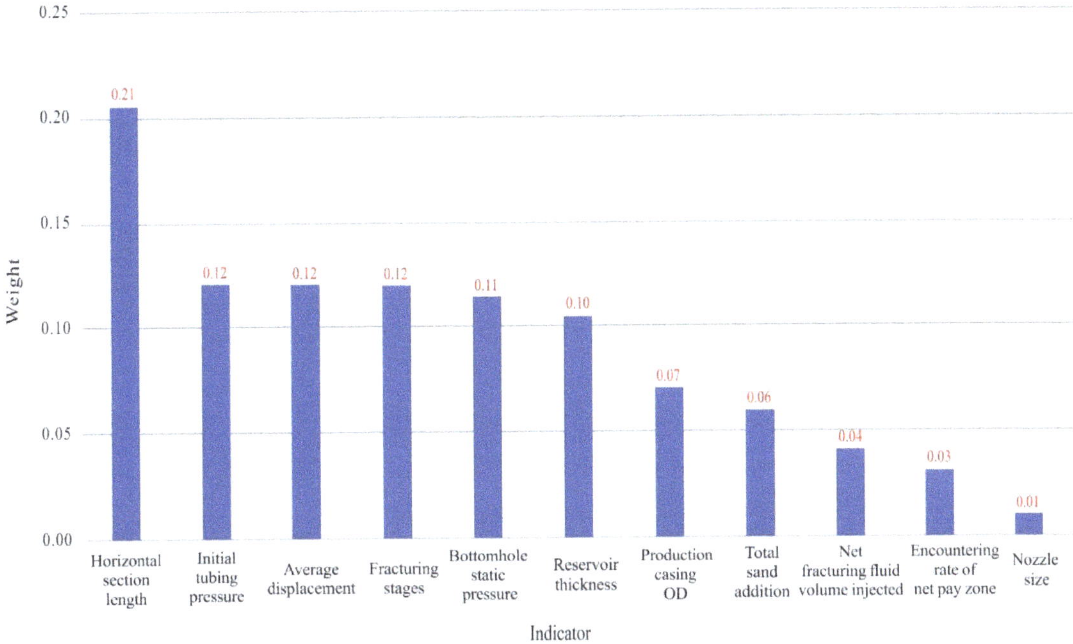

Figure 10. Weights of CPII indicators for fractured horizontal wells.

In terms of development factors, initial tubing pressure and bottomhole static pressure had the greatest weights. Reservoir pressure is an important factor affecting production, as mentioned by many scholars (Hamm and Struyk, Chaikine et al.). The production data of horizontal wells in this block also confirmed the significant impact of formation pressure. The section above described the difference in performance of the same fracturing technology among wells, such as W19, W20, and W21. According to statistics, the horizontal wells in this study were all located in the enrichment zone of S2, with similar bottomhole static pressure (15.07–23.16 MPa), but greatly variable initial tubing pressure (2.28–17.76 MPa). Typically, W19, W20, and W21 had similar horizontal section lengths and reservoir thicknesses, and they adopted the same fracturing technology, but their tubing pressures were quite different: only 9.22 MPa for W21, in contrast to about 13 MPa for W19 and W20. The production of W21 was also much lower than that of W19 and W20. Similarly, due to the influence of tubing pressure, the production performance of W24 (tubing pressure of 7.43 MPa) was far lower than that of W22 and W23 (tubing pressure of 16.31–17.76 MPa), as shown in Table 5. These cases demonstrated the reliability of the computation results of the entropy method, revealing pressure as an important factor that significantly affects the well production.

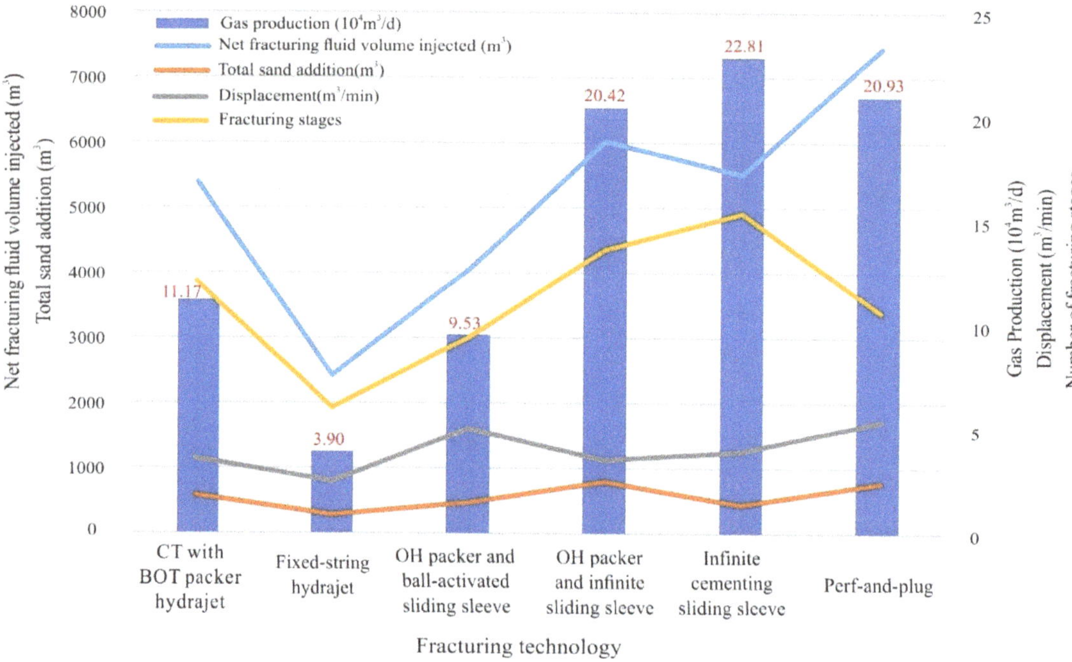

Figure 11. Well production and parameters according to multistage fracturing technologies.

In terms of geological factors, horizontal section length and reservoir thickness had the greatest weights. Analytical results noted that wells in different areas seemed to benefit from a longer overall length and reservoir thickness (Omatsone, Galas, et al.). In Block M, most of the wells had basically similar horizontal section length, ranging from 652 m to 1399 m. However, the reservoir thickness varied greatly, ranging from 4 to 4–12 m. For W4 and W5, given the similarity of other parameters, the reservoir thickness was only 4 m in W4 and 10 m in W5, resulting in lower production of W4 than W5. Meanwhile, fixed-string hydra-jet multistage fracturing was adopted in both wells, with horizontal section lengths of 652 m and 933 m, respectively, which were relatively small in this block. It was, thus, inferred that the actual performance of fixed-string hydra-jet multistage fracturing should be better than the recorded data of the above two wells, although it is less applicable to Block M than other technologies, due to the limitation of displacement and number of fracturing stages.

Furthermore, field data confirmed that the well production is more affected by a comprehensive influence of multiple factors; the productivity index ratio is a function of variables including reservoir permeability, reservoir thickness, fracture half-length, fracture spacing along the horizontal, and completion method [19]. For example, the production difference between W1 and W2/W3 was greatly dependent on tubing pressure and reservoir thickness. The production of W1 (tubing pressure of 3.9 MPa, reservoir thickness of 4.5 m) was lower than that of W2 and W3 (tubing pressure of 11.1–16.3 MPa, reservoir thickness of 7.5–9.5 m). The production of W14 and W15 (tubing pressure of 8.9–9.5 MPa, reservoir thickness of 6.5 m) was lower than that of W16, W17, and W18 (tubing pressure of 16.3–17.8 MPa, reservoir thickness of 11–9.5 m). Similarly, due to the relatively small horizontal section length (834 m) and low initial tubing pressure (7.4 MPa), W24 revealed lower production when using the perf-and-plug multistage fracturing technology, which was considered highly applicable to this block, than other wells adopting the same fracturing technology. In contrast, W13 had a large horizontal section length (1200 m) and high initial tubing pressure (16.9 MPa), but it achieved higher production than other wells, ranking among the top three fractured wells in terms of overall performance, although it adopted the open-hole packer and ball-activated sliding sleeve multistage fracturing, which features a small scale. This also proves from another aspect that the impact of fracturing technology on horizontal well production is restricted by the geological reservoir conditions of horizontal wells.

4.3. Comparison of Application Effects of Fracturing Technologies

The above analysis suggests that the conclusion drawn in Section 3.2 only by the classification of fracturing technologies but ignoring the influence of other factors is not accurate. To identify the influence of fracturing technologies on horizontal well production and their applicability in Block M, some Class I horizontal wells with similar horizontal section length, reservoir thickness, and pressure but adopting different fracturing technologies were selected for comparison, as shown in Table 7 and Figure 12.

From Table 7 and Figure 12, it can be seen that, for Class I horizontal wells, infinite sliding sleeve multistage fracturing exhibited the best performance, followed by perf-and-plug and traditional open-hole packer multistage fracturing, while CT with BOT packer hydra-jet showed poor performance. The performance of fixed-string hydra-jet technology needs to be further evaluated due to the lack of samples. This conclusion is basically in line with historical studies; that is, completion methods do have a significant impact on production. However, this study suggests a much smaller difference between open-hole and cemented completions than illustrated in previous articles. As mentioned above, the productivity of horizontal wells in this area was comprehensive affected by multiple factors, especially length of horizontal section, fracturing construction parameters, and reservoir pressure.

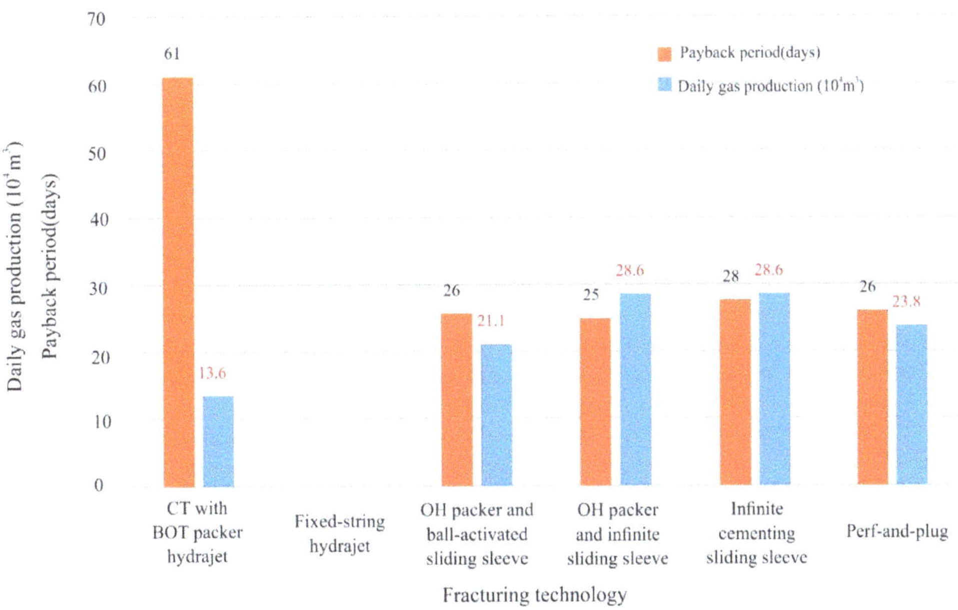

Figure 12. Production and payback period of Class I horizontal wells by multistage fracturing technology.

Table 7. Parameters and production of Class I horizontal wells in Block M.

Well	Horizontal Section Length (m)	Reservoir Thickness (m)	OD. of Production Casing (mm)	Multistage Fracturing Technology	Number of Fracturing Stages	Net Fracturing Fluid Volume Injected (m^3)	Displacement (m^3/min)	Total Sand Addition (m^3)	Bottomhole Static Pressure (MPa)	Initial Tubing Pressure, (MPa)	Production (10^4 m^3/day)
W3	1175	9.5	114.3	CT with BOT packer Hydra-jet	12	3768.4	2.5–3.9	508.5	22.5	16.3	13.6
W13	1200	12.0	88.90	OH packer and ball-activated sliding sleeve	10	3872.4	4.5–5.5	324.4	22.9	16.9	21.1
W16	1246	11.0	114.3	OH packer and infinite sliding sleeve	16	7261.8	3.0–3.9	960.3	22.9	17.0	31.2
W17	1212	12.0	114.3	OH packer and infinite sliding sleeve	13	5872.6	2.7–3.8	830.0	23.2	16.8	26.0
W19	1192	8.0	114.3	Infinite cementing sliding sleeve	15	5897.5	3.4–3.9	434.6	19.6	13.0	29.0
W20	1193	8.0	114.3	Infinite cementing sliding sleeve	15	5209.5	3.9–4.2	510.3	20.9	13.1	28.3
W22	1230	9.5	114.3	Perf-and-plug	10	7154.0	4.0–7.0	765.5	22.5	17.8	29.3
W23	1201	10.0	114.3	Perf-and-plug	10	6377.9	3.0–7.0	798.8	22.3	16.3	18.2

5. Comprehensive Productivity Impact Index (CPII)

The results demonstrated that horizontal well production was affected by a combination of factors, but there is no method available to characterize this comprehensive index. In this paper, in order to explore the comprehensive impact of indicators on the production of horizontal wells, on the basis of the factors weights calculated above, a comprehensive productivity impact index (CPII) was constructed in Equation (7).

$$W_{ij} = \sum_{i=1}^{n} Y_i(j) W_i \tag{7}$$

The correlation between the production of the fractured horizontal well and CPII is plotted in Figure 13.

It can be seen from Figure 13 that the CPII constructed has a good positive correlation with the well production. Without considering the economic and technical constraints, a greater CPII denotes greater production of the fractured horizontal well. For a newly drilled horizontal well, where horizontal section length, reservoir thickness, and initial tubing pressure are given, the only way to improve well production is to optimize the fracturing technology. According to Figure 13 and the weight ranking of the three-level indicators mentioned above, it can be seen that the well production of horizontal wells can be increased as much as possible by increasing the number of fracturing stages and displacement. This also further demonstrates the reason why the infinite sliding sleeve multistage fracturing and perf-and-plug multistage fracturing technologies with unlimited fracturing stages and larger fracturing scale have relatively better performance in Block M.

Meanwhile, according to the curve matching relation between well production and CPII in Figure 13, the production of fractured horizontal wells in Block M can be roughly predicted by using Equation (8), so as to provide reference for selection of fracturing technology and optimization of parameters for fracturing design of new horizontal wells for the purpose of better fracturing performance and higher economic benefit of wells.

$$Q = 0.1532e^{4.0449W_{ij}} \tag{8}$$

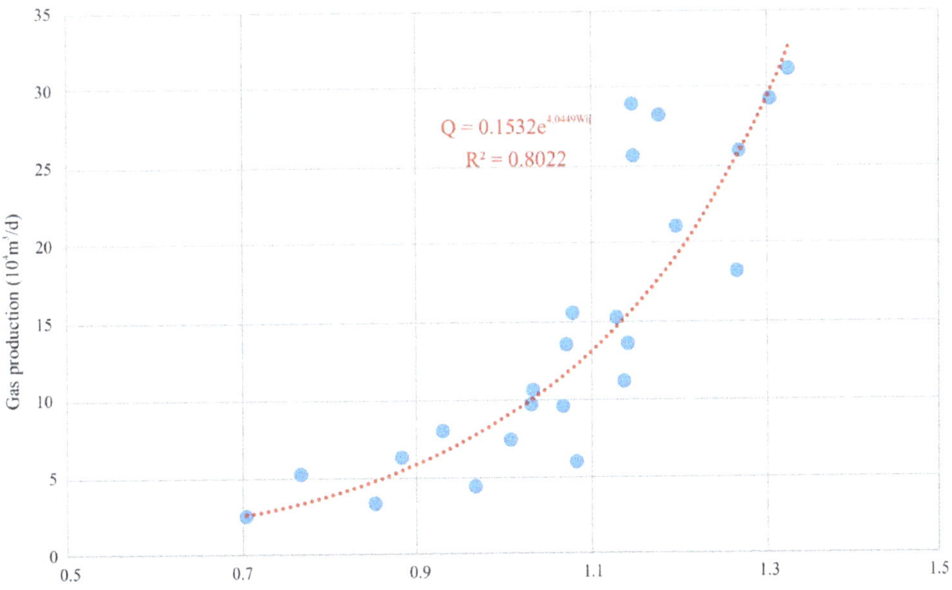

Figure 13. Well production vs. CPII.

6. Conclusions

The following conclusions can be drawn in this study:

(1) All horizontal well multistage fracturing technologies can achieve the purpose of gas reservoir stimulation, production enhancement, and improved economics, but each technology has its own advantages, disadvantages, and applicability. Considering technical and economic benefits, infinite sliding sleeve multistage fracturing and perf-and-plug multistage fracturing are preferred for Class I wells in Block M and horizontal wells in similar gas reservoirs.

(2) The comprehensive productivity impact index (CPII) constructed in this paper shows a good positive correlation with the production of fractured horizontal wells. Without considering the economic and technical constraints, a greater CPII, denotes greater well production. This relation can be used to predict the production of newly drilled horizontal wells and guide fracturing design and optimization of fracturing parameters.

(3) For Block M, from the perspective of Level-2 indicators, the production of fractured horizontal wells is comprehensively controlled by geological, engineering, and development factors. From the perspective of Level-3 indicators, horizontal section length, initial tubing pressure, displacement, number of fracturing stages, bottom hole static pressure, and reservoir thickness are the main indicators with the largest weights influencing the production of fractured horizontal wells.

(4) Horizontal well fracturing technology should be selected with consideration of the geological reservoir parameters of horizontal wells, as well as the fracturing design and parameters to be optimized according to specific well conditions. For newly drilled horizontal wells, where the geological reservoir conditions are given, selection of fracturing technology and optimization of parameters are critical for production enhancement, because the fracturing technology largely determines the range of parameters, which further affects the fracturing performance and well production.

(5) The study method in this paper, that is, determining the main factors dominating the production of horizontal wells and constructing the comprehensive productivity impact index (CPII) on the basis of the data from Block M, is universal and can be easily applied to other similar reservoirs.

7. Research Limitations

This paper may be deficient in several aspects. The horizontal wells involved in this study were all located in the reserve enrichment zone of S2 in Block M, with similar reservoir physical properties and fluid properties; hence, the influence of reservoir physical properties and fluid properties was ignored when constructing the CPII. Meanwhile, the effect of inter-well interference was ignored in this paper because of the short production time of Block M. It is recommended to consider these indicators when using the proposed method to analyze the performance of fracturing technologies in similar reservoirs. Secondly, some viewpoints in this paper need to be further quantitatively demonstrated using more samples. For instance, the fixed-string hydra-jet multistage fracturing technology is undoubtedly less applicable in this block due to the limitation of displacement and number of fracturing stages, but its performance needs further quantification, since the two horizontal wells involved in the analysis were affected by horizontal section length, tubing pressure, and type of fracturing technology. Moreover, the applicability of fracturing technologies in Class I horizontal wells was verified in this paper, while the applicability in Class II and Class III horizontal wells needs to be further confirmed with more samples.

Author Contributions: Conceptualization, L.W.; methodology, J.Z. and L.W.; software, L.W. and Y.Y.; validation, L.W., Y.Y. and S.W.; formal analysis, J.Z.; investigation, L.W. and Y.Y.; resources, J.Z. and D.J.; data curation, L.W., Y.Y. and J.Z. writing—original draft preparation, L.W.; writing—review and editing, S.W. and J.Z.; visualization, L.W. and J.Z.; supervision, S.W.; project administration, D.J.; funding acquisition, D.J. All authors have read and agreed to the published version of the manuscript.

Funding: This work was supported by the National Natural Science Foundation of China (Grant Number 52074345).

Institutional Review Board Statement: Not applicable.

Informed Consent Statement: Not applicable.

Data Availability Statement: Not applicable.

Conflicts of Interest: The authors declare no conflict of interest.

References

1. Xin, T. *Research on Fracturing Stimulation Method and Technology of Horizontal Well in Tight Sandstone Reservoir*; China University of Geosciences: Beijing, China, 2016.
2. Snyder, D.J.; Seale, R.; Hollingsworth, R. SPE 139370: Optimization of completions in unconventional reservoirs for ultimate recovery. In Proceedings of the Society of Petroleum Engineers SPE Latin American and Caribbean Petroleum Engineering Conference, Lima, Peru, 1 December 2010.
3. Ding, Q.X.; Hou, S.H.; Du, X.F.; Wang, W.M. Research progress of horizontal well fracturing technology in China. *Pet. Mach.* **2016**, *44*, 78–82. [CrossRef]
4. Zhang, H.-Z.; He, Y.-H.; Liu, J.; Zhang, H.-Z. Development status and trend of staged fracturing technology for horizontal wells abroad. *Pet. Sci. Technol. Forum* **2012**, *31*, 47–52.
5. Cipolla, C.L.; Fitzpatrick, T.; Williams, M.J.; Ganguly, U.K. Seismic-to-simulation for unconventional reservoir development. In Proceedings of the SPE Reservoir Characterisation and Simulation Conference and Exhibition, Abu Dhabi, United Arab Emirates, 9–11 October 2011. [CrossRef]
6. Ma, Y.Z. *Unconventional Oil and Gas Resources Handbook*; Elsevier: Amsterdam, The Netherlands, 2016.
7. Rahman, M.M.; Hossain, M.M.; Crosby, D.G.; Rahman, M.K.; Rahman, S.S. Analytical, numerical and experimental investigations of transverse fracture propagation from horizontal wells. *J. Pet. Sci. Eng.* **2002**, *35*, 127–150. [CrossRef]
8. East, L.E.; William, G.; McDaniel, B.; Bill, J.; Randy, J.; Kevin, F. Successful application of hydrajet fracturing on horizontal wells completed in a thick shale reservoir. In Proceedings of the SPE Eastern Regional Meeting, Charleston, WV, USA, 15–17 September 2004.
9. Mcdaniel, B.W.; Surjaatmadja, J.B. *Hydrajetting Applications in Horizontal Completions to Improve Hydraulic Fracturing Stimulations and Improve ROI*; Massachusetts Institute of Technology: Cambridge, MA, USA, 2009.
10. Lu, X.; Ye, D.; Zhu, J.; Song, D.; Yin, C.; Bin, G.; Wang, G. Applications of new coiled tubing multi-staged fracturing technology. In Proceedings of the IADC/SPE Asia Pacific Drilling Technology Conference and Exhibition, Tianjin, China, 9–11 July 2012.
11. Xue, J.R.; Guo, J.C.; Fan, W.T.; Zhang, P.D. Study and application on hydrajet multistage fracturing technology. *Appl. Mech. Mater.* **2013**, *433–435*, 1965–1968. [CrossRef]
12. Jiang, T.; Zhao, X.; Yin, F.; Qu, H. *Application of Multistage Hydrajet-Fracturing Technology in Horizontal Wells with Slotted Liner Completion in China*; Tottori University: Tottori, Japan, 2014.
13. Li, Q.; Xing, H.; Liu, J.; Liu, X. A review on hydraulic fracturing of unconventional reservoir. *Petroleum* **2015**, *1*, 8–15. [CrossRef]
14. Qin, J.L.; Chen, Z.; Yang, T.Y. Multistage sliding sleeve multistage fracturing technology for horizontal wells in ordos basin. *Oil Drill. Technol.* **2015**, *43*, 7–12.
15. Parshin, N.; Burdin, K.; Starodubtseva, K.; Novikov, M.; Demkovich, M. Experience and optimization of essential multistage fracturing technology with coiled tubing at the well for the vinogradova oilfield. In Proceedings of the SPE Russian Petroleum Technology Conference, Moscow, Russia, 15–17 October 2018.
16. Lohoefer, D.S. *Long-Term Comparison of Production Results from Open Hole and Cemented Multi-Stage Completions in the Barnett Shale*; Society of Petroleum Engineers: London, UK, 2010.
17. Edwards, W.; Braxton, D.K.; Smith, V. Tight gas multistage horizontal completion technology in the granite wash: A case study. In Proceedings of the Tight Gas Completions Conference, San Antonio, TX, USA, 2–3 November 2010.
18. Wilson, B.; Lui, D.; Kim, J.; Kenyon, M.; Mccaffrey, M. Comparative study of multistage cemented liner and openhole system completion technologies in the montney resource play. In Proceedings of the Canadian Unconventional Resources Conference, Calgary, AB, Canada, 15–17 November 2011.
19. Augustine, J.R. *Openhole versus Cemented Completions for Horizontal Wells with Transverse Fractures: An Analytical Comparison*; Society of Petroleum Engineers: London, UK, 2011.
20. Theppornprapakorn, V. Computational Fluid Dynamics Flow Comparison between Openhole Sleeve and Plug-and-Perf Completion in a Hydraulic Fractured Horizontal Well. Master's Thesis, Missouri University of Science and Technology, Rolla, MO, USA, 2013.
21. Vasudevan, H. Computational Fluid Dynamic Analysis in High Permeability Hydraulic Fractured Horizontal Gas Wells. Master's Thesis, Missouri University of Science and Technology, Rolla, MO, USA, 2016.
22. Lingchuan, L. Practice and understanding of full-diameter staged fracturing stimulation of horizontal wells in tight sandstone gas reservoir. *Drill. Prod. Technol.* **2019**, *42*, 62–65.
23. Thompson, D.; Rispler, K.; Stadnyk, S.; Hoch, O.; Mcdaniel, B. Operators evaluate various stimulation methods for multizone stimulation of horizontals in north east british columbia. In Proceedings of the SPE Hydraulic Fracturing Technology Conference, The Woodlands, TX, USA, 19–21 January 2009.

24. Mcdaniel, B.W.; Rispler, K.A. SPE 125903: Horizontal wells with multistage fracs prove to be best economic completion for many low-permeability reservoirs. In Proceedings of the SPE Eastern Regional Meeting, Charleston, WV, USA, 23–25 September 2009.
25. Stanojcic, M.; Jaripatke, O.A.; Sharma, A. Pinpoint fracturing technologies: A review of successful evolution of multistage fracturing in the last decade. In Proceedings of the SPE/ICoTA Coiled Tubing and Well Intervention Conference and Exhibition, The Woodlands, TX, USA, 23–24 March 2010.
26. Jingbao, X. Research and application of staged fracturing technology for horizontal wells in daniudi gas field. *Drill. Prod. Technol.* **2011**, *34*, 113–114.
27. Kennedy, R.L.; Knecht, W.N.; Georgi, D.T. Comparisons and Contrasts of Shale Gas and Tight Gas Developments, North American Experience and Trends. In Proceedings of the SPE Saudi Arabia Section Technical Symposium and Exhibition, Al-Khobar, Saudi Arabia, 8–11 April 2012.
28. Burton, W.A. Unconventional Completions: Which One is Right for Your Application? In Proceedings of the SPE Annual Technical Conference and Exhibition, New Orleans, LA, USA, 30 September–2 October 2013.
29. Salah, M.; Gabry, M.; El-Sebaee, M. Evaluation of multistage fracturing stimulation horizontal well completion methods in western desert, Egypt. In Proceedings of the SPE Middle East Oil & Gas Show & Conference, Manama, Bahrain, 6–9 March 2017.
30. Prudskiy, M.Y.; Khaliulin, R.R.; Shakirov, R.A. Application of the multi-stage hydraulic fracturing technology in horizontal wells with large vertical outflow to improve the development efficiency of reservoirs with deteriorated filtration-capacitance properties at the sakhalin shelf field. In Proceedings of the SPE Russian Petroleum Technology Conference, Moscow, Russia, 26–29 October 2020.
31. Xianwen, L.; Zhe, L.; Yuanxiang, X.; Baochun, C.; Yanming, Z.; Hua, S. Comparative study on completion and fracturing technology of sulige tight gas horizontal wells. *Oil Drill. Prod. Technol.* **2021**, *43*, 48–53. [CrossRef]
32. Omatsone, E.N.; Bagheri, M.; Galas, C.M.F.; Curtis, B.; Frankiw, K. Redevelopment of the cardium formation using fractured horizontal wells: Reservoir engineering perspectives and early case histories. In Proceedings of the Canadian Unconventional Resources and International Petroleum Conference, Calgary, AB, Canada, 19–21 October 2010.
33. Rankin, R.R.; Thibodeau, M.; Vincent, M.C.; Palisch, T. Improved production and profitability achieved with superior completions in horizontal wells: A bakken/three forks case history. In Proceedings of the SPE Annual Technical Conference and Exhibition, Florence, Italy, 19–22 September 2010.
34. Taylor, R.S.; Barree, R.; Aguilera, R.; Hoch, O.; Storozhenko, K. CSUG/SPE 148680: Why not to base economic evaluations on initial production alone. In Proceedings of the Canadian Unconventional Resources Conference, Calgary, AB, Canada, 15–17 November 2011.
35. Hamm, B. CSUG/SPE 149000: Quantifying the results of horizontal multistage development in tight oil reservoirs of the western canadian sedimentary basin: Technical and economic case studies from a reservoir evaluators perspective. In Proceedings of the Canadian Unconventional Resources Conference, Calgary, AB, Canada, 15–17 November 2011.
36. Al-Ghazal, M.; Al-Driweesh, S.; Al-Ghurairi, F.; Al-Sagr, A.; Al-Zaid, M. Assessment of Multistage Fracturing Technologies as Deployed in the Tight Gas Fields of Saudi Arabia. In Proceedings of the International Petroleum Technology Conference, Beijing, China, 26–28 March 2013.
37. Alekseev, A. Methods for forecasting and evaluation of reservoir properties under conditions of its development using hydraulic fracturing technology. In Proceedings of the SPE Russian Petroleum Technology Conference, Moscow, Russia, 15–17 October 2018.
38. Chodzicki, J. SPE 153064: Reserves and performance comparisons of multi-fractured spearfish horizontal wells, to various fracturing techniques and well spacing configurations. In Proceedings of the SPE/EAGE European Unconventional Resources Conference and Exhibition, Vienna, Austria, 20–22 March 2012.
39. Sun, C.C.; Tan, C.D.; Song, J.; Song, W.R.; Gao, D. Overview of horizontal well fracturing technology abroad. *China Pet. Chem. Ind.* **2014**, 50–52. [CrossRef]
40. Qu, H.; Li, G.S.; Huang, Z.W.; Li, X.W.; Zhao, W.; Bu, X.Q.; Wang, X.D. Sealing mechanism of hydraulic jet staged fracturing. *J. Pet.* **2011**, *32*, 514–517.
41. Li, X.W.; Zhao, W.Z. Application of hydraulic perforating jet fracturing technology in Changqing oilfield. *Pet. Drill. Prod. Technol.* **2008**, 67–70. [CrossRef]
42. Zhong, S.; Lin, L.; Fang, X.; Qu, J. Research and application of hydraulic jet staged fracturing technology with fixed- string. *Oil Gas Reserv. Eval. Dev.* **2012**, *2*, 50–52. [CrossRef]
43. Qin, F.X. Research and application of hydraulic jet staged fracturing completion technology for horizontal wells with fixed-string. *China Pet. Chem. Stand. Qual.* **2016**, *36*, 121–122.
44. Ling, Y.; Li, X.W.; Mu, L.J.; Ma, X. New progress in fracturing technology for tight sandstone gas reservoirs in Sulige gas field. *Nat. Gas Ind.* **2014**, *34*, 66–72. [CrossRef]
45. Zhao, G.M.; Zhao, M.Q.; Guo, S.W.; Shao, Y.; Zhang, W.X. Application of open hole packer staged fracturing technology in changqing oil and gas field. In Proceedings of the International Conference on Oil and Gas Field Exploration and Development 2016 IFEDC, Beijing, China, 11–12 August 2016; pp. 233–237.
46. Zhang, J.; Ren, J.M.; Tian, H.R.; Li, Y.B. Application of openhole packer staged fracturing technology in Shengli oilfield. *Pet. Mach.* **2013**, *41*, 102–105.
47. Teng, C.M.; Du, Z.H.; Shi, L.L.; Hu, J.L.; Yu, P.; Li, H.C. Application of multistage staged fracturing technology with openhole packer in horizontal wells in Su 75 block. *Drill. Prod. Technol.* **2012**, *35*, 31–33.

48. Yuan, F.; Blanton, E.; Convey, B.A.; Palmer, C. Unlimited multistage frac completion system: A revolutionary ball—Activated system with single size balls. In Proceedings of the SPE Annual Technical Conference and Exhibition, New Orleans, LA, USA, 30 September–2 October 2013.
49. Liu, J.B.; Huang, Q.; Yang, M.; Wang, M. Current situation and prospect of horizontal well staged fracturing tool technology. *Pet. Mach.* **2021**, *49*, 110–119. [CrossRef]
50. Yao, B.; Ding, Q.; Hou, Y.; Liu, S.; Zhang, S. A new openhole multistage hydraulic fracturing system and the ball plug motion in a horizontal pipe. *J. Nat. Gas Sci. Eng.* **2018**, *50*, 11–21. [CrossRef]
51. Li, M.; Liu, Z.B.; Lu, H.; Cheng, Z.Y.; Wang, Q.J.; Zheng, W.J. Application of coiled tubing infinite sliding sleeve multistage fracturing technology in Sulige. *Pet. Mach.* **2015**, *43*, 40–43. [CrossRef]
52. Yang, W.B. Application of casing cementing sliding sleeve multistage fracturing technology in Dongsheng gas field. *Xinjiang Pet. Nat. Gas* **2016**, *12*, 57–59.
53. Wang, H.D.; Wang, Q.; Li, R.; Zhang, L.; Li, T.; Geng, M.L. Application of the combined technology of soluble bridge plug and cluster perforation in shale gas horizontal well. *Drill. Prod. Technol.* **2019**, *42*, 113–114.
54. Ren, Y.; Ye, D.S.; Li, J.Q.; Jiang, H. Practice of perf-and-plug combined operation in horizontal well staged fracturing. *Oil Drill. Prod. Technol.* **2013**, *35*, 90–93. [CrossRef]
55. Liu, T.L.; Shi, J.G.; Feng, D.; Tian, Y. Staged fracturing technology and development trend of soluble bridge plugs in horizontal wells. *Pet. Mach.* **2020**, *48*, 103–110. [CrossRef]
56. Yuan, F.; Blancton, E.; Palmer, C.; Andersen, C.R.; Grogan, A. Is it possible to do unlimited multistage fracturing economically? In Proceedings of the SPE/EAGE European Unconventional Resources Conference and Exhibition, Vienna, Austria, 25–27 February 2014.
57. Oberhofer and Rob. Application of ball-drop technology to improve efficiency and stimulation of limited entry completion systems. In Proceedings of the Abu Dhabi International Petroleum Exhibition & Conference, Abu Dhabi, United Arab Emirates, 7–10 November 2016.
58. Joshi, S.D. A review of horizontal well and drainhole technology. In Proceedings of the SPE Technical Conference & Exhibition, Lafayette, LA, USA, 22–23 February 1987.
59. Yin, S.; Han, C.; Wu, Z.; Li, Q. Developmental characteristics, influencing factors and prediction of fractures for a tight gas sandstone in a gentle structural area of the ordos basin, China. *J. Nat. Gas Sci. Eng.* **2019**, *72*, 103032. [CrossRef]
60. Sun, Y.; Wang, Y.; Wang, P.; Cao, Z. Research on contribution analysis method of lubricating oil performance index. *IOP Conf. Ser. Earth Environ. Sci.* **2021**, *651*, 022031. [CrossRef]
61. Yin, D.; Wu, T. Optimizing well for fracturing by fuzzy analysis method of applying computer. In Proceedings of the 2009 First International Conference on Information Science and Engineering, Nanjing, China, 26–28 December 2009.

energies MDPI

Article

Geochemical Characteristics and Process of Hydrocarbon Generation Evolution of the Lucaogou Formation Shale, Jimsar Depression, Junggar Basin

Wenjun He [1,2], Yin Liu [3,*], Dongxue Wang [2], Dewen Lei [1,2], Guangdi Liu [1], Gang Gao [1], Liliang Huang [1,2] and Yanping Qi [2]

1 College of Geosciences, China University of Petroleum (Beijing), Beijing 102249, China; fchwj@petrochina.com.cn (W.H.); leidw@petrochina.com.cn (D.L.); lgd@cup.edu.cn (G.L.); gaogang2819@sina.com (G.G.); hlil@petrochina.com.cn (L.H.)
2 Research Institute of Exploration and Development, Xinjiang Oilfield Company, PetroChina, Urumqi 830013, China; fcwdx@petrochina.com.cn (D.W.); qiyanp@petrochina.com.cn (Y.Q.)
3 Key Laboratory of Deep Oil and Gas, School of Geosciences, China University of Petroleum (East China), Qingdao 266580, China
* Correspondence: liuyin@upc.edu.cn; Tel.: +86-18366266387

Abstract: Lacustrine shale, represented by the Middle Permian Lucaogou Formation in the Jimsar Depression in the eastern Junggar Basin, has become one of the main areas of shale oil exploration in China. In this study, we used 137 samples of shale from the Lucaogou Formation, drawn from 14 wells in the Jimsar Depression, to investigate their characteristics of pyrolysis, organic carbon and soluble organic matter content, biomarkers, organic microscopic composition, and vitrinite reflectance. Basin simulation and hydrocarbon generation thermal simulation experiments were also conducted in a closed system. The results of this study indicate that the input of an algae source was dominant in the source rocks of the Lucaogou Formation, that the water in which the rocks were deposited had high salinity and strong reducibility, and that the source rocks were oil-prone. The Lucaogou source rocks generally had good hydrocarbon generation capability, but showed significant heterogeneity. At the end of the Cretaceous period, the shales in the Lucaogou Formation entered the oil-generation window as a whole. Currently, the shales of the Lucaogou Formation are generally in the high-maturity stage in the deep part of the depression, producing a large amount of high-maturity oil and condensate gas, while those in the shallow part have relatively low maturity and can only produce a large amount of conventional crude oil. The maximum crude oil generation rate of the Lucaogou Formation shale obtained from the thermal simulation results was 220.2 mg/g of the total organic carbon (TOC), and the maximum hydrocarbon expulsion efficiency was estimated to be 59.3–76.4%.

Keywords: Junggar Basin; Jimsar Depression; shale oil; thermal simulation; source rock evaluation; hydrocarbon generation evolution

Citation: He, W.; Liu, Y.; Wang, D.; Lei, D.; Liu, G.; Gao, G.; Huang, L.; Qi, Y. Geochemical Characteristics and Process of Hydrocarbon Generation Evolution of the Lucaogou Formation Shale, Jimsar Depression, Junggar Basin. *Energies* **2022**, *15*, 2331. https://doi.org/10.3390/en15072331

Academic Editors: Hossein Hamidi

Received: 6 February 2022
Accepted: 21 March 2022
Published: 23 March 2022

1. Introduction

Crude oil and gas exploration in the Junggar Basin began in 1909 and has a history of more than 100 years. As of 2013, the proven crude oil and natural gas reserves in the Junggar Basin were 23.34×10^8 t and 1972×10^8 m^3, respectively [1]. The Junggar Basin is one of the most important onshore oil and gas production sites in China [1,2]. Although the current oil and gas exploration in the Junggar Basin is dominated by conventional reservoirs, the exploration of unconventional shale oil, represented by the Middle Permian Lucaogou Formation (P_2l) in the Jimsar Depression, the Middle Permian Pingdiquan Formation (P_2p) in the Wucaiwan–Shishugou area, and the Lower Permian Fengcheng Formation (P_1f) in Fengcheng area of the northwestern margin, has shown promising results [3–5].

Compared with conventional oil reservoirs, shale oil reservoirs are significantly different, and the main differences are as follows: (1) source–reservoir integration and near-source or intra-source accumulation; (2) extremely low porosity and permeability; (3) the ambiguous boundary conditions of the trap; and (4) Darcy's law of seepage flow not being met during the secondary migration of oil and gas [6–8]. At present, this type of integrated source–reservoir shale oil and gas exploration is still in its infancy in China, and the predicted resource amount is around 11×10^9 t, of which Junggar Basin alone accounts for 29×10^8 t (26.4% of the total) [6]. Unlike foreign shale oil, which is dominated by marine deposits, the shale oil in China is dominated by lacustrine deposits. The source rocks in the lake are typically thick, with a small distribution, and their porosity (<10%) and permeability (<0.1 mD) are lower than those in the maritime environment [6,9]. The P_2l lacustrine source rocks in the Jimsar Depression, located in the eastern part of the Junggar Basin, can be regarded as the world's largest lacustrine mudstone in terms of its thickness and organic matter abundance (Figure 1a,b) [10]. At present, this area is one of the main areas of shale oil exploration in China, with a predicted resource amount of 7.02×10^8 t [1,6].

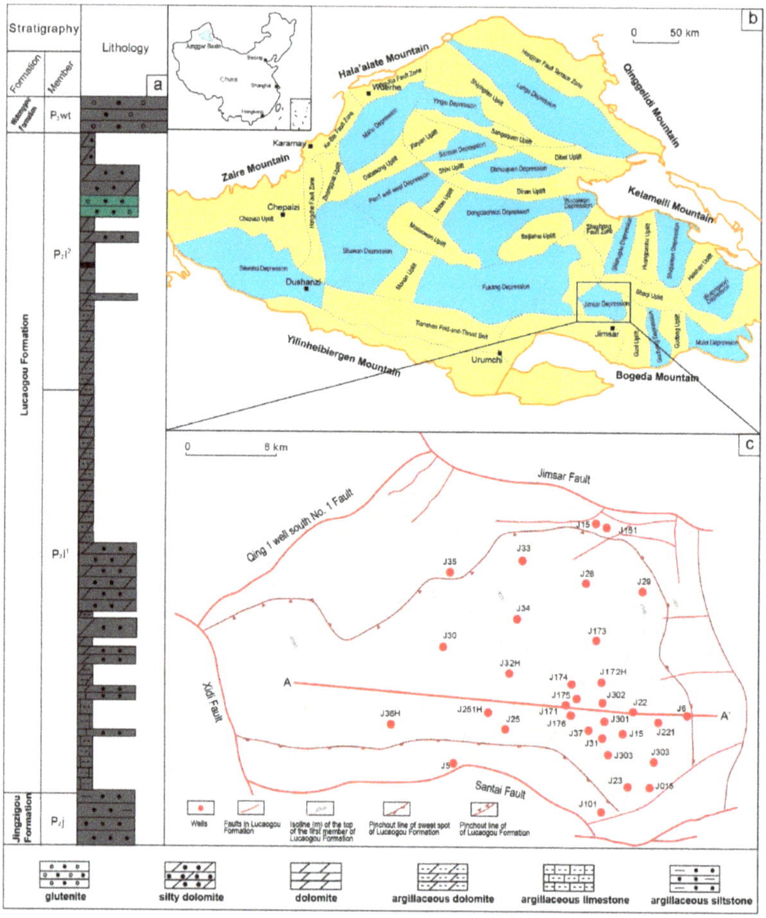

Figure 1. Overview of the petroleum geology of the Jimsar Depression. (**a**) The lithology column of Lucaogou Formation; (**b**) The location of Jimsar Depression in the Junggar Basin; (**c**) Detailed structural map of the Jimsar Depression.

The organic matter abundance and hydrocarbon generation capacity of source rocks are the most important factors used for evaluating the quality of shale oil [8]. Previous researchers had studied P_2l shale oil in the Jimsar Depression, mainly focusing on the sedimentary facies and lithology of the P_2l shale [11], oil and gas source comparison [12], the heterogeneity of the source rocks [13], and the hydrocarbon accumulation process [14,15]. These studies have greatly enriched research on unconventional oil and gas, and promoted the exploration and development of shale oil in the Jimsar Depression. However, current research on the characteristics of hydrocarbon generation and the evolution of P_2l source rocks in the study area is still in a relatively early stage. Based on the analysis of the static hydrocarbon generation capacity of the P_2l source rocks, in this study a gold-tube thermal simulation experiment was conducted in a closed system. Based on this and the two-dimensional basin simulation, the hydrocarbon generation mechanism, hydrocarbon expulsion efficiency, and the evolution process were systematically studied, providing strong support for the evaluation of continental shale oil and gas.

2. Geologic Background

The Junggar Basin is located in the Xinjiang Uygur Autonomous Region in northwestern China, with an area of approximately 1.3×10^5 km^2. It can be divided into a total of six primary structural units, including two major depressions (the Central Depression and the Wulungu Depression), three major uplifts (the Luliang Uplift, the Western Uplift, and the Eastern Uplift), and a thrust-fold belt (the North Tianshan Thrust-Fold Belt) (Figure 1b) [16,17]. The study area is a secondary structural unit in the Eastern Uplift in the Jimsar Depression, bounded by the Shaqi Uplift to the north, the Santai Uplift to the west, the Fukang Fault Zone to the south, and the Guxi Uplift to the east, and has an area of around 1500 km^2 (Figure 1c) [1,18]. The depression has undergone multiple tectonic movements (Hercynian, Indosinian, Yanshanian, and Himalayan movements). There are huge, thick Paleozoic–Cenozoic sedimentary rocks overlying the Precambrian crystalline basement and Paleozoic metamorphic basement. At their thickest, the rocks exceed 5000 m, and the deposits in the depression gradually become thinner from west to east (Figure 1c) [1,17].

The Middle Permian Lucaogou Formation in the Jimsar Depression was formed in a saline lacustrine basin environment after the relict sea was closed [1,19], and it is widely distributed in the depression, with an area of approximately 1278 km^2, a thickness of 200–350 m, and a maximum thickness of >500 m (Figure 1a) [20]. The Lucaogou Formation is a set of lacustrine deposits that has recorded multiple complete cycles of lake water depth fluctuations from shallow to deep, and can be further divided into the upper sub-member (P_2l^2) and the lower sub-member (P_2l^1) (Figure 1a). Previous studies have shown that the Lucaogou Formation contains a set of lacustrine source rocks that are rich in organic matter [15,18,21]. Their lithology includes gray–black mudstone, dolomitic mudstone, siltstone, and dolomite [15,21]. The main lithology of the reservoir includes dolomite, dolomitic siltstone, sandy dolomite, diabase, and a small amount of tuff [11,17]. The storage space of the reservoir is dominated by nanopores and microfractures, the pore throat radius is 100–500 nm, the porosity is 2.5–16.27%, and the permeability is less than 0.1 mD [6].

3. Samples and Analytical Methods

3.1. Samples

In this study, a total of 137 samples of the P_2l source rocks were collected from 14 wells in the Jimsar Depression, including 40 samples from the P_2l^1 member and 97 samples from the P_2l^2 member. The locations of the wells are shown in Figure 1c. Based on these samples, various analytical measurements were carried out on more than 700 samples, including the pyrolysis, extraction, and quantification of the organic carbon and soluble organic matter; the analysis of biomarker compounds; the determination of the whole-rock organic microscopic composition; and the determination of the vitrinite reflectance (Ro).

3.2. Analytical Methods

3.2.1. Evaluation of Source Rocks

The source rock samples were cut into thin sections to observe the lithology of the source rocks and the characteristics of the organic detritus, and determine the Ro. The "Method for Isolation of Kerogen in Sedimentary Rocks", published and implemented in 2003, was used to isolate the kerogen, and the component identification, quantification, and the classification of the kerogen types of the isolated kerogen were carried out based on the "Methods for Microscopic Composition Identification and Type Classification by Transmitted Light-Fluorescence" industry standard (SY/T5125-2014).

The pyrolysis of the source rocks was conducted using a Rock-Eval VI standard pyrolysis instrument (France). The sample was crushed, and 100 mg of the sample was heated in helium gas. The hydrocarbons released during the pyrolysis were monitored using a hydrogen flame ionization monitor, and the CO and CO_2 generated by the heating and oxidation of the residual organic matter after the pyrolysis were detected using a thermal conductivity detector. Following the rock pyrolysis analysis standard (GB/T18602-2012), a constant temperature of 300 °C was set for 3 min to obtain S_1; the temperature was increased from 300 °C to 650 °C at a rate of 25 °C for 1 min to obtain S_2. Then the highest pyrolysis peak temperature (T_{max}), Hydrocarbon Index (HCI), Hydrogen Index (HI), Oxygen Index (OI), Hydrocarbon Generation Potential ($S_1 + S_2$), and Production Index (PI) were obtained through equations for calculating the pyrolysis parameters.

3.2.2. Simulation of Two-Dimensional Basin

The simulation of a two-dimensional basin was carried out using the PetroMod software. The data of heat flows of present and past, and the electrical conductivity of the source rocks, were obtained from previous studies that had carried out simulations [22,23]. The solution (EasyRo) given by Sweeney and Burnham (1990) [24] was used to convert the result to source rock maturity, and the applicable maturity interval was 0.3–4.6% Ro.

3.2.3. Thermal Simulation Experiment in a Closed System

The source rock sample was sealed in a gold tube under the protection of an argon gas atmosphere. The gold tube was placed in an autoclave, and the autoclave was filled with water using a high-pressure pump. The high-pressure water caused the gold tube to be flexibly deformed, thereby exerting pressure on the sample. The sample was sealed under the protection of argon to ensure that there was no air contamination. The gold tube was sealed via arc welding. The samples were heated at rates of 20 °C/min and 2 °C/min. The temperature difference between each autoclave was less than 1 °C. The pressure was 50 MP, and the pressure fluctuation was less than 1 MP. The temperature range was 150–456 °C, and the temperature fluctuation was less than 1 °C. After heating, the gold tube was removed from the autoclave. The detected contents were gas (C_1–C_4), light hydrocarbons (C_6–C_{14}), and heavy hydrocarbons (C_{14+}).

4. Results and Discussion

4.1. Depositional Environment of Source Rocks

4.1.1. Normal Alkanes and Isoprenoid Alkanes

Isoprenoid alkane compounds are commonly used as indicators of the sedimentary environment [25,26]. The pristane/phytane (Pr/Ph) ratios of the P_2l source rocks in the Jimsar Depression were 0.81–2.62, with an average of 1.37. Most of the values were in the range of 0.75–1.5, but quite a few samples had values of >2 (Table 1). Peters et al. (2005) [25] pointed out that when the Pr/Ph ratio is 0.8–3.0, there is great uncertainty in using this ratio to explain the depositional environment of the source rock. Therefore, the ratio of isoprenoid to normal alkanes [27] was used in this study. As is shown in Figure 2, the Pr/n-C_{17} and Ph/n-C_{18} values were 0.14–6.40 (average 1.54) and 0.09–4.33 (average 1.43), respectively, indicating a reducing environment overall, with bacteria and algae as the main sources of organic matter. The contribution of land-based organic matter was low, and the maturity was not high. It should be noted that the distribution characteristics of the isoprenoid alkanes and n-alkanes are also controlled by their maturity [25,28]. The maturity of the samples from the study area was relatively low, so the impact of the maturity could generally be ignored (Table 1, Figure 2).

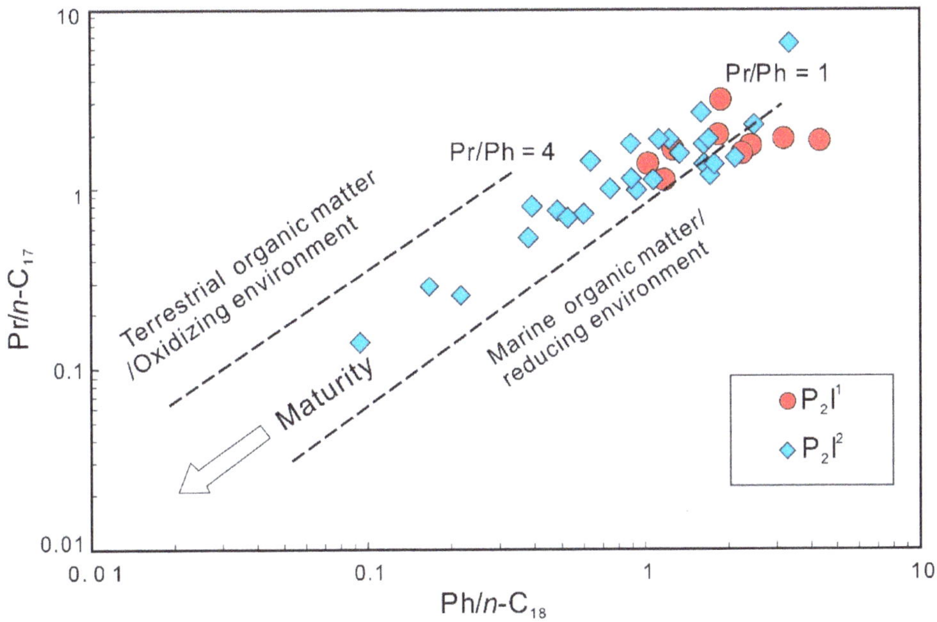

Figure 2. Plot of Pr/n-C_{17} vs. Ph/n-C_{18} showing the depositional environment of the P_2l source rocks in the Jimsar Depression (modified after Shanmugam, 1985) [27] (terrestrial organic matter/oxidation environment; maturity; marine organic matter/reducing environment).

Table 1. Biomarker fingerprints of P_2l shale in the Jimsar Depression.

Well	Depth(m)	Strata	Lithology	C_{19}/C_{21}-Tricyclic Terpane	Tricyclic Terpane/Pentacyclic Terpane	C_{24}-Tetracyclic Terpane/C_{26}-Tricyclic Terpane	$C_{35}/(C_{31}-C_{35})$ Homohopane %	$C_{30}\alpha\beta$-Hopane/$C_{29}\alpha\alpha\alpha$,(R + S)-Sterane	Gammacerane Index	C_{31R}/C_{30}-Hopane	Ts/(Ts + Tm)	C_{27}-Sterane (%)	C_{28}-Sterane (%)	C_{29}-Sterane (%)
Ji15	2280.63	P_2l^1	Dark gray fluorescent mudstone	0.41	0.29	1.12	7.07	6.79	0.23	0.10	0.43	23.78	23.77	52.46
Ji174	3218.97	P_2l^1	Dark gray calcareous mudstone	0.12	0.12	1.95	2.72	4.85	0.28	0.12	0.05	13.80	37.24	48.96
Ji174	3227.14	P_2l^1	Dark gray dolomitic mudstone	0.13	0.13	1.55	2.66	6.59	0.31	0.17	0.04	14.81	37.11	48.09
Ji24	1691.75	P_2l^1	Gray black mudstone	0.08	0.07	0.62	2.87	5.58	0.31	n.d.	0.27	18.41	32.77	48.83
Ji17	3138.01	P_2l^2	Black gray mudstone	0.10	0.14	1.06	4.16	8.20	0.28	0.11	0.09	14.99	35.39	49.62
Ji17	3138.01	P_2l^2	Black gray mudstone	0.10	0.05	0.82	4.01	12.73	0.36	0.16	0.32	34.81	26.13	39.06
Ji17	3135.33	P_2l^2	Gray black mudstone	0.13	0.06	1.40	3.79	9.88	0.39	0.22	0.33	36.51	26.17	37.33
Ji17	3135.33	P_2l^2	Gray black mudstone	0.07	0.04	0.71	4.05	19.28	0.35	0.15	0.29	42.15	22.16	35.70
Ji174	3135.31	P_2l^2	Dark gray mudstone	0.12	0.15	1.28	2.40	4.62	0.34	0.12	0.07	20.25	34.09	45.66
Ji174	3113.3	P_2l^2	Dark gray mudstone	0.10	0.16	1.46	2.57	4.08	0.42	0.22	0.10	18.88	33.39	47.73
Ji174	3134.05	P_2l^2	Black gray mudstone	0.08	0.10	1.38	2.70	5.44	0.35	0.20	0.10	22.94	31.56	45.50
Ji174	3146.16	P_2l^2	Dark gray mudstone	0.12	0.18	1.65	2.90	3.39	0.19	n.d.	0.12	18.53	41.45	40.01
Ji174	3158.88	P_2l^2	Dark gray mudstone	0.32	0.04	8.93	2.56	6.01	0.21	0.18	0.23	19.84	39.07	41.10
Ji174	3166.74	P_2l^2	Dark gray mudstone	0.11	0.21	2.42	2.23	3.02	0.27	0.17	0.16	21.47	40.29	38.24
Ji174	3177.55	P_2l^2	Dark gray mudstone	0.16	0.09	3.62	2.13	6.44	0.12	0.12	0.25	22.08	37.87	40.05
Ji174	3217.51	P_2l^2	Dark gray calcareous mudstone	0.12	0.17	1.30	2.80	4.89	0.31	0.11	0.06	13.10	35.79	51.11
Ji174	3130.76	P_2l^2	Black gray mudstone	0.12	0.12	2.29	2.32	8.37	0.26	0.16	0.14	25.81	30.83	43.36
Ji174	3114.73	P_2l^2	Dark gray mudstone	0.11	0.16	1.37	2.39	7.26	0.39	0.20	0.14	31.83	28.72	39.45
Ji174	3117.75	P_2l^2	Dark gray mudstone	0.13	0.15	1.96	2.45	11.13	0.41	n.d.	0.10	20.59	32.54	46.88
Ji174	3118.78	P_2l^2	Dark gray mudstone	0.14	0.19	2.15	2.04	7.08	0.38	0.19	0.09	17.65	34.22	48.13
Ji174	3122.14	P_2l^2	Dark gray mudstone	0.18	0.09	5.37	3.39	4.14	0.31	0.19	0.17	15.76	31.31	52.93

Table 1. *Cont.*

Well	Depth(m)	Strata	Lithology	C_{19}/C_{21}-Tricyclic Terpane	Tricyclic Terpane/Pentacyclic Terpane	C_{24}-Tetracyclic Terpane/C_{26}-Tricyclic Terpane	$C_{35}/(C_{31}-C_{35})$ Homohopane %	$C_{30}\alpha\beta$-Hopane/$C_{29}\alpha\alpha\alpha$,(R + S)-Sterane	Gammacerane Index	C_{31R}/C_{30}-Hopane	Ts/(Ts + Tm)	C_{27}-Sterane (%)	C_{28}-Sterane (%)	C_{29}-Sterane (%)
Ji174	3146.19	P_2l^2	Dark gray mudstone	0.12	0.16	1.11	3.66	2.80	0.12	0.20	0.11	19.22	44.60	36.19
Ji174	3152.98	P_2l^2	Dark gray mudstone	0.18	0.21	4.31	2.94	6.27	0.27	0.18	0.12	14.69	52.95	32.36
Ji174	3155.32	P_2l^2	Dark gray dolomitic mudstone	0.22	0.10	8.54	1.46	5.91	0.26	0.21	0.15	14.69	57.26	28.05
Ji174	3162.02	P_2l^2	Dark gray mudstone	0.22	0.03	7.27	2.34	7.27	0.25	0.18	0.22	18.76	41.56	39.68
Ji174	3196.3	P_2l^2	Dark gray dolomitic mudstone	0.09	0.16	1.55	2.44	7.47	0.27	0.16	0.18	20.70	37.28	42.02
Ji22	2543.06	P_2l^2	Dark gray mudstone	0.10	0.15	1.69	1.75	5.25	0.35	0.13	0.12	23.97	33.63	42.40
Ji22	2554.54	P_2l^2	Dark gray dolomitic mudstone	0.06	0.13	0.87	2.44	6.56	0.40	0.08	0.08	23.69	30.96	45.35
Ji22	2552.67	P_2l^2	Dark gray dolomitic mudstone	0.09	0.09	2.39	1.51	5.97	0.25	0.08	0.21	35.35	32.11	32.54
Ji22	2554.8	P_2l^2	Dark gray dolomitic mudstone	0.08	0.07	1.76	2.35	10.36	0.55	0.22	0.10	33.07	33.13	33.80
Ji23	2296.12	P_2l^2	Gray black calico mudstone	0.05	0.07	0.81	2.59	8.57	0.51	0.21	0.21	31.11	30.44	38.46
Ji23	2296.04	P_2l^2	Gray black mudstone	0.04	0.06	0.61	2.48	8.32	0.50	0.21	0.22	30.02	30.41	39.57
Ji5	3535.56	P_2l^2	Gray black mudstone	1.43	0.08	n.d.	n.d.	4.73	0.37	0.25	0.52	17.43	32.51	50.06
Ji7	2059	P_2l^2	Gray black mudstone	0.16	0.05	1.26	4.76	1.72	0.16	0.21	0.30	17.42	17.36	65.22
Ji7	1949.61	P_2l^2	Gray black mudstone	0.05	0.08	1.19	6.70	1.26	0.24	0.17	0.24	18.73	31.91	49.36
Ji7	2069.99	P_2l^2	Dolomitic mudstone	0.29	0.19	n.d.	n.d.	2.10	0.13	0.23	0.09	32.46	31.15	36.39
Ji7	2160.2	P_2l^2	Dolomitic mudstone	0.13	0.12	3.62	3.62	2.50	0.24	0.25	0.34	24.89	31.31	43.79
Ji7	2287.45	P_2l^2	Dolomitic mudstone	0.09	0.17	1.06	2.75	1.29	0.14	n.d.	0.35	22.83	17.69	59.47
Ji7	1949.61	P_2l^2	Dark gray dolomitic mudstone	0.00	0.08	1.17	7.19	2.16	0.25	0.21	0.26	19.69	33.79	46.52
Ji7	2160.2	P_2l^2	Dark gray dolomitic mudstone	0.24	0.05	10.90	n.d.	8.14	0.23	0.13	0.37	29.38	30.63	39.99
Ji7	2287.45	P_2l^2	Dark gray dolomitic mudstone	0.33	0.06	2.80	3.06	9.21	0.20	n.d.	0.10	32.70	26.05	41.25
Ji7	2069	P_2l^2	Black mudstone	0.07	0.08	0.86	n.d.	1.51	0.27	n.d.	0.21	17.61	28.06	54.33
Ji7	2160	P_2l^2	Black dolomitic mudstone	0.15	0.06	1.86	5.31	4.14	0.61	n.d.	0.28	16.64	23.63	59.73

4.1.2. Terpanes

Tricyclic terpanes are abundant in the P_2l shale of the study area, mainly in the range of C_{19}–C_{28} (Figure 3), indicating that their formation was closely related to the input of an algae source [29]. The baseline of the GC–MS spectra is nearly horizontal, with no UCM "hump", suggesting that there was nearly no influence of biodegradation. The source rocks are rich in hopanes. The $C_{30}\alpha\beta$-hopane/$C_{29}\alpha\alpha\alpha$,(R + S)-sterane ratios are 1.26–19.28, with an average of 6.12 (Figure 3, Table 1), suggesting that bacteria reproduced rapidly in a lake environment and provided a large amount of hopane precursors. The ratios of some individual samples were relatively low, which may be related to a certain amount of terrestrial organic matter input. C_{31}–C_{35} homohopanes are closely related to prokaryotic microorganisms, and high abundance of C_{35} homohopane often indicates a strongly reducing depositional environment [25,30]. The C_{35} homohopane index (C_{35}/(C_{31}–C_{35}) homohopane%) of the P_2l shale is relatively high, with a range of 1.46–7.19 and an average of 3.17, indicating that the sedimentary water body had strong reducibility (Figure 3, Table 1). The C_{31}R homohopane/C_{30} hopane ratios are 0.08–0.25, with an average of 0.17, also exhibiting the characteristics of lacustrine crude oil (Figure 3, Table 1). Gammacerane can effectively reflect the stratification of sedimentary water bodies, which in turn is related to the salinity of the water body [31]. The abundance of gammacerane in coal-measure source rocks formed in a swamp environment is very low, and their gammacerane index (Gammacerane/(Gammacerane + $C_{30}\alpha\beta$-hopane)) is usually less than 0.05 (or even less than 0.01). The abundance of gammacerane in lacustrine sedimentary mudstone is relatively high, and the gammacerane index is greater than 0.05 in most cases [31,32]. The gammacerane index values of P_2l shale source rocks in the study area were 0.12–0.61, with an average of 0.30 (Figure 3, Table 1), indicating that evaporation was very strong and the sedimentary water body had a fairly high salinity. In other words, in the Middle Permian, the Jimsar Depression was a saline lake basin.

Table 2. The definition of the biomarker code.

Code	Name	Code	Name
S1	5α(H),14β(H)-pregnane	T6	C24,13β(H),14α(H)-tricyclic terpane
S2	C22-homopregnane	T7	C25,13β(H),14α(H)-tricyclic terpane
S3	20S-ααα-sterane	T8	C26-tricyclic terpane
S4	20R-αββsterane	T9	C26-tricyclic terpane
S5	20S-αββ-sterane	T10	C24-tetracyclic terpane
S6	20R-ααα-sterane	T11	C28-tricyclic terpane
S7	20R-24-ethyl-10α13β17α-rearranged sterane	T12	C28-tricyclic terpane
S8	20S-24-ethyl-10α13α17β-rearranged sterane	T13	18α,-22,29,30-trisnorhopane(Ts)
S9	20S-24-methyl-ααα-sterane	T14	17α,-22,29,30-trisnorhopane(Tm)
S10	20R-24-methyl-αββ-sterane	T15	17α,21β-30-norhopane
S11	20S-24-methyl-αββ-sterane	T16	17α(H),21β(H)-hopane
S12	20R-24-methyl-ααα-sterane	T17	17β(H),21α(H)-moretane
S13	20S-24-ethyl-ααα-sterane	T18	22S-17α(H),21β(H)-homohopane
S14	20R-24-ethyl-αββ-sterane	T19	22R-17α(H),21β(H)-homohopane
S15	20S-24-ethyl-αββ-sterane	T20	Gammacerane
S16	20R-24-ethyl-ααα-sterane	T21	22R-17β(H),21α(H)-homohopane

Table 2. *Cont.*

Code	Name	Code	Name
T1	C19,14β(methyl)-tricyclic terpane	T22	22S-17α(H),21β(H)-double homohopane
T2	C20,13β(H),14α(H)-tricyclic terpane	T23	22R-17α(H),21β(H)-double homohopane
T3	C21,13β(H),14α(H)-tricyclic terpane	T24	22S-17α(H),21β(H)-trishomohopane
T4	C22,13β(H),14α(H)-tricyclic terpane	T25	22R-17α(H),21β(H)-trishomohopane
T5	C23,13β(H),14α(H)-tricyclic terpane		

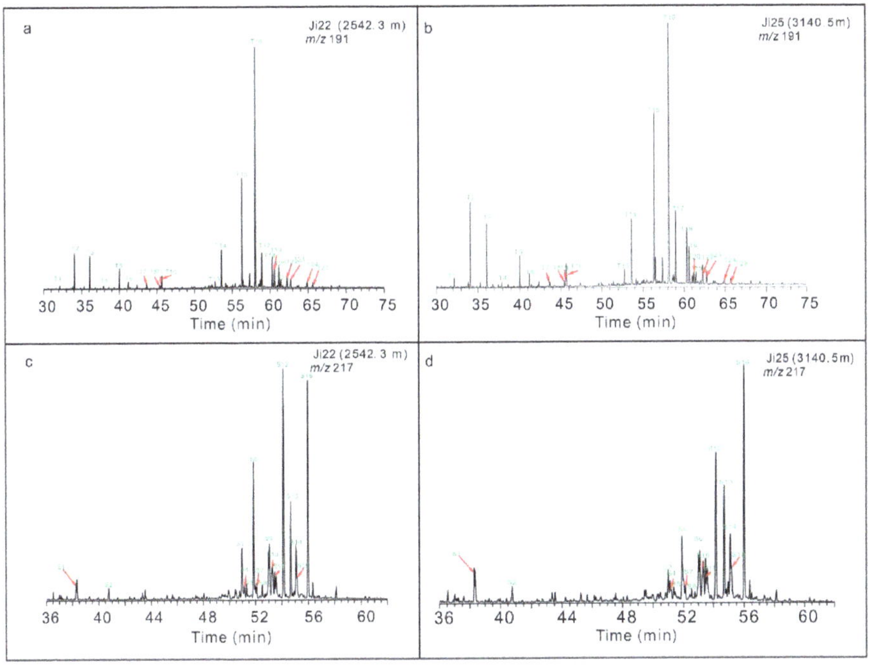

Figure 3. Gas chromatography–mass spectroscopy (GC–MS) spectra of the biomarkers of the (**a**,**b**) terpanes (m/z 191) and (**c**,**d**) steranes (m/z 217) in the P_2l source rocks in the Jimsar Depression (the compounds corresponding to all of the labels are shown in Table 2).

4.1.3. Sterane

The data suggest that the Lucaogou source rocks were rich in C_{27} and C_{28} steranes, implying the existence of large quantities of algae organic matter [33]. The contents of the C_{27}, C_{28}, and C_{29} regular steranes in the P_2l shale in the study area were 13.1–42.2% (average 22.9%), 17.4–57.3% (average 33.0%), and 28.1–57.3% (average 44.1%), respectively (Table 1, Figures 3 and 4). The total content of the C_{27} and C_{28} regular steranes in the source rock samples exceeded the content of C_{29} regular steranes. This indicates that algae organic matter was dominant; however, it also indicates that there may have been some contribution from terrestrial organic matter to source rocks in the study area. The relatively high C_{29} sterane content was not just related to the source of the organic matter; it was also affected by the depositional environment of the source rocks. Generally, lacustrine source rocks dominated by input of algae sources have higher relative C_{29} sterane content than marine carbonate/shale [34–36]. Overall, the P_2l shale had relatively high C_{27} and C_{28} sterane content (Figure 4).

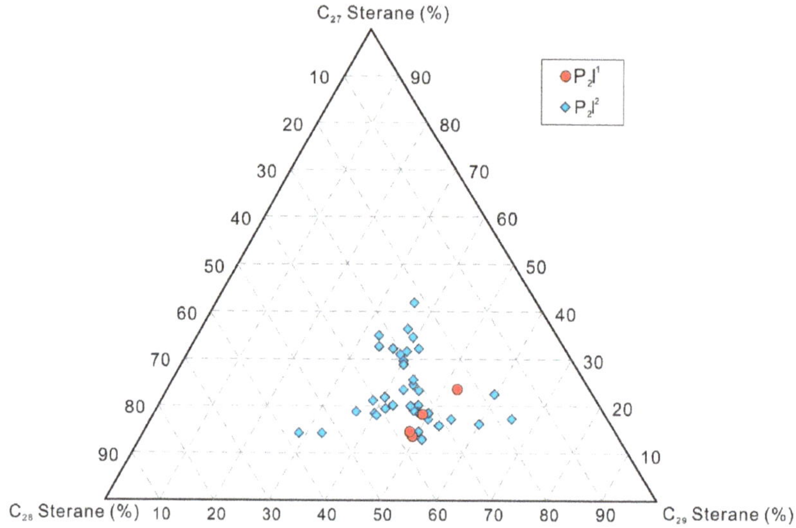

Figure 4. Ternary diagram of the relative percentages of the C_{27}, C_{28}, and C_{29} regular steranes in the P_2l source rocks in the Jimsar Depression.

4.2. Type and Abundance of Source Rocks

4.2.1. Types of Organic Matter

The elemental analysis revealed that the H/C atomic ratios of the P_2l^1 member in the study area were 0.79–1.33 (average 1.06); the O/C atomic ratios were 0.04–0.08 (average 0.06); the H/C atomic ratios of the P_2l^2 member were 0.82–1.60 (average 1.23); the O/C atomic ratios were 0.03–0.30 (average 0.08); and the two sets of source rocks did not exhibit significant differences, mainly exhibiting the characteristics of type II kerogen (Figure 5a). Hunt (1996) [37] pointed out that when the H/C atomic ratio is >0.8, organic matter starts to have the capability to generate oil. According to this standard, the P_2l shale had a relatively high H/C atomic ratio and was a set of oil-prone source rocks. It should be noted that large differences in the H/C atomic ratios of the different samples indicated that this set of source rocks was strongly heterogeneous.

Based on the pyrolysis HI values of >600, 300–600, 200–300, 50–200, and <50 mg hydrocarbons/g TOC, the organic matter can be divided into five types: type I (extremely oil-prone type), type II (oil-prone type), II/III type (oil/gas-prone type), type III (gas-prone type), and type IV (non-source rock) (Figure 5b) [38,39]. The HI values of the P_2l^1 member in the study area were 12–781 mg hydrocarbons/g TOC (average of 367 mg hydrocarbons/g TOC), the HI values of the P_2l^2 member were 72–808 mg hydrocarbons/g TOC (average of 390 mg hydrocarbons/g TOC), and there was no large difference between the two sets of source rocks (Figure 5b). The type I, type II, type II/III, type III, and type IV samples accounted for 14.5%, 47.1%, 21.0%, 15.2%, and 2.0% of the total samples, respectively, indicating that the P_2l source rocks were dominated by type II kerogen, and that type I kerogen did not account for a large proportion. In addition, this set of source rocks was very strongly heterogeneous, and the types of parent materials involved were very different. This heterogeneity was very obvious even in samples from the same well. As an example, for Well Ji174, from which a large number of samples were collected, the HI values of the P_2l^1 and P_2l^2 members were 48–294 and 72–808 mg hydrocarbons/g TOC, respectively.

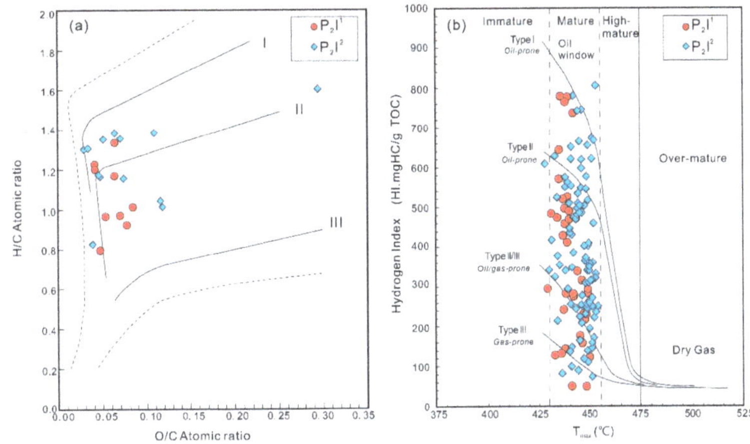

Figure 5. Plots of (**a**) O/C vs. H/C and (**b**) atomic ratio of HI vs. T_{max} showing the organic matter types of the P_2l source rocks in the Jimsar Depression (H/C, atomic ratio; O/C, atomic ratio; immature; mature; oil generation window; highly mature; condensate oil-wet gas; over-mature; dry gas generation).

The analysis of the organic microscopic compositions of whole-rock samples revealed that the contents of the vitrinite group and exinite group of the P_2l source rocks were low (<20%), while the contents of the sapropelinite group + exinite group were very high (62.3–78.7%). In general, the sapropelinite group component mainly generated crude oil, the vitrinite group component mainly generated natural gas, the exinite group component generated both, and the inertinite group component had almost no potential for hydrocarbon generation. Therefore, the P_2l source rocks in the study area should mainly be oil-generating. Figure 6 shows the results of the whole-rock organic microscopic composition analysis of a core sample (2321.4 m) of P_2l^1 from Well Ji15. Under reflected fluorescence, the clay (Cl) substrate exhibited a strong fluorescence, and, among them, the parallel-distributed microsporinite (MiS), resinite (R), liptodetrinite (Ld), and vitrodetrinite (Cd) were also widely distributed (Figure 6). Under reflected plane-polarized light (oil immersion), the clay (Cl) minerals exhibited a granular structure in which Cd and semifusinite (SF) fragments were distributed and the latter's structure was broken. Pyrite (Py) was widely distributed, reflecting the strong reducibility of the sedimentary environment and the good organic matter preservation conditions (Figure 6).

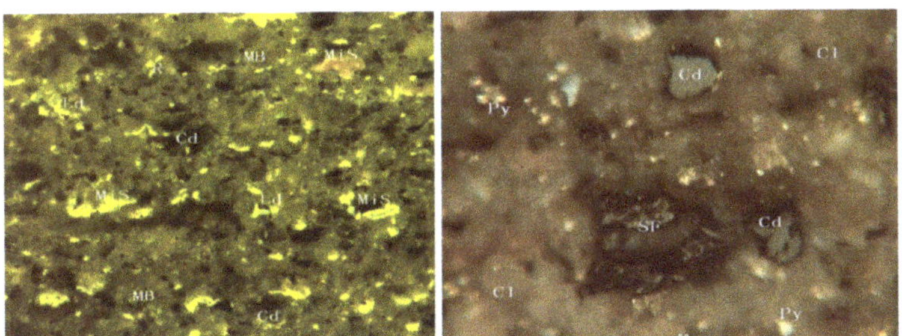

Figure 6. Analysis of the organic microscopic composition of the whole-rock P_2l^1 sample (2321.4 m) from Well Ji15 (left image: reflected fluorescence; right image: reflected plane-polarized light). Cl is clay; MiS is microsporinite; R is resinite (R); Ld is liptodetrinite; Cd is vitrodetrinite; SF is semifusinite; MB is mineral bitumen; Py is pyrite.

4.2.2. Abundance of Organic Matter

Huang et al. (1982) [40] summarized the abundance of organic matter in the source rocks of the main petroliferous basins in China, and proposed corresponding evaluation criteria for lacustrine source rocks deposited in freshwater and brackish water environments (Figure 7). The TOC values of the P_2l^1 and P_2l^2 source rocks were 0.38–7.55% (average 3.30%) and 0.34–12.45% (average 3.80%), respectively. From the perspective of organic carbon, the proportions of good and extremely good source rocks in the P_2l^1 source rocks were 15.0% and 70.0%, respectively, while the proportions of good and extremely good source rocks in the P_2l^2 source rocks were 14.3% and 75.5%, respectively. The source rock $S_1 + S_2$ values of the P_2l^1 and P_2l^2 members were 0.26–39.44 mg hydrocarbons/g rock (average of 15.21 mg hydrocarbons/g rock) and 0.47–78.96 mg hydrocarbons/g rock (average of 17.60 mg hydrocarbons/g rock), respectively (Figure 7). In terms of $S_1 + S_2$, the proportions of the good and extremely good source rocks in the P_2l^1 member were 40% and 30%, respectively, while the proportions of the good and extremely good source rocks in the P_2l^2 source rocks were 37.8% and 34.7%, respectively (Figure 7).

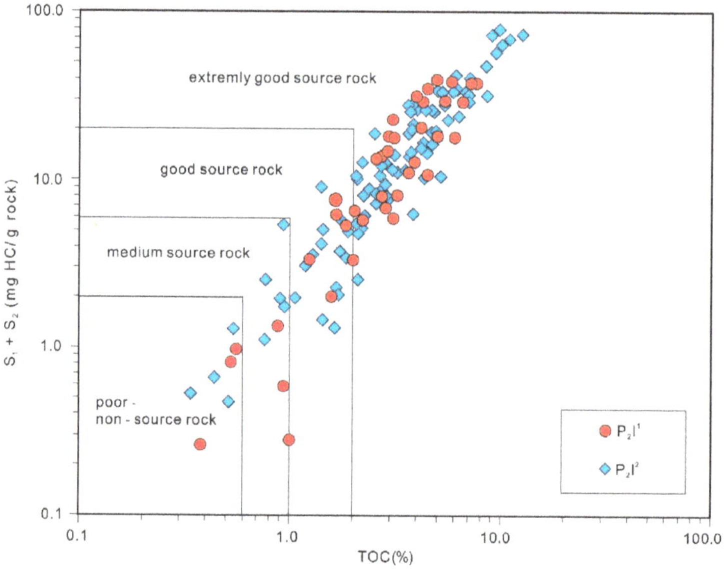

Figure 7. TOC vs. $S_1 + S_2$ diagram showing the abundance of organic matter in the P_2l source rocks in the Jimsar Depression (extremely good source rock; good source rock; medium-grade source rock; poor-non-source rock).

From the perspective of the TOC and $S_1 + S_2$, the P_2l source rocks were generally good to extremely good source rocks, but they also exhibited a strong heterogeneity (Figure 7). It should be noted that when the TOC was used as the evaluation standard, the quality of the source rocks was considerably better than the results obtained using the hydrocarbon generation potential as the standard. For example, when TOC was used as the evaluation criterion, the proportion of extremely good source rocks out of the total number of samples was twice as high as the results obtained based on $S_1 + S_2$. This is mainly due to the difference in the hydrogen content of the source rocks. In thermochemical reactions, carbon must be combined with hydrogen to generate hydrocarbons. If the hydrogen content is low, even if the organic carbon content is high, the result is often ineffective. The heterogeneity of the hydrogen content of the source rocks, which is reflected in the analysis of organic elements and the pyrolysis hydrogen index, also confirms this view (Figure 5). When evaluating the P_2l source rocks, the results obtained using the pyrolysis hydrocarbon generation potential as the standard may be more reliable.

4.3. Thermal Evolution of the Source Rocks

4.3.1. Maturity of the Source Rocks

The vitrinite reflectance values of the P_2l source rocks in the study area were 0.52–1.24%, with an average of 0.71%, meaning they were within the main oil generation window. It is generally believed that when S_2 is less than 0.2 mg hydrocarbons/g rock, T_{max} is unreliable [41]. Except for the two samples from Well Ji5, the S_2 values of the samples from the study area were greater than 0.2 mg hydrocarbons/g rock. The T_{max} values obtained based on these values were generally reliable, ranging from 428 °C to 454 °C, with an average of 445 °C. The samples were essentially in a mature stage, which is consistent with the situation reflected by the vitrinite reflectance.

The gas chromatographic analysis of the extracted source-rock bitumen revealed that the carbon preference index (CPI) values of the P_2l source rocks in the Jimsar Depression were 1.18–2.0 (average 1.41), and the odd-to-even preference (OEP) was 1.14–1.71 (average 1.30), exhibiting a significant odd–even predominance, which indicates that the maturity of the source rocks was not high. The $C_{29}20S/(S + R)$-sterane and $C_{29}\beta\beta/(\alpha\beta\beta + \alpha\alpha\alpha)$-sterane ratios of the source rock extracts were 0.09–0.46 and 0.11–0.45, respectively, making them lower than the equilibrium values (0.67–0.71 and 0.52–0.55), though they were still within the oil generative window (Figure 8, Table 1) [25].

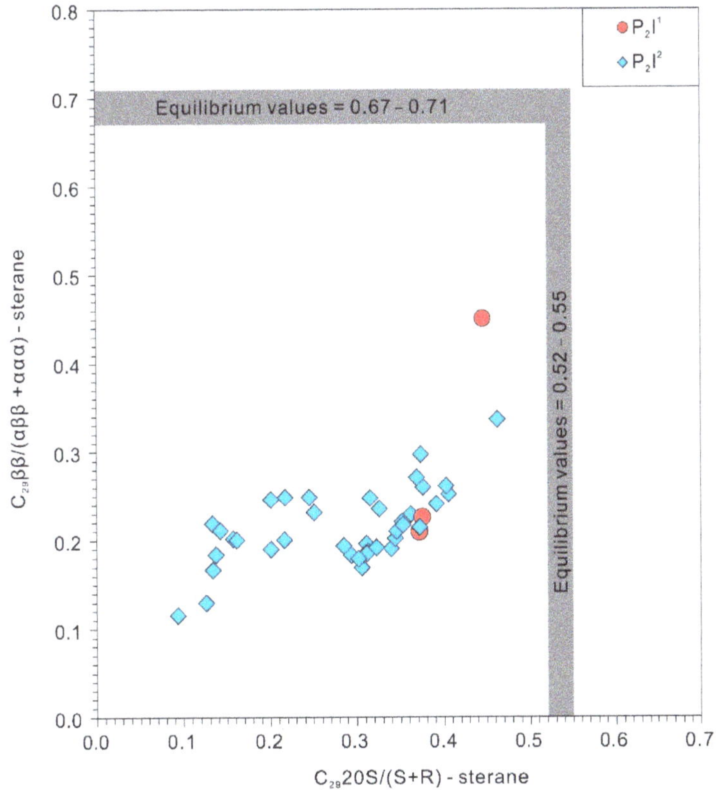

Figure 8. Use of sterane isomer maturity parameters to determine the organic matter maturity of the P_2l source rocks in the Jimsar Depression.

4.3.2. Thermal Evolution of the Source Rocks

Based on a two-dimensional seismic profile crossing the Jimsar Depression in the east–west direction, we restored the hydrocarbon generation and expulsion history and burial history of the P_2l source rocks (Figure 9; the location of the profile is shown in Figure 1c). At the end of the Triassic, the maturity of the P_2l shale was low, EasyRo < 0.5%, and it had not yet reached the oil generation threshold (Figure 9a). The maturity of the P_2l shale increased significantly at the end of the Cretaceous, with an EasyRo > 0.7%, and it entered the main oil generation window, producing a large amount of normal crude oil. The highly mature stage (1.0–1.3% EasyRo) was reached in the deep part of the Depression, and certain amounts of highly mature crude oil and condensate gas were generated (Figure 9b). After the Cretaceous deposition, the formation was uplifted and denuded, and the hydrocarbon generation process in the P_2l source rocks stalled (Figure 9c). In the Paleogene, P_2l continued to settle (Figure 9d). Currently, the deep part of the depression has generally reached the high-maturity stage, producing a large amount of highly mature oil and condensate gas (1.0–1.3% Ro). The shallow part had relatively low maturity and could only generate a large amount of normal crude oil (0.7–1.0% EasyRo) (Figure 9e).

4.4. *Thermal Simulation Experiment on Source Rocks in a Closed System*

In this study, a P_2l calcareous and siliceous black shale sample (core, 2321.4 m; sampling location is shown in Figure 1c) from Well Ji15 in the eastern margin of the Jimsar Depression was selected. Its kerogen was extracted, and a gold-tube thermal simulation experiment was carried out in a closed system. The total organic carbon content of this sample was 2.46%, the hydrocarbon generation potential (S_1 + S_2) was 13.36 mg hydrocarbons/g rock, and the HI was 491 mg/g TOC. Generally speaking, the abundance index was in the middle of the range of the P_2l shales in the Jimsar Depression, and was representative. In addition, the maturity of the sample was relatively low, the measured vitrinite reflectance was 0.54%, and the sample was determined to be an ideal thermal simulation sample.

When the gas leaves the gold tube under the internal pressure of 1×10^2 Pa and enters the vacuum equipment for gaseous compound analysis, large quantities of C_6 and C_7 compounds evaporate and are lost. Therefore, when the total amount of liquid hydrocarbon compounds produced was determined, only the amount of C_{8+} compounds was measured (Figure 10). At 2 °C/h and 20 °C/h, the maximum amounts of oil generated by the P_2l shale in the experiment were 209.2 mg/g (0.96EASY%Ro) and 172.3 mg/g (1.08EASY%Ro), respectively (Figure 10). There was no large difference in the maximum amount of oil generated by the sample at the heating rates of 2 °C/h and 20 °C/h. When EASY%Ro > 1.0, the maximum amount of oil generated at a heating rate of 2 °C/h was slightly higher than that generated at a heating rate of 20 °C/h (Figure 10). The experimental results show that the heating rate was not the key factor controlling the crude oil yield in this study. During the thermal simulation process, when the kerogen was cracked to produce crude oil for the first time, part of the crude oil was cracked to generate natural gas at the same time. Therefore, we could assume that the measured maximum amount of oil generated represented a conversion rate of 95% for the kerogen sample [42]. Based on this assumption, the maximum amount of oil generated by the sample in this study was 220.2 mg/g TOC (209.2 mg/g TOC/0.95).

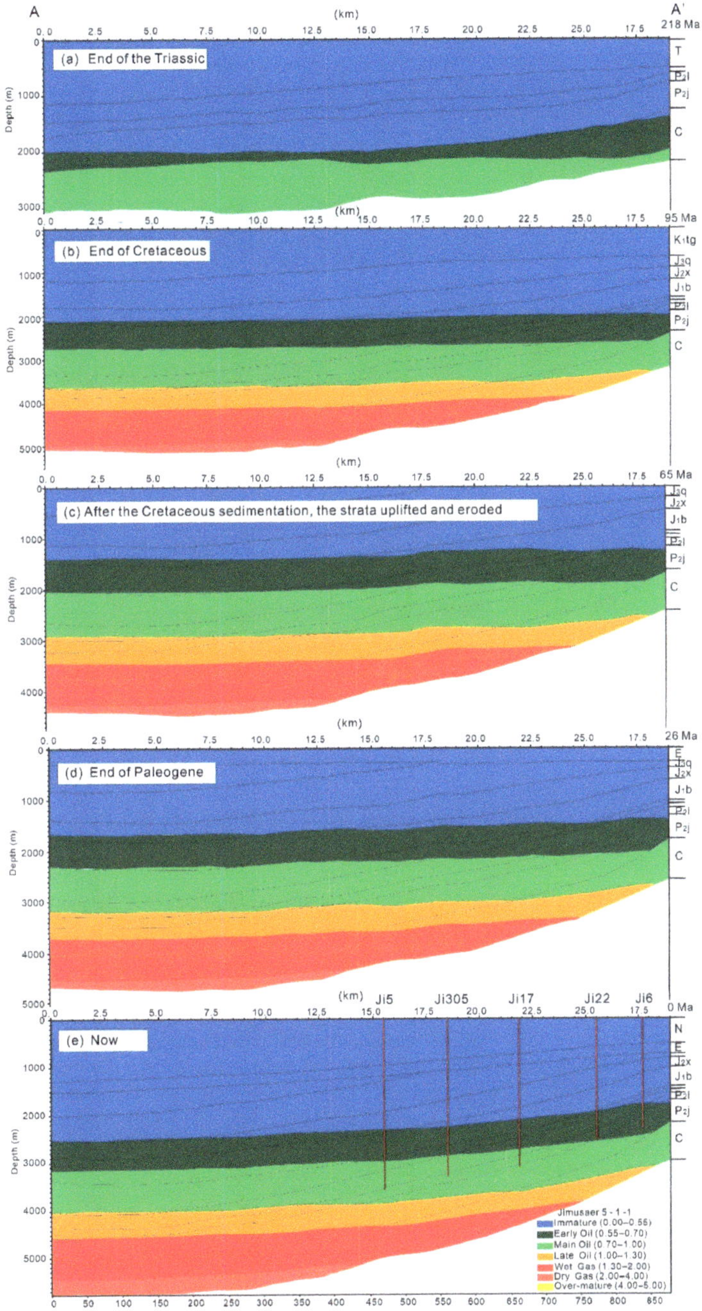

Figure 9. Two-dimensional simulation profile of the thermal evolution and burial history of the P_2l source rocks in the Jimsar Depression (Late Triassic (**a**); Late Cretaceous (**b**); after Cretaceous deposition, uplift and denudation of formations (**c**); Late Paleogene (**d**); present day(**e**)). The location of this section can be found in Figure 1c.

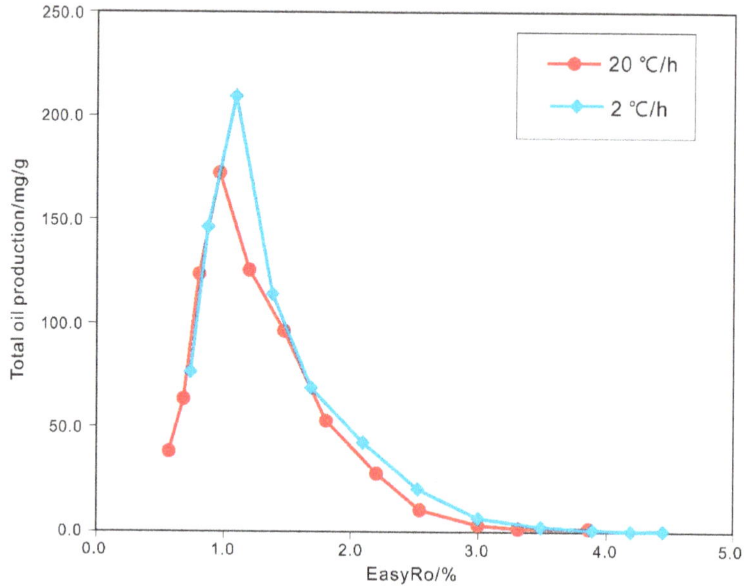

Figure 10. Simulation results of the oil production rate under different heating rates of the P_2l source rocks in the Jimsar Depression.

The organic adsorption model proposed by Pepper (1992) and Sandvik et al. (1992) [43,44] can be used to estimate the amount of oil expelled from the source rocks. We defined R as the amount of residual oil in g/g TOC, and we defined M and W as the final amount of oil generated and the final amount of oil expelled per gram of initial organic carbon in the source rocks, respectively [42]. Assuming that the carbon content of the expelled crude oil was 80% after the oil expulsion was completed, the amount of residual organic carbon and residual oil per gram of original organic carbon in the source rocks were (1 − 0. 8 × W) and R × (1 − 0.8 × W), respectively; thus, the following equations could be established [42]:

$$M = W + R \times (1 - 0.8 \times W), \quad (1)$$

$$W = (M - R)/(1 - 0.8 \times R), \quad (2)$$

where the units of M and W are g/g TOC. As previously mentioned, according to the pyrolysis experiment conducted in the closed system, the maximum amount of oil generated (M) by the sample was 220.2 mg/g TOC. The content (R) of bitumen extracted from the P_2l source rocks in the Jimsar Depression was mainly in the range of 60–100 mg/g, the maximum amount of expelled crude oil (W) was 130.7–168.2 mg/g TOC, and the maximum hydrocarbon expulsion efficiency was 59.3–76.4% [42,43,45].

5. Conclusions

Based on data from the hydrocarbon pyrolysis, the contents of organic carbon and soluble organic matter, the biomarkers, the organic microscopic composition, the vitrinite reflectance, the basin simulation and hydrocarbon generation experiments, the sedimentary environment, hydrocarbon generation capacity, quality, and maturity of the Lucaogou Formation were discussed. Four main conclusions could be drawn.

(1) The majority of the organic matter in the Lucaogou Formation came from algae, and the sedimentary water environment had high reducibility and salinity.
(2) The organic microscopic components of the Lucaogou source rocks were dominated by the sapropelinite group and exinite group, and they were a set of oil-prone source

rocks. Most of them were rated as good and extremely good, but their heterogeneity was strong.

(3) At the end of the Cretaceous period, the Lucaogou Formation source rocks entered the main oil generation window and began to generate a large amount of normal crude oil. Currently, the deep part of the depression has generally reached the high-maturity stage and is generating a large amount of highly mature oil and condensate gas. The maturity of the shallow part is relatively low, and only a large amount of conventional crude oil can be generated.

(4) The thermal simulation experiment conducted in a closed system suggests that the Lucaogou source rocks have great hydrocarbon generation capacity and expulsion efficiency, and the heating rate did not have a significant impact on the maximum amount of oil generated.

Author Contributions: Conceptualization, W.H., D.L. and G.L.; methodology, W.H.; software, D.W. and Y.L.; validation, Y.L.; formal analysis, G.L.; investigation, G.G.; resources, W.H. and L.H.; data curation, L.H. and Y.Q.; writing—original draft preparation, W.H. and Y.L.; writing—review and editing, Y.L.; visualization, D.W. and Y.L.; supervision, D.L. and G.L.; project administration, W.H.; funding acquisition, W.H. and Y.L. All authors have read and agreed to the published version of the manuscript.

Funding: This research was funded by the Strategic Priority Research Program of the Chinese Academy of Sciences (Grant No. XDA14010301), National Natural Science Foundation of China (Grant No. 41702138).

Institutional Review Board Statement: Not applicable.

Informed Consent Statement: Not applicable.

Data Availability Statement: Not applicable.

Acknowledgments: We want to thank the Xinjiang Oilfield Company for the support in preparing this manuscript. We also thank the editors and anonymous reviewers for their helpful suggestions and comments.

Conflicts of Interest: The authors declare no conflict of interest.

References

1. Bai, H.; Pang, X.Q.; Kuang, L.C.; Pang, H.; Wang, X.L.; Jia, X.Y.; Zhou, L.M.; Hu, T. Hydrocarbon expulsion potential of source rocks and its influence on the distribution of lacustrine tight oil reservoir, Middle Permian Lucaogou Formation, Jimsar Sag, Junggar Basin, Northwest China. *J. Pet. Sci. Eng.* **2017**, *149*, 740–755. [CrossRef]
2. Chen, Z.H.; Cao, Y.C.; Ma, Z.J.; Zhen, Y.S. Geochemistry and origins of natural gases in the Zhongguai area of Junggar Basin, China. *J. Pet. Sci. Eng.* **2014**, *119*, 17–27. [CrossRef]
3. Su, Y.; Zha, M.; Liu, K.Y.; Ding, X.J.; Qu, J.X.; Jin, J.H. Characterization of Pore Structures and Implications for Flow Transport Property of Tight Reservoirs: A Case Study of the Lucaogou Formation, Jimsar Sag, Junggar Basin, Northwestern China. *Energies* **2021**, *14*, 1251. [CrossRef]
4. Jiang, Y.H.; Hou, D.J.; Li, H.; Zhang, Z.M.; Guo, R.B. Impact of the Paleoclimate, Paleoenvironment, and Algae Bloom: Organic Matter Accumulation in the Lacustrine Lucaogou Formation of Jimsar Sag, Junggar Basin, NW China. *Energies* **2020**, *13*, 1488. [CrossRef]
5. Tang, Y.; Cao, J.; He, W.J.; Shan, X.; Liu, Y.; Zhao, K.B. Development Tendency of Geological Theory of Total Petroleum System: Insights from the Discovery of Mahu Large Oil Province. *Xinjiang Pet. Geol.* **2021**, *42*, 1–9, (In Chinese with English abstract).
6. Jia, C.Z.; Zou, C.N.; Li, J.Z.; Li, D.H.; Zheng, M. Assessment criteria, main types, basic features and resource prospects of the tight oil in China. *Acta Pet. Sin.* **2012**, *33*, 343–350, (In Chinese with English abstract).
7. Zou, C.N.; Yang, Z.; Tao, S.Z.; Yuan, X.J.; Zhu, R.K.; Hou, L.H.; Wu, S.T.; Sun, L.; Zhang, G.S.; Bai, B.; et al. Continuous hydrocarbon accumulation over a large area as a distinguishing characteristic of unconventional petroleum; the Ordos Basin, north-central China. *Earth-Sci. Rev.* **2013**, *126*, 358–369. [CrossRef]
8. Zou, C.N.; Zhu, R.K.; Bai, B.; Yang, Z.M.; Hou, L.H.; Zha, M.; Fu, J.H.; Shao, Y.; Liu, K.Y.; Cao, H.; et al. Significance, Geologic Characteristics, Resource Potential and Future Challenges of Tight Oil and Shale Oil. *Bull. Mineral. Petrol. Geochem.* **2015**, *34*, 3–17, (In Chinese with English abstract).
9. Katz, B.; Lin, F. Lacustrine basin unconventional resource plays: Key differences. *Mar. Pet. Geol.* **2014**, *56*, 255–265. [CrossRef]

10. Carroll, A.R. Upper Permian lacustrine organic facies evolution, southern Junggar Basin, NW China. *Org. Geochem.* **1998**, *28*, 649–667. [CrossRef]
11. Jiang, Y.Q.; Liu, Y.Q.; Yang, Z.; Nan, Y.; Wang, R.; Zhou, P.; Yang, Y.J.; Kou, J.Y.; Zhou, N.C. Characteristics and origin of tuff-type tight oil in Jimusar Depression, Junggar Basin, NW China. *Pet. Explor. Dev.* **2015**, *42*, 741–749, (In Chinese with English abstract). [CrossRef]
12. Wang, Y.T.; Yang, Z.M.; Ma, W.C.; Pan, C.C.; Wang, F. Geochemical Characteristics and Genesis of Tight Oil in Lucaogou Formation of Jimsar Sag. *Xinjiang Pet. Geol.* **2017**, *38*, 379–384, (In Chinese with English abstract).
13. Hu, T.; Pang, X.Q.; Jiang, S.; Wang, Q.F.; Zheng, X.W.; Ding, X.G.; Zhao, Y.; Zhu, C.X.; Li, H. Oil content evaluation of lacustrine organic-rich shale with strong heterogeneity: A case study of the Middle Permian Lucaogou Formation in Jimusaer Sag, Junggar Basin, NW China. *Fuel* **2018**, *221*, 196–205. [CrossRef]
14. Kuang, L.C.; Wang, X.T.; Guo, X.G.; Chang, Q.S.; Jia, X.Y. Geological Characteristics and Exploration Practice of Tight Oil of Lucaogou Formation in Jimsar Sag. *Xinjiang Pet. Geol.* **2015**, *36*, 629–634, (In Chinese with English abstract).
15. Gao, G.; Xiang, B.L.; Li, T.T.; Ren, J.L.; Kong, Y.H. Tight Oil System Particularity of Lucaogou Formation in Jimusaer Sag, Junggar Basin. *Acta Sedimentol. Sin.* **2017**, *35*, 824–833, (In Chinese with English abstract).
16. Wang, X.L.; Zhi, D.M.; Wang, Y.T.; Chen, J.P.; Qin, Z.J.; Liu, D.G.; Xiang, Y.; Lan, W.F.; Li, N. *Organic Geochemistry of Source Rocks and Hydrocarbons in the Junggar Basin*; Petroleum Industry Press: Beijing, China, 2013; pp. 18–73. (In Chinese)
17. Wu, H.G.; Hu, W.X.; Tang, Y.; Cao, J.; Wang, X.L.; Wang, Y.C.; Kang, X. The impact of organic fluids on the carbon isotopic compositions of carbonate-rich reservoirs: Case study of the Lucaogou Formation in the Jimusaer Sag, Junggar Basin, NW China. *Mar. Pet. Geol.* **2017**, *85*, 136–150. [CrossRef]
18. Shao, Y.; Yang, Y.Q.; Wan, M.; Qiu, L.W.; Cao, Y.C.; Yang, S.C. Sedimentary Characteristic and Facies Evolution of Permian Lucaogou Formation in Jimsar Sag, Junggar Basin. *Xinjiang Pet. Geol.* **2015**, *36*, 635–641, (In Chinese with English abstract).
19. Hou, L.H.; Zou, C.N.; Liu, L.; Wen, B.H.; Wu, X.Z.; Wei, Y.Z.; Mao, Z.G. Geologic essential elements for hydrocarbon accumulation within Carboniferous volcanic weathered crusts in northern Xinjiang, China. *Acta Pet. Sin.* **2012**, *33*, 533–540, (In Chinese with English abstract).
20. Kuang, L.C.; Hu, W.X.; Wang, X.L.; Wu, H.G.; Wang, X.L. Research of the Tight Oil Reservoir in the Lucaogou Formation in Jimusar Sag: Analysis of Lithology and Porosity Characteristics. *Geol. J. China Univ.* **2013**, *19*, 529–535, (In Chinese with English abstract).
21. Cao, Z.; Liu, G.D.; Kong, Y.H.; Wang, C.Y.; Niu, Z.C.; Zhang, J.Y.; Geng, C.B.; Shan, X.; Wei, Z.P. Lacustrine tight oil accumulation characteristics: Permian Lucaogou Formation in Jimusaer Sag, Junggar Basin. *Int. J. Coal Geol.* **2016**, *153*, 37–51. [CrossRef]
22. Wang, S.J.; Hu, S.B.; Li, T.J.; Wang, J.Y.; Zhao, W.Z. Terrestrial heat flow in Junggar Basin, Northwest China. *Chin. Sci. Bull.* **2000**, *45*, 1327–1332, (In Chinese with English abstract). [CrossRef]
23. Wang, S.J.; Hu, S.B.; Wang, J.Y. The Characteristics of heat flow and geothermal fields in Junggar Basin. *Chin. J. Geophys.* **2000**, *43*, 771–779, (In Chinese with English abstract). [CrossRef]
24. Sweeney, J.J.; Burnham, A.K. Evaluation of a Simple Model of Vitrinite Reflectance Based on Chemical Kinetics. *AAPG Bull.* **1990**, *74*, 1559–1570.
25. Peters, K.E.; Walters, C.C.; Moldowan, J.M. *The Biomarker Guide: Interpreting Molecular Fossils in Petroleum and Ancient Sediments*; Cambridge University Press: Cambridge, UK, 2005; pp. 1–1155.
26. Powell, T.G. Pristane/phytane ratio as environmental indicator. *Nature* **1988**, *333*, 604. [CrossRef]
27. Shanmugan, G. Significance of coniferous rain forests and related oil, Gippsland Basin, Australia. *AAPG Bulletin* **1985**, *69*, 1241–1254.
28. Vu, T.T.A.; Zink, K.G.; Mangelsdorf, K.; Sykes, R.; Wilkes, H.; Horsfield, B. Changes in bulk properties and molecular compositions within New Zealand Coal Band solvent extracts from early diagenetic to catagenetic maturity levels. *Org. Geochem.* **2009**, *40*, 963–977. [CrossRef]
29. Azevedo, D.A.; Aquino Neto, F.R.; Simoneit, B.R.T.; Pinto, A.C. Novel series of tricyclic aromatic terpanes characterized in Tasmanian tasmanite. *Org. Geochem.* **1992**, *18*, 9–16. [CrossRef]
30. Ourisson, G.; Albrecht, P.; Rohmer, M. The hopanoids, paleochemistry and biochemistry of a group of natural products: Pure Applied Chemistry. *Pure Appl. Chem.* **1979**, *51*, 709. [CrossRef]
31. Damsté, J.S.S.; Duin, A.; Hollander, D.; Kohnen, M.E.L.; Leeuw, J.W.D. Early diagenesis of bacteriohopanepolyol derivatives: Formation of fossil homohopanoids. *Geochim. Cosmochim. Acta* **1995**, *59*, 5141–5157. [CrossRef]
32. Ten Haven, H.L.; De Leeuw, J.W.; Sinninghe, D.J.S.; Schenck, P.A.; Palmer, S.E.; Zumberge, J.E. Application of Biological Markers in the Recognition of Palaeohypersaline Environments. In *Lacustrine Petroleum Source-Rocks*; Fleet, A.J., Kelts, K., Talbot, M.R., Eds.; Geological Society Special Publication: Oxford, UK, 1988; Volume 40, pp. 123–130.
33. Schwark, L.; Empt, P. Sterane biomarkers as indicators of palaeozoic algal evolution and extinction events. *Palaeogeogr. Palaeoclimatol. Palaeoecol* **2006**, *240*, 225–236. [CrossRef]
34. Moldowan, J.M.; Seifert, W.K.; Gallegos, E.J. Relationship between Petroleum Composition and Depositional Environment of Petroleum Source Rocks1. *AAPG Bull.* **1985**, *69*, 1255–1268.
35. Dai, J.X. *Large Coal-Formed Gas Fields and Gas Sources in China*; Science Press: Beijing, China, 2017; pp. 1–454. (In Chinese)
36. Gong, D.Y.; Lan, W.F.; Xiang, H.; Ding, J.; Wu, W.A.; Hu, Z.L. Genetic types and origins of natural gases from the eastern Junggar basin. *J. China Univ. Min. Technol.* **2019**, *48*, 142–152, (In Chinese with English abstract).

37. Hunt, J.M. *Petroleum Geology and Geochemistry*; Freeman Company: San Francisco, CA, USA, 1996.
38. Tissot, B.P.; Welte, D.H. *Petroleum Formation and Occurrence*; Springer: Berlin/Heidelberg, Germany; New York, NY, USA, 1984; pp. 1–699.
39. Bordenave, M.L. *Applied Petroleum Geochemistry*; Editions Technip: Paris, France, 1993; pp. 344–353.
40. Huang, D.F.; Li, J.C. *Continental Hydrocarbon Generation in China*; Petroleum Industry Press: Beijing, China, 1982; pp. 1–355. (In Chinese)
41. Peters, K.E. Guidelines for Evaluating Petroleum Source Rock Using Programmed Pyrolysis1. *AAPG Bulletin* **1986**, *70*, 318–329.
42. Xiang, B.L.; Li, E.T.; Gao, X.W.; Wang, M.; Wang, Y.; Xu, H.; Huang, P.; Yu, S.; Liu, J.Z.; Zou, Y.R.; et al. Petroleum generation kinetics for Permian lacustrine source rocks in the Junggar Basin, NW China. *Org. Geochem.* **2016**, *98*, 1–17. [CrossRef]
43. Pepper, A.S. Estimating of Petroleum Expulsion Behavior of Source Rocks: A Novel Quantitative Approach. In *Petroleum Migration*; England, A.J., Fleet, A.J., Eds.; Geological Society Special Publication: London, UK, 1992; Volume 59, pp. 9–31.
44. Sandvik, E.I.; Young, W.A.; Curry, D.J. Expulsion from hydrocarbon sources: The role of organic absorption. *Org. Geochem.* **1992**, *19*, 77–87. [CrossRef]
45. Cooles, G.P.; Mackenzie, A.S.; Quigley, T.M. Calculation of petroleum masses generated and expelled from source rocks. *Org. Geochem.* **1986**, *10*, 235–245. [CrossRef]

energies

Article

Characteristics and Origins of the Difference between the Middle and High Rank Coal in Guizhou and Their Implication for the CBM Exploration and Development Strategy: A Case Study from Dahebian and Dafang Block

Fuping Zhao [1,2,3], Shuxun Sang [1,4,5,*], Sijie Han [4,5,*], Zhangli Wu [2,3], Jinchao Zhang [1], Wenxin Xiang [1] and Ang Xu [1]

1 School of Resources and Geosciences, China University of Mining and Technology, Xuzhou 221116, China; db20010022b4cx@cumt.edu.cn (F.Z.); jinchaozhang@cumt.edu.cn (J.Z.); xiangwx@cumt.edu.cn (W.X.); xuang_xa@cumt.edu.cn (A.X.)
2 Key Laboratory of Unconventional Natural Gas Evaluation and Development in Complex Tectonic Areas, Ministry of Natural Resources, Guiyang 550081, China; lilybrbr@163.com
3 Guizhou Engineering Research Institute of Oil & Gas Exploration and Development, Guiyang 550081, China
4 Carbon Neutrality Institute, China University of Mining and Technology, Xuzhou 221008, China
5 Jiangsu Key Laboratory of Coal-Based Greenhouse Gas Control and Utilization, China University of Mining and Technology, Xuzhou 221008, China
* Correspondence: shxsang@cumt.edu.cn (S.S.); hsj_1991@cumt.edu.cn (S.H.)

Citation: Zhao, F.; Sang, S.; Han, S.; Wu, Z.; Zhang, J.; Xiang, W.; Xu, A. Characteristics and Origins of the Difference between the Middle and High Rank Coal in Guizhou and Their Implication for the CBM Exploration and Development Strategy: A Case Study from Dahebian and Dafang Block. *Energies* **2022**, *15*, 3181. https://doi.org/10.3390/en15093181

Academic Editors: Shu Tao, Dengfeng Zhang, Huazhou Huang, Shuoliang Wang, Yanjun Meng and Manoj Khandelwal

Received: 1 April 2022
Accepted: 25 April 2022
Published: 27 April 2022

Abstract: The coalbed methane (CBM) geology in Guizhou is characterized by a high gas content, pressure and resource abundance, indicating superior CBM resource potential. However, there are also many unfavorable factors, such as complex structure geology, significant regional differences in CBM geology, the widespread development of tectonically deformed coal, and the unclear understanding of the configuration of geological factors for CBM enrichment and high yield, which restrict the increase in CBM production and a large-scale development. Taking the Dahebian Block in Liupanshui coal field and the Dafang Block in Qianbei coal field as examples, this study presented the CBM geological differences between middle- and high-rank coals; their origins were analyzed and the effect of depth on gas content and permeability was discussed. A CBM enrichment and high-yield model was illustrated, and the geologic fitness-related exploration and development methods for Guizhou CBM were finally proposed. The results show that (1) significant differences between the middle- and high-rank coals occur in coal occurrence and distribution, coal qualities, and coal reservoir properties. Compared to Dahebian coal, Dafang coal has a higher coal rank, vitrinite content, and gas content, but a lower number of coal layers and permeability. (2) The sedimentary–tectonic evolution of the Longtan coal-bearing sequence is the fundamental reason for CBM geological differences between the Dadebian Block and Dafang Block, consisting of coal occurrence, qualities, maceral, rank, structure, and their associated reservoir properties. (3) The coordinated variation of gas content and permeability contributes to a greater depth for CBM enrichment and a high yield of the middle-rank coal. It is suggested that the best depths for CBM enrichment and high yield in Guizhou are 600–800 m for the middle-rank coal and 500 m for the high-rank coal, respectively. (4) Considering the bottleneck of inefficient CBM development in Guizhou, we proposed three CBM assessment and development technologies, including the CBM optimization of the classification–hierarchical optimization–analytical hierarchy, multiple coal seams commingling production with the pressure relief of tectonically deformed coal, and surface–underground CBM three-dimensional drainage development. The aim of this study was to provide new insights into the efficient exploration and development of CBM in Guizhou.

Keywords: different CBM geology; middle and high rank coal; tectonism and sedimentation; enrichment and high yield model; exploration and development strategy

1. Introduction

Guizhou Province is rich in coalbed methane (CBM) resources, accounting for 10% of the total CBM geological resources in China. The CBM in Guizhou is characterized by a large reserve, a concentrated distribution, and a high quality, and has high potential for large-scale exploration and development. In the 14th Five-year Plan of Guizhou Province, it has been clearly stated that "accelerating the exploration and development of CBM, promoting the Bijie-Liupanshui-Xingren CBM industrial base and approaching CBM annual output of 4×10^8 m^3" [1]. The government also vowed to support and promote the construction of the Zhijin–Panzhou CBM industrial base. The above policies and advantages reinforce the exploration and development of CBM in Guizhou Province. The coal-bearing sequence in Guizhou was formed during the Late Permian, and the tectonic evolutions of the coal-bearing basin are characterized by the multistage, strong different lifting-subsidence and late shaping. The sedimentary setting of the Longtan coal-bearing sequence showed a strong change between marine and land facies across the surface and at the vertical sequence [2]. There are various coal types, and the Guizhou coals show a wide distribution of coal rank, ranging from high volatile bituminous to anthracite. The genetic types of coal both contain regional magmatic thermal metamorphism and plutonic metamorphism [3,4]. The gas-bearing property of coal is highly related to the structural style and depth, and its control via the coupling of multiple factors is the fundamental reason for the diversity of CBM-rich types in Guizhou [5]. The attributes of CBM, characterized by "small but fat" (a small area but a high abundance of CBM resources), were significantly affected by the sedimentation and tectonic [6]; namely, coal-bearing syncline is the enrichment location of CBM, and there is a positive correlation between depth and gas content. The general characteristics of the coal reservoir in Guizhou consist of low porosity, permeability and high gas saturation, in situ stress, and damage degree; meanwhile, there are significant differences among the coal-bearing tectonic units [7–9]. On the one hand, strong reservoir heterogeneity results in the morphology of pores and fractures, especially nano-pores, and then affects the gas flow and efficient CBM production; on the other hand, various pore structures also contribute to different methane adsorption, which will enlarge the difference between excess and absolute adsorption, and finally affect the accurate evaluation of gas content [10–12]. The extensive occurrence of tectonically deformed coal is an important factor restricting the large-scale development of CBM in Guizhou [13,14]. Significantly, extremely thick coal usually shows a higher deformation and failure degree of the coal body, such as No. 11 coal in Dahebian Block and No. 6 coal in Wenjiaba Block. The CBM geological conditions, associated with its enrichment and accumulation, vary in Guizhou. The lithofacies paleogeography, differential tectonic subsidence and abnormal thermal evolution resulted in the observation that the occurrence and physicochemical properties of coal show a regional and heterogeneous distribution in Guizhou [15–17]. Affected by the sequence stratigraphic framework and hydrogeology, the gas-bearing property, gas-bearing gradient, and pressure coefficient of coal are independently segmented at the vertical sequence without a direct fluid exchange, resulting in the formation of the unattached multiple superposed CBM system [18–20]. Some advances have been made in the exploration and development of the CBM of middle- and high-rank coals in Guizhou in recent years. For example, well Yangmeican 1 (middle-rank coal) and well Wencong 1 (high-rank coal) both show breakthroughs in technology and gas production [21,22]. However, most CBM wells still present a low gas production and short time of stable gas production. The significant geological and regional difference in CBM in different coal-bearing units resulted from their tectonic–sedimentary differentiations, which poses challenges to increasing CBM production and conducting large-scale development in Guizhou.

The representative coal-bearing blocks in Guizhou, the Dahebian Block (middle-rank coal), and Dafang Block (high-rank coal) were used as the cases in this study. The differences in the occurrence, maceral, coal quantity, and reservoir physical properties were analyzed, and their origins of sedimentation and tectonic were interpreted. Then, the effect of depth on the gas content and permeability of coal was identified, and the geological model of the CBM enrichment and high yield in Guizhou was established. Finally, based on the bottleneck induced by the geological differences in CBM in Guizhou, the exploration and development methods and technologies with geological suitability were put forward. This study aims to provide a geological theory and new insights into the exploration and development of CBM in complex geological settings.

2. Materials and Method

2.1. Samples and Data Sources

The samples and data in this study are composed of two parts: one is the coal samples collected from boreholes and underground coal mines, and the other is the logging interpretation data of boreholes in study area. The pulverized coals included Nos. 1, 4, 7, 8, 11, 12, 13, 14, 17, and 30 coals, which are collected from boreholes in the Dahebian Block, and Nos. 6_1, 6_2, 6_3, 7, 10, 14, 23, 33, and 34 coals, which were collected from Dafang Block. The pulverized coals were applied for maceral identification, and the determination of the total sulfur, ash and major element contents. The block coals are No. 11 coal from Wangjiazhai mine in the Dahebian Block and No. 6 coal from the Lvtang mine in the Dafang Block. The block coals were made into a cylinder with a size of 2.5 cm × 2.5 cm × 5 cm for the porosity and permeability measurement of the core in net confining stress. The wells used for the logging interpretation of gas content, porosity and permeability included Zhong-1, Zhong-1-1, Zhong-1-4, Zhong-1-6, Zhong-1-8, CK2302, CK2402, and CK2602 in the Dahebian Block and Da-201, Da-2, and Da-203 in the Dafang Block.

2.2. Methods

The pulverized coals were crushed and reduced into the required size and qualities for maceral identification and determination of the total sulfur, ash, and major element contents. These experiments were conducted according to Chinese National Standard GB/T 8899-2013 "Determination of maceral group composition and minerals in coal", GB/T 212-2008 "Proximate analysis of coal", GB/T 214-2007 "Determination of total sulfur in coal" and GB/T 14506. 28-2010 "Methods for chemical analysis of silicate rocks".

High-pressure nitrogen was used to provide the confining pressure, and high-purity helium was used as the test gas. The test temperature was the indoor temperature, and its changes were controlled within 1 °C. The test pressure was designed as 3, 6, 9, 12, and 15 MPa, and the equilibrium time for each pressure was 30 min. The test procedures were conducted according to the Chinese oil and gas industry standard "The porosity and permeability measurement of core in net confining stress".

3. Results and Discussion

3.1. Different Characteristics between the Middle and High-Rank Coals

The Dahebian Block in the northwest of the Liupanshui coal field and the Dafang Block in the south of the Qianbei coal field develop the middle- and high-rank coal, respectively. There are obvious differences between them regarding tectonic characteristics. The Dahebian Block is located in the western part of the Yangtze platform with a NNW–SEE direction. There is only one large, wide, and gentle syncline, named Dahebian syncline, with relatively few faults and a simple structure. The Dafang Block is located in the Zhunyi Fault-Arch zone of Qianbei Uplift with a NE–SW direction. The Luojiaohe syncline and Dafang anticline control the structure of the block, and their related faults and sub-folds are widely distributed. There are also great differences in coal occurrence, qualities and properties between these two blocks. Importantly, a stronger tectonic deformation results in Dafang Block having a wider distribution of tectonically deformed coal, accounting for approximately 40% of the total coal distribution. In contrast, this proportion is 10% in the Dahebian Block. The thickness of the coal-bearing sequence in the Dahebian Block ranges from 200.70 m to 257.28 m, with an average of 232.97 m. This sequence contains 14–29 layers of coal, and the thickness of the coal ranges from 14.24 m to 17.54 m, with an average of 16.08 m. Dahebian low-middle volatile bituminous is mainly composed of semidull and semibright–semidull coal. The ratio of coal maceral is 2:1:1 in Dahebian coal, and vitrinite is the primary maceral. The thickness of the coal-bearing sequence in the Dafang Block ranges from 166.14 m to 261.5 m, with an average of 204.14 m. This sequence contains 23–50 layers of coal, and the thickness of the coal ranges from 3.97 m to 13.75 m, with an average of 9.5 m. Dafang anthracite is mainly composed of bright coal and vitrinite content, which accounts for approximately 80% and inertinite content for approximately 20%. Previous studies have demonstrated that there are significant differences between these two blocks in terms of coal occurrence, coal maceral, coal quality, and reservoir properties [7,23,24]. Therefore, the key geological parameters of the two blocks are summarized here, as shown in Table 1, and the CBM geological conditions are comprehensively compared from the perspectives of the occurrence, composition and physical properties of coal.

Table 1. Comparison of the representative CBM geological parameters between Dahebian block and Dafang block.

Geological Parameters		Dahebian Block	Dafang Block
Occurrence	Total thickness of coal/m	14.24–17.54/16.08	3.97–13.75/9.50
	Depth/m	600–1200/800	400–1000/600
	Number of coal layer	14–29/18	23–50/35
	Average coal interval/m	4.79–108.71/25.78	11.64–57.84/34.74
	Total thickness of coal measure/m	200.70–257.28/232.97	166.14–261.50/204.14
Coal properties	Vitrinite reflectance/%	0.73–0.97/0.82	3.03–3.69/3.32
	Vitrinite content/%	38.53–63.95/53.81	79.74–92.84/83.96
	Ash content/%	11.27–39.05/25.16	23.75–28.63/26.05
	Sulfur content/%	0.24–5.91/3.1	1.32–2.76/1.94
Coal reservoir parameters	Gas content/$m^3 \cdot t^{-1}$	8.39–14.92/11.66	17.84–23.75/20.8
	Permeability/mD	0.03–0.66/0.3	0.016–0.43/0.17
	Gas saturability/%	70.45–102.42/83.34	75.59–84.8/78.33
	Langmuir volume/$cm^3 \cdot g^{-1}$	10.57–18.72/14.52	28.18–35.14/32.04
	Langmuir pressure/MPa	2.48–3.09/2.71	2.28–2.51/2.38
	Tectonically deformed coal development	Weak	Medium–well developed
	Pressure gradient/MPa·100 m^{-1}	0.87–1.32/1	0.82–0.93/0.89

Annotation: Min–Max/Average.

As shown in Table 1, the typical differential CBM geological characteristics between Dahebian and Dafang Blocks include coal layer number; vitrinite reflectance; maceral; coal quality; coal structure; and their highly related reservoir parameters, such as gas content and permeability. First, in terms of coal occurrence, the coals in both blocks have low to medium thickness, but they have a higher thickness and a more concentrated distribution in the Dahebian Block. In the Dafang Block, coals thinner than 0.5 m are more developed, suggesting the differences in the stability of coal environment. Second, in terms of coal properties, there are significant differences in the maceral between the middle- and high-rank coals. The vitrinite content of Dafang high-rank coal is dominant, concentrated at approximately 80%, while that of Dahebian middle-rank coal is generally low and varies widely, mainly in the range of 35–65% (Figure 1a). Although the overall average value of the ash and sulfur content of these two blocks is similar, the variation range of Dahebian coal is larger, which also implies the difference in the sedimentary environment. Third, on the one hand, the gas content of Dahebian coal is generally lower than that of Dafang coal, which is related to the coal rank. The scattered distribution of gas content and the occurrence of an abnormally high gas content of Dafang coal suggest the development of tectonically deformed coal (Figure 1b). On the other hand, the porosity of coals in these two blocks is similar, but Dafang coals show more scattered porosity (Figure 2a). The permeability of Dafang coals (<0.1 mD) is generally lower than that of Dahebian coals (0.2–0.3 mD), and the values are more concentrated (Figure 2b). These reservoir characteristics also suggest that Dafang high-rank coals have a larger development of tectonically deformed coal. Tectonic compression promotes the increase in porosity, especially the micropores that affect adsorption capacity; however, the destruction of connectivity of the original pore network finally decreases the permeability.

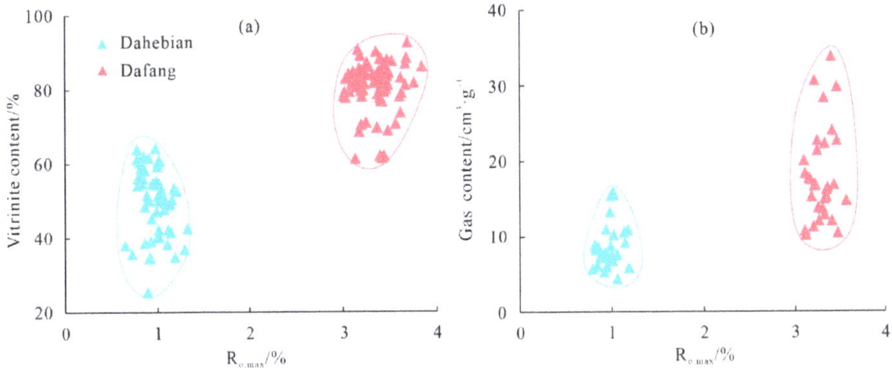

Figure 1. Relationship between vitrinite reflectance with vitrinite content (**a**) and gas content (**b**) of the primary mineable coals.

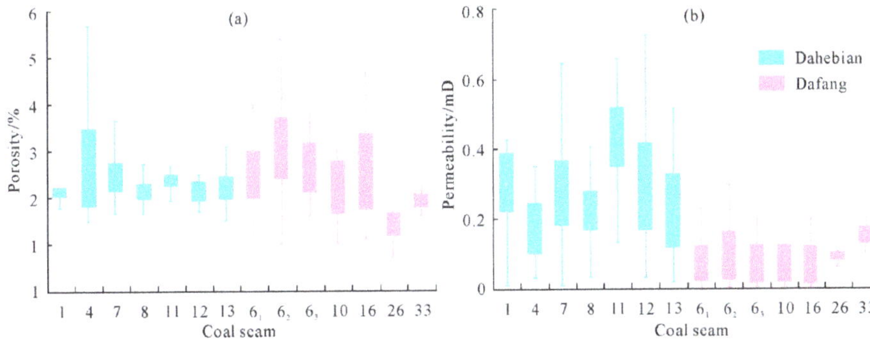

Figure 2. Distribution of porosity (**a**) and permeability (**b**) in various coal seams obtained from the log interpretation.

Significantly, the geological characteristics of typical medium–high-rank CBM in Guizhou are significantly different in most geological parameters, such as the number of coal layer, vitrinite content, coal structure, gas content, and permeability. These key geological characteristics are the results of sedimentation and tectonic. Gas content and permeability, which are the indictors of CBM enrichment and high yield, are not only related to the above key geological parameters, but are also controlled by depth. Therefore, the origins of different geological characteristics between middle- and high-rank coal were discussed based on their sedimentary, tectonic, and depth dependences.

3.2. Origins of Different Geological Characteristics between Middle- and High-Rank Coals

3.2.1. Different Sedimentary Environment

The sedimentary environment, REDOX conditions, terrigenous supply, and seawater activities play important roles in maceral and coal qualities, and the variation of maceral and coal qualities can also indicate the evolution of the sedimentary environment [23,25]. Although the Dahebian and Dafang Blocks are geographically neighbors and have similar transitional sedimentary facies settings and coal-forming plants, there are obvious differences in sedimentary subfacies, microfacies, and seawater activities intensity. This results in the change in the invasion of exogenous components and gelatinization during the peat accumulation, further contributing to the variation of maceral and coal qualities.

The total sulfur content and ash index (Fe_2O_3 + CaO + MgO/SiO_2 + Al_2O_3) in Dahebian and Dafang Blocks both showed a general increase from bottom to top coal (Figure 3a,b), indicating that the REDOX condition of the sedimentary environment changed from weak oxidation–weak reducibility to strong reducibility, which is consistent with the seawater activities in Guizhou during Late Permian [2]. The difference is that the variation in coal qualities in the Dahebian Block is more significant, suggesting stronger seawater activities during peat accumulation. Significantly, the total sulfur content of No. 12, 13, 14, 17, and 30 coals are lower than 1 %, and the ratio between organic sulfur and total sulfur approach 42.42–68.75%, suggesting that these coals were formed in a terrestrial environment and the organic sulfur mainly came from the absorption of terrestrial plants. The triangular diagram of ash compositions also indicated that the Dafang Block had a relatively stable and weak reducibility environment (Figure 3c). However, Fe_2O_3 + SO_3 gradually increased from bottom to top in Dahebian coals, suggesting that this area evolved from near the provenance to the environment with deep overlying water, weak hydrodynamics, and strong reducibility. Therefore, according to the total sulfur content and ash compositions, it can be inferred that the Dahebian Block experienced the evolution from the upper delta plain fluvial environment to the tidal flat-lagoon environment, while the Dafang Block mainly developed the tidal lower delta plain environment. The sedimentary environment indicated by maceral content showed a similar result with coal qualities. In the Dafang Block, the vitrinite content ranges from 70.84% to 86.36%, with an average of 78.16%, and the inertinite content ranges from 13.64% to 29.16%, with an average of 21.19%. In the Dahebian Block, the vitrinite varies, with the content ranging from 35.37% to 72.4%, and the inertinite content ranges from 11.07% to 52.41%, with an average approximately 25%. However, coals above No. 12 coal have a significantly higher vitrinite content, and the inertinite content is opposite. The results of maceral also suggested that the sedimentary environment in the Dafang Block was relatively stable with a weak hydrodynamic condition and weak to strong reduction environment, but the Dahebian Block underwent a significant change from a weak oxidation to a strong reduction environment.

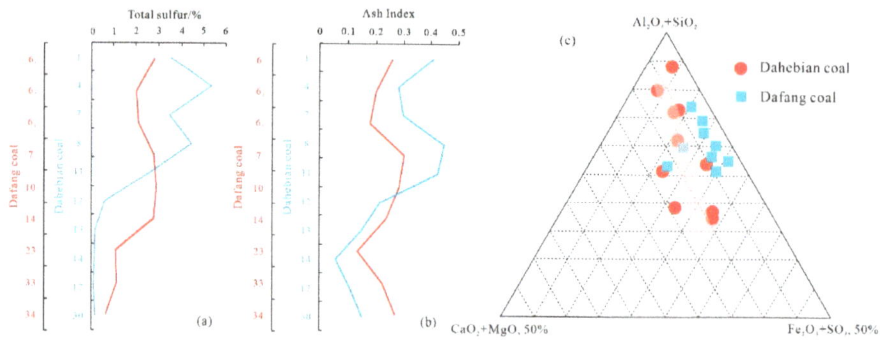

Figure 3. Sedimentary environment indication using different coal qualities parameters. (**a**) Total sulfur; (**b**) Ash index; and (**c**) Triangular diagram of ash compositions.

3.2.2. Tectonic Effects on Coal Rank and Structure

Tectonic and its associated magmatic activities have important effects on the coal rank and structure. The coal rank is the product of the coupling effects of temperature, pressure, and time, and the coal structure indicates the process of in situ stress transformation. Plutonic thermal metamorphism in different geothermal environments during tectonic subsidence and high-temperature thermal anomalies caused by regional magmatic activities are the fundamental driving forces of coalification.

It is well known that the thermal evolution of Longtan coals in Guizhou suffered plutonic thermal metamorphism superimposed by high-temperature metamorphism during the age of the Yanshan period [26]. Therefore, the uneven tectonic subsidence and plutonic magmatic intrusion during the Yanshan period caused the uneven distribution of temperature gradient and further resulted in a significant difference in coal rank. The coal rank in Guizhou presents an X-type distribution with a low coal rank in the east and west, but a high coal rank in the middle north and south of Guizhou [27,28]. The plane distribution of coal rank shows a high relationship with deep faults in Guizhou. Affected by the NW-oriented Ziyun–Shuicheng, NE-oriented Zongshi–Guiyang, and NNE-oriented Zunyi–Huishui faults, anthracites are distributed in the north and south blocks, while bituminous coals often occur in the east and west blocks (Figure 4). Meanwhile, the coal rank of bituminous coals showed a decrease from these deep faults. The middle-rank coals in the Liupanshui coal field are the results of plutonic thermal metamorphism, and they are controlled by the depth, with increasing magmatic thermal effects from west to east [29]. The uneven tectonic subsidence during the evolution of the Ziyun–Shuicheng and Zongshi–Guiyang faults resulted in the Qianzhong Uplift in the northern Zhina coal field. The formation of the Qianzhong Uplift indicated the thinning of the lithosphere and, therefore, the heating induced by the latent magmatic intrusion was most obvious in this area [3]. In fact, the coal rank had a high affinity with the distance of the Qianzhong Uplift. A previous study on the inclusions demonstrated that the highest temperature in the Zhijin area approached 244 °C, but the highest temperature in Panxian only reached 150 °C [4], which verified that the different tectonic thermal events are the governing control on the various coal ranks in Guizhou.

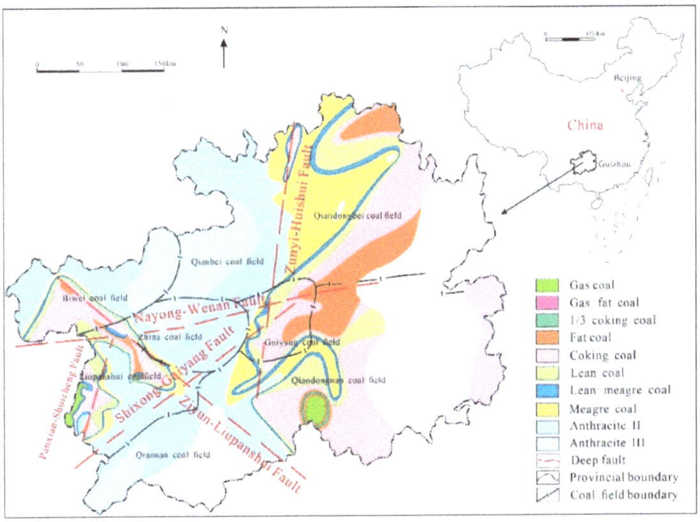

Figure 4. Relationship between deep faults and coal rank in Guizhou. Adapted from [27].

The effect of tectonic activities on coal is also indicated by the deformation and damage of the coal structure, and the development of tectonically deformed coal is its typical feature. Subsequent to the formation of the Longtan coal-bearing sequence in Guizhou, coals suffered a long-term compression during the Yanshan and Himalayan periods, resulting in serious damage to the coal structure and the wide development of tectonically deformed coal. Meanwhile, the effect of structure on tectonically deformed coals is reflected in their distribution and damage degree [30]. The development of tectonically deformed coal is often accompanied by large compressional faults, and the distribution of tectonically deformed coal has a high affinity with Late Permian synsedimentary faults. During the Yanshan period, the squeeze thrust and interlayer slip of these faults reinforced the deformation and damage to the coal structure and, therefore, the closer to the faults, the more broken the coal structure was [31]. A previous study indicated that the Ziyun–Shuicheng, Shizong–Guiyang, and Zunyi–Huishui faults were not only the main areas for tectonically deformed coal, but also the boundary of the high gas content area, suggesting the tectonic effect on gas content and coal structure [32]. In addition to the obvious structural control, thick coal is also the key to searching for tectonically deformed coal in Guizhou. This is because thick coals have a lower mechanical strength compared with their surrounding rocks and, therefore, they are more prone to crumpling and interlayer slipping under the influence of squeezing and shear stress, resulting in greater damage to thick coals. For example, No. 6 coal in the Dafang Block is composed of granulated coal and mylonitized coal, whereas No. 16 and 27 coals, located below No. 6 coal, are composed of primary and cataclastic coal with a minor granulated coal. In addition, the hierarchical structure of thick coal will also produce a stress concentration area in some soft layers, contributing to a typical sandwich-like structure. The classical case is No.11 coal in the Dahebian Block, which has a thickness of 6–8 m; such thick coal is extremely rare in Guizhou and even in southwest China. Primary coal, schistose coal (Figure 5a) or mylonitized coal, granulated coal (Figure 5b), and cataclastic coal/lenticular coal (Figure 5c) occur at their ascend sequence. The small crumpled and sliding surface are often discerned in the middle tectonically deformed coal zone (Figure 5d).

Figure 5. Main tectonically deformed coal types of thick coal (11#) in Dahebian Block. (**a**) Schistose coal (2 m below the roof); (**b**) Granulated coal (2.5 m below the roof); (**c**) Cataclastic coal/lenticular coal (0.5 m above the floor); and (**d**) Schistose coal with a small fold (1.5 m above the floor).

3.2.3. Effect of Depth on Gas Content and Permeability between the Middle- and High-Rank Coal

Gas content and permeability, which show high affinities with CBM enrichment and high yield, have different dependences of depth. They show an opposite change against depth at a low depth; namely, as depth increases, gas content linearly increases, but permeability shows a significant decrease. However, with the increase in depth, both gas content and permeability decrease, but their decrease in amplitude is significantly different. Due to the completely different control effects of depth on gas content and permeability, there is a balance depth that enables these two key factors to meet the requirements of CBM enrichment and high yield [33]. It should be noted that there is a difference in the balance depth between middle- and high-rank coal. This is because (1) the high-rank coal has a higher gas content due to its high Langmuir volume at the same depth; and (2) the permeability of high-rank coal has a stronger stress sensitivity. On the one hand, a previous study showed that the methane adsorption capacity has a maximum value of depth profile, and this value is closely related to temperature and pressure [34]. The gas content of Guizhou coal also follows the above law, but the difference is that the gas content of middle-rank coal is smaller than that of high-rank coal at the same depth, and the critical depth of the maximum gas content of high-rank coal is deeper than that of middle-rank coal (Figure 6a,b), which is related to the control effect of coal rank on gas adsorption capacity [35]. On the other hand, in fact, the control effect of depth on permeability is the modification of effective stress on the gas migration channel. The increase in depth indicates an increase in overburden pressure, as well as effective stress. The stronger effective stress on coal results in the compression or even closure of efficient gas migration channels, such as cleat and fracture, which eventually contributes to the decrease in permeability. The results of the change in the porosity and permeability of Guizhou medium- and high-rank coals under the loading of effective stress show that the porosity and permeability of high-rank coal are lower than those of middle-rank coal under the same effective stress, and the permeability attenuation of high-rank coal is more obvious at a high pressure (Figure 7). This demonstrates that the negative effect of the porosity and permeability of the high-rank coal is stronger than that of the middle-rank coal. The permeability of high-rank coal is lower than that of middle-rank coal due to the extinction of macropores and the decrease in cleat density. The dependence of fracture on the permeability of high-rank coal results in its stress sensitivity being stronger than middle-rank coal [36,37].

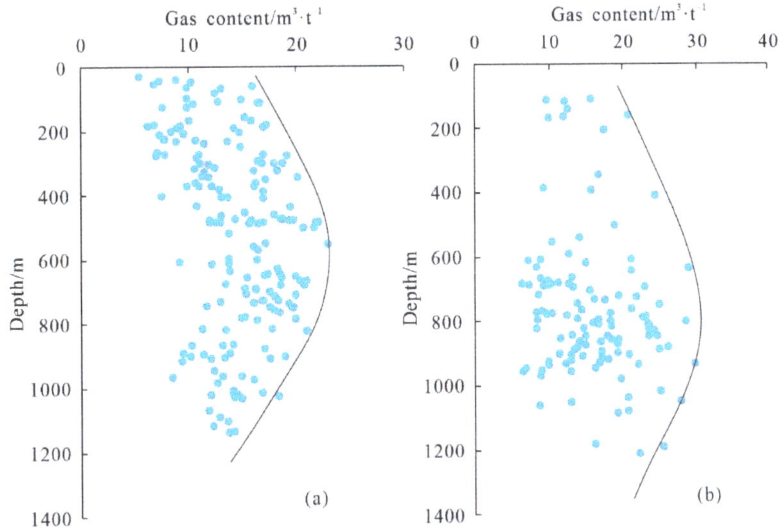

Figure 6. Relationship between gas content obtained from log interpretation and depth of the middle and high rank coals. (**a**) Dahebian Block; (**b**) Dafang Block.

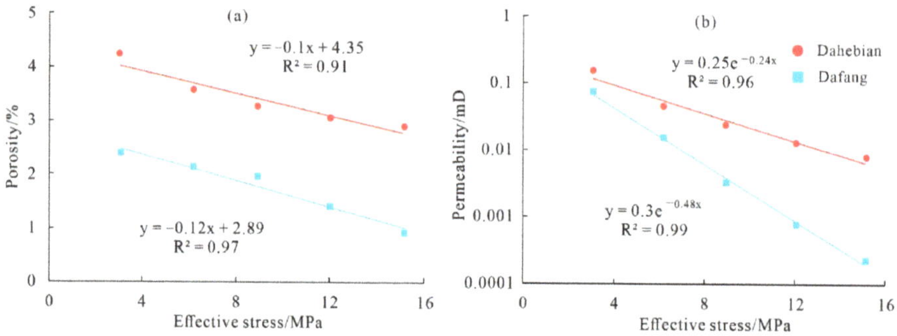

Figure 7. Relationship between effective stress with porosity (**a**) and permeability (**b**).

In general, the depths of coal with a high gas content and permeability are different for middle- and high-rank coals. At the same depth, middle-rank coal has a lower gas content than high-rank coal, but its permeability is higher than that of high-rank coal. The increase in depth has a negative effect on the permeability of high-rank coal, but has a positive effect on the gas content of middle-rank coal before its transition depth. Therefore, the depth of high-rank coal with a high gas content and permeability is shallower than that of middle-rank coal, which is consistent with the results of the CBM exploration and development in Guizhou.

3.3. Model of Enrichment and High Yield of CBM in Guizhou

By 2020, more than 300 CBM wells had been constructed in Guizhou, with a production capacity of 1×10^8 m^3 per year and an output of 0.24×10^8 m^3 per year. However, the vast majority of wells have common problems, such as low single well output and short stable production time. The characteristic of gas-rich coal in Guizhou is significantly superior to that in north and northwest China. The high CBM resource abundance and gas saturation are the original driving forces behind CBM exploration and development in Guizhou. In fact, the key reasons for the low production of CBM wells are the low permeability, development of tectonically deformed coal and weak water content in coals. As a result, the hydrophobic depressed CBM development model has been blocked from widespread application in Guizhou. Therefore, in the current era of low-permeability coal, in which especially tectonically deformed coals have not been broken through, it is an inevitable choice to find high-permeability primary or weakly deformed coals controlled by the configuration of key geological factors.

The enrichment and high yield of CBM is the result of the coupling controlling of gas content and permeability via key geological factors [33]. Guizhou has complex geological conditions, and the coal-bearing basins are divided into uneven tectonic units. The sealed syncline is a potential position for CBM enrichment. On the one hand, the overlying Feixianguan Formation is developed with waterproof mud shale with low porosity and permeability, which provides a good lithologic seal for the CBM reservoir. In addition, the hydrodynamic condition of the Longtan Formation coal-bearing sequence is weak, and the weak runoff area is only developed near the outcrop layer of the wing. Underground water bedding supply plays a hydrodynamic sealing role in deep CBM, and gas content is positively correlated with depth, contributing to gas enrichment in the deep syncline. On the other hand, the synclinal axial is under a compressive stress environment, and the permeability of coal here decreases significantly under the joint action of depth and tectonic stress, which is not conducive to CBM production. Therefore, under the combined control of structure, hydrodynamic sealing, and depth, primary coals in the monocline near the syncline axis can easily form coalbed methane enrichment areas (Figure 8). However, it should be noted that due to the stronger adsorption capacity and higher permeability stress sensitivity of high-rank coals, their depth for CBM enrichment and high yield is usually lower than that of middle-rank coal. The high-yield CBM development project in Guizhou also shows that the main producing coals in well Zhong 1 in the Dahebian Block is at the depth of 700–800 m, with a gas content of 11.5–17.66 cm^3/g and permeability of 0.05–0.53 mD, while in the Wenjiaba Block, the producing high-rank coals in well Wen 1 are at the depth of 300–500 m with a gas content of 12.72–13.34 cm^3/g and permeability of 0.18–0.92 mD. This indicated that the coals of different coal ranks in these two blocks have a similar gas content and permeability. Therefore, it can be speculated that the depth of high-rank coal for CBM enrichment and high yield is less than 500 m, while the depth of middle-rank coal is expected to be 600–800 m. The key geological factors of CBM enrichment and high yield of different coal ranks in Guizhou have the common characteristics of structure–hydrodynamic coupling control. However, the effect of synergistic depth control on gas-bearing and permeability properties is considerably more important.

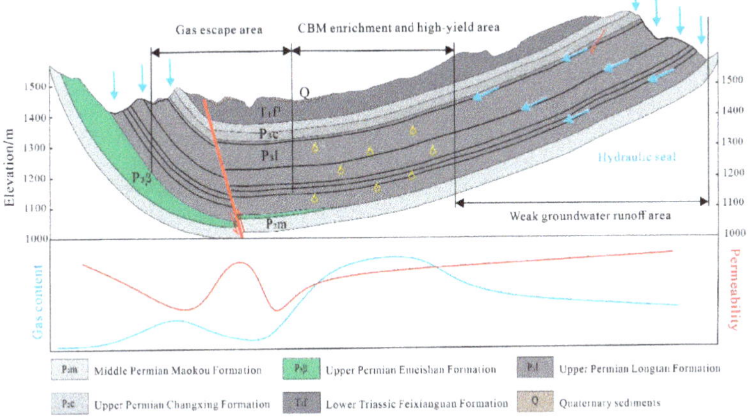

Figure 8. High rank coal CBM enrichment and high-yield model in Guizhou.

3.4. *Geological Suitable CBM Exploration and Development Strategies in Guizhou*

Various sedimentations and tectonics constitute an essential reason for the different geological characteristics of CBM in Guizhou, and they are also the key geological factors controlling gas content and permeability. Such complex geological conditions also pose challenges to the exploration and development of CBM, so it is necessary to propose specific technical methods according to the different coal reservoirs and the coupling law between gas content and permeability.

3.4.1. Optimization of Block and Coal Seam for CBM Exploration and Development Based on Coal Rank Classification

The geological characteristics of CBM vary greatly under complex geological conditions and, therefore, coal reservoirs have strong heterogeneity. There are differences in the collaborative configuration of key geological factors for CBM enrichment and high yield under the coupling effects of gas content and permeability between different coal ranks. As a result, the influence of some parameters on the evaluation results is exaggerated under the same standards. For example, although the gas content of low-rank coal is low, the macropore and microfracture for gas seepage are more developed. In contrast, the high-rank coal has a high gas content, but low permeability, and the attenuation of permeability is more obvious under the action of increased stress. Accordingly, China's energy industry standards NB/T 10013-2014 "CBM Geological Selection Evaluation" stated that there are different quantitative classification indexes of key geological parameters for the CBM geological selection evaluation of different coal ranks. For example, the best gas content and permeability are ≥ 6 m^3/t and ≥ 3 mD for the low rank coal and ≥ 15 m^3/t and ≥ 1 mD for the middle–high-rank coal. Therefore, in view of the differences in CBM geological characteristics in a complex geological setting, a classification of coal rank is suggested before the original optimization for CBM geological selection evaluation. The main coal-bearing areas in Guizhou, such as Liupanshui, Qianbei, and Zhina coal fields, have a wide distribution of coal rank, ranging from high volatile bituminous coal to anthracite. The coals of different ranks have significantly different gas contents and permeabilities. However, under the same evaluation standard, it is easy to omit middle-rank coals with a low gas content or high-rank coals with a low permeability during optimization for CBM geological selection evaluation. In fact, the CBM exploration and development in China have shown that there are high-yield CBM wells in both middle- and high-rank coal areas [38,39]. At present, there are few high-yielding CBM wells, and there is also no commercialized CBM blocks. However, the high-yielding performance of well Zhong 1 in the Dahebian Block, Liupanshui coal field, and well group Wencong 1 in the Wenjiaba Block, Zhina coal field, both indicate that CBM blocks of different coal ranks in Guizhou both have high production potential under the reasonable configuration of geological factors. Therefore, CBM geological selection evaluation based on the classification of key geological parameter is especially suitable for the geological optimization of CBM blocks and coal seams in a complex geological setting. According to the previous CBM geological optimization used in eastern Yunnan and western Guizhou [40,41], we proposed a modified method for the optimization of block and coal seam for CBM exploration and development in Guizhou, as shown in Figure 9. First, the blocks are classified into two groups according to the coal rank (middle- and high-rank coals). Second, the analytic hierarchy process (AHP) is used to establish the evaluation system. By comparing the importance of geological parameters, the discriminant matrix is established, and then the weight of the geological parameters is calculated for the system. Third, the membership function of each geological parameter is established according to the fuzzy mathematics principle. The actual weight of each geological parameter is calculated, and then the order of the blocks is obtained through the accumulation of the actual weights. The evaluation of the sweet point areas and sections have the same evaluation process as that of blocks. The aim of this process is to obtain the potential block and coal seam for CBM exploration and development.

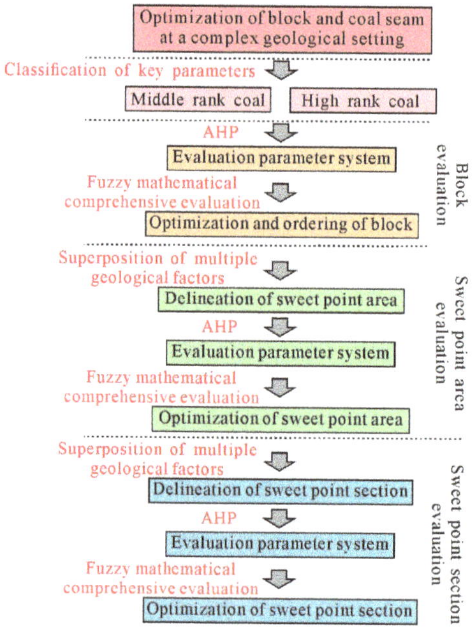

Figure 9. Flow diagram of optimization of block and coal seam for CBM exploration and development.

3.4.2. Multiple Coal Seams Commingling Production with Pressure Relief of Tectonically Deformed Coal

Tectonically deformed coal is characterized by a high gas content, low permeability, and soft structure. Reservoir reconstruction technology based on hydraulic fracturing has little effect on the permeability improvement of these coals, and even easily leads to hole collapse and blocking. Therefore, the CBM development system based on hydrophobic depressurization and desorption theory of primary coal or weakly deformed coal is not suitable for the efficient development of tectonically deformed coal. Surprisingly, the coal-bearing basins in Guizhou were strongly deformed and underwent multi-stage tectonic deformations, resulting in the modification of the coal-bearing sequence. As a result, tectonically deformed coals were widely distributed in Guizhou coal-bearing basins, especially in some complex tectonic areas, such as Zhaozihe syncline, Tudiya syncline, and other tectonic transition zones. How to enhance the CBM recovery of tectonically deformed coal is the key to increasing CBM reserves and yields. According to the positive results of the surface gas drainage technology in the coal mine goaf and open hole cave completion technology in a straight well [42,43], Sang et al. [44] innovatively proposed a CBM development model of tectonically deformed coal stress relief. Based on the volume expansion, increase in permeability and decrease in pressure induced by stress relief of tectonically deformed coal, a large-range efficient desorption-seepage area was formed, resulting from the formation of the horizontal well cavern and significantly enhanced CBM recovery. More importantly, due to the deformation and fracture of the overlying strata, longitudinal and stratified fractures of different sizes were produced, which provides favorable conditions for the development of the overlying multi-coal seam group (Figure 10). The multi-coal seam groups widely occur in Guizhou and southwest China, and the commingling production of CBM has been successfully applied to these areas. According to the permeability improvement of the overlying coals via tectonically deformed coal development, the synthesis of a CBM development technique combined with the stress relief of tectonically deformed coal and the commingling production of multiple coal seams are recommended. A combined implementation of tectonically deformed coal in situ CBM recovery via horizontal well cavern completion and stress relief and hydraulic fracturing in overlying primary coal can further improve the drainage efficiency in multiple coal seam commingling production. This technology transforms the development of tectonically deformed coal CBM into a high-yield advantage, and promotes the efficient exploration and development of CBM under the complex geological condition in Guizhou Province.

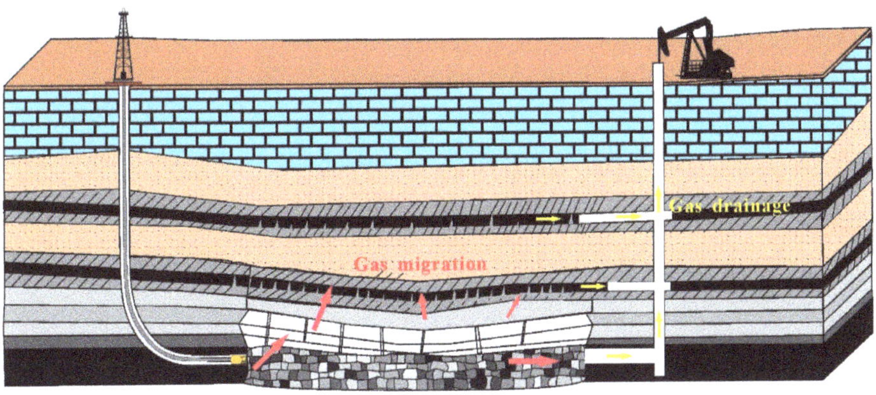

Figure 10. The multiple coal seams commingling production with tectonically deformed coal in-situ CBM recovery by horizontal well cavern completion and stress relief.

3.4.3. Surface–Underground CBM Three-Dimensional Drainage Development in Coal Mine Area

Due to the imperfect policies and regulations in China's energy industry, the mining rights affiliation of coal and CBM are often different, resulting in a conflict of mining rights between coal and CBM development. As a result, it is difficult for CBM enterprises to enter coal mining areas for CBM exploration and development. However, the CBM projects indicated that the depth of CBM enrichment and high-yield is lower than 800 m, and even 500 m, for high-rank coal. These depths basically overlap with the coal mining depth, which greatly restricts the development of the CBM industry in Guizhou. Significantly, most of the coal mines in Guizhou belong to high gas mines, and the high concentration of gas seriously threatens the safety production of coal mines, which creates an opportunity for CBM extraction in coal mine areas. Surface–underground gas pre-drainage can effectively reduce gas outbursts whilst making full use of CBM, which has great development potential. The driving forces of surface–underground CBM three-dimensional drainage are gas control and coal safety mining. Therefore, this technology needs to be combined with coal production in time and space and make full use of the mining effect on the pressure relief and permeability of coal. The three-dimensional highly efficient migration channel network of CBM was produced via surface vertical well fracturing and downhole long borehole and eventually realized the efficient development of CBM in coal mining areas (Figure 11) [45–47]. Surface–underground CBM three-dimensional drainage development has multiple advantages, such as ensuring coal production safety, effectively utilizing CBM resources and reducing methane emissions, which is expected to be an important and suitable technology for CBM exploration and development in Guizhou. In recent years, scholars and engineers have begun to explore the surface–underground CBM three-dimensional drainage in coal mine areas, and the first demonstration project was launched in the Xintian coal mine in 2020. Other similar demonstration projects are also being advanced in Qianbei and Zhina coal fields.

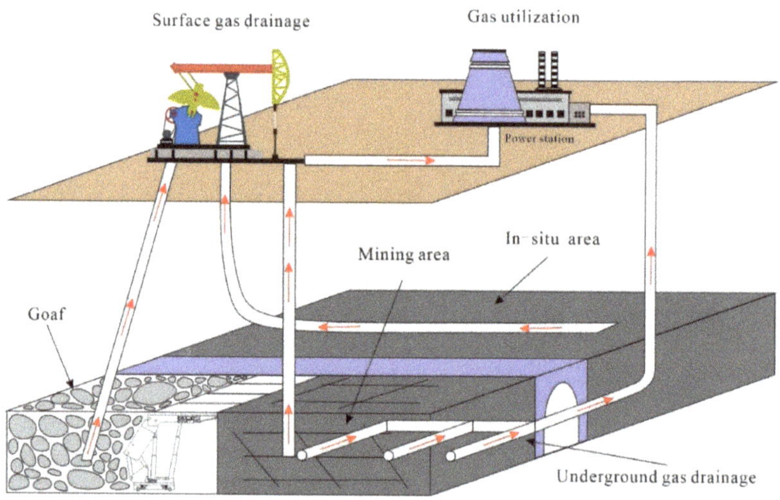

Figure 11. Integrated development model of the surface–underground CBM three-dimensional drainage in coal mine area.

4. Conclusions

The middle-rank coal-rich area—Dahebian Block—and the high-rank coal-rich area—Dafang Block—were taken as examples to compare coal occurrence, coal maceral, and properties. The effects of sedimentation, tectonic, and depth on the CBM geological differences between the Dahebian and Dafang Blocks were interpreted. The Guizhou CBM enrichment and high-yield model and its different depths for the middle- and high-rank coals were shown. Geologically suitable CBM exploration and development technologies were proposed for geological selection, development technology, and modeling in Guizhou. The main conclusions are summarized as follows:

(1) The geological differences in the middle- and high-rank coal CBM in Guizhou are mainly reflected in the number of coal layers, maceral, coal structure, and their influence on gas-bearing capacity and permeability. Compared to the middle-rank coal in the Dahebian Block, the Dafang high-rank coal shows more coal layers (30–66), and a higher vitrinite (2.94–3.42%) and gas content (17.84–23.75 m^3/t), but a lower thickness (9.5 m), a lower permeability (0.02–0.64 mD), and a wider tectonically deformed coal distribution;

(2) Although the sedimentary environment was the transitional sedimentary facies system in most of Guizhou during the Late Permian, the frequent transgression–regression and terrigenous source input resulted in a significant difference in coal occurrence, maceral, and qualities. The distribution of coal rank shows a high affinity with deep faults, and the uneven tectonic subsidence and its related thermal anomaly result in an X-type distribution of coal rank. The distribution and damage degree of the tectonically deformed coal are also related to these faults;

(3) The favorable configuration of structure, hydrology, and depth promotes enrichment and high CBM yield in Guizhou. However, there are differences in depth for CBM enrichment and high yield between middle-rank coal (<500 m) and high-rank coal (600–800 m). The coupling relationship between permeability and gas content controlled by depth is the fundamental reason for the difference in depth for CBM enrichment and high yield between the middle- and high-rank coal;

(4) We provided three new concepts for the efficient exploration and development of CBM in Guizhou; namely, (a) the optimization of block and coal seam for CBM exploration and development (classification–successive optimization–analytical hierarchy process), (b) the multi-layer CBM development with the pressure relief of tectonically deformed coal, and (c) surface–underground CBM three-dimensional drainage development in coal mines.

Author Contributions: Conceptualization, F.Z., S.H. and S.S.; methodology, F.Z.; validation, S.S. and S.H.; formal analysis, J.Z., A.X. and W.X.; investigation, F.Z., S.H. and J.Z.; resources, F.Z. and Z.W.; data curation, F.Z. and Z.W.; writing—original draft preparation, F.Z.; writing—review and editing, F.Z., S.H. and S.S.; visualization, J.Z., A.X. and W.X.; supervision, S.H. and S.S.; project administration, F.Z. and Z.W.; funding acquisition, F.Z., S.H. and S.S. All authors have read and agreed to the published version of the manuscript.

Funding: This research was funded by [National Natural Science Foundation of China] grant number [No. 42030810, 41727801, 42102207] and [Geological Exploration Foundation of Guizhou Province] grant number [No. 208-9912-JBNUTS0, 52000021MGQSE7S7K6PRP]. And The APC was funded by [Geological Exploration Foundation of Guizhou Province].

Acknowledgments: The authors would like to acknowledge Guizhou Natural Gas Energy Investment Co., Ltd. and Guizhou Shuicheng Coal Mining & Orion Clean Energy LLC for assisting in logging data collection.

Conflicts of Interest: The authors declare no conflict of interest.

References

1. Guizhou Daily. Available online: http://szb.gzrbs.com.cn/pc/cont/202102/27/content_6542.html (accessed on 27 February 2021).
2. Shao, L.; Hua, F.; Yi, T.; Guo, L.; Wang, X. Sequence-paleogeography and coal accumulation of Lopingian in Guizhou Province. *Coal Geol. Explor.* **2021**, *49*, 45–56.
3. Chen, Y.; Qin, Y.; Li, Z.; Chang, H.; Luo, Z. Magma thermal metamorphism of the Longtan formation coals in Zhijin Guizhou. *J. China Univ. Min. Technol.* **2012**, *41*, 406–414.

4. Tang, S.; Tang, D.; Xu, H.; Tao, S.; Li, S.; Geng, Y. Geological mechanisms of the accumulation of coalbed methane induced by hydrothermal fluids in the western Guizhou and eastern Yunnan regions. *J. Nat. Gas Sci. Eng.* **2016**, *33*, 644–656. [CrossRef]
5. Li, M.; Jiang, B.; Lin, S.; Lan, F.; Wang, J. Structural controls on CBM reservoirs in Faer coal mine, Southwest China. *J. Earth Sci.* **2013**, *24*, 437–448. [CrossRef]
6. Zhang, Z.; Qin, Y.; You, Z.; Yang, Z. Distribution characteristics of in situ stress field and vertical development unit division of CBM in Western Guizhou, China. *Nat. Resour. Res.* **2021**, *30*, 3659–3671. [CrossRef]
7. Song, L.; Tang, D.; Xu, H.; Yang, Z. The pore-fracture system properties of CBM reservoirs in the Panguan Syncline, Guizhou, China. *Geosci. Front.* **2012**, *3*, 853–862.
8. Li, S.; Tang, D.; Pan, Z.; Xu, H.; Guo, L. Evaluation of CBM potential of different reservoirs in western Guizhou and eastern Yunnan, China. *Fuel* **2015**, *139*, 257–267. [CrossRef]
9. Chen, S.; Tang, D.; Tao, S.; Chen, Z.; Xu, H.; Li, S. Coal Reservoir Heterogeneity in Multicoal Seams of the Panguan Syncline, Western Guizhou, China: Implication for the Development of Superposed CBM-Bearing Systems. *Energy Fuels* **2018**, *32*, 8241–8253. [CrossRef]
10. Tan, Y.; Pan, Z.; Liu, J.; Zhou, F.; Connell, L.D.; Sun, W.; Haque, A. Experimental study of impact of anisotropy and heterogeneity on gas flow in coal. Part II: Permeability. *Fuel* **2018**, *230*, 397–409. [CrossRef]
11. Jia, B.; Xian, C. Permeability measurement of the fracture-matrix system with 3D embedded discrete fracture model. *Pet. Sci.* **2022**. [CrossRef]
12. Jia, B.; Chen, Z.; Xian, C. Investigations of CO_2 storage capacity and flow behavior in shale formation. *J. Pet. Sci. Eng.* **2022**, *208*, 109659. [CrossRef]
13. Li, M.; Jiang, B.; Lin, S.; Wang, J.; Ji, M.; Qiu, Z. Tectonically deformed coal types and pore structures in Puhe and Shanchahe coal mines in western Guizhou. *Min. Sci. Technol.* **2011**, *21*, 353–357. [CrossRef]
14. Ju, W.; Yang, Z.; Qin, Y.; Yi, T.; Zhang, Z. Characteristics of in-situ stress state and prediction of the permeability in the Upper Permian CBM reservoir, western Guizhou region, SW China. *J. Pet. Sci. Eng.* **2018**, *165*, 199–211. [CrossRef]
15. Shen, Y.; Qin, Y.; Guo, Y.; Yi, T.; Yuan, X.; Shao, Y. Characteristics and sedimentary control of a CBM-bearing system in lopingian (late permian) coal-bearing strata of western Guizhou Province. *J. Nat. Gas Sci. Eng.* **2016**, *33*, 8–17. [CrossRef]
16. Ren, P.; Xu, H.; Tang, D.; Li, Y.; Chen, Z. Pore structure and fractal characterization of main coal-bearing synclines in western Guizhou, China. *J. Nat. Gas Sci. Eng.* **2019**, *63*, 58–59. [CrossRef]
17. Ju, W.; Yang, Z.; Shen, Y.; Yang, H.; Wang, G.; Zhang, X.; Wang, S. Mechanism of pore pressure variation in multiple coal reservoirs, western Guizhou region, South China. *Front. Earth Sci.* **2021**, *15*, 770–789. [CrossRef]
18. Qin, Y.; Xiong, M.; Yi, T.; Yang, Z.; Wu, C. On unattached multiple superposed coalbed-methane system: In a case of the Shuigonghe Syncline, Zhijin-Nayong coalfield, Guizhou. *Geol. Rev.* **2008**, *54*, 65–70.
19. Shen, Y.; Qin, Y.; Guo, Y.; Yi, T.; Shao, Y.; Jin, H. Sedimentary controlling factor of unattached multiple superimposed coalbed-methane system formation. *Earth Sci.-J. China Univ. Geosci.* **2012**, *37*, 573–579.
20. Zhou, P.; Jin, J.; Luo, K.; Gao, W.; Chen, J. Unattached multiple layer superimposed coalbed methane system in Songhe mine, west Guizhou. *Coal Geol. Explor.* **2017**, *45*, 66–69.
21. Shan, Y.; Bi, C.; Chi, H.; Wang, F.; Li, H. Geological characteristics of CBM and optimization for favorable productive intervals of Yangmeishu syncline in Liupanshui area. *Nat. Gas Geosci.* **2018**, *29*, 122–129.
22. Wang, S.; Gao, W.; Guo, T.; Bao, S.; Jin, J.; Xu, Q. The discovery of shale gas, coalbed gas and tight sandstone gas in Permian Longtan Formation, northern Guizhou Province. *Geol. China* **2020**, *47*, 249–250.
23. Cheng, W.; Yang, R.; Cui, Y.; Fu, H.; Zhang, T. Characteristic of Late Permian coal quality from Bijie, Guizhou Province, SW China, and its significance for paleoenvironment. *Acta Geol. Sin.* **2013**, *87*, 1763–1777.
24. Fang, X.; Wu, C.; Jiang, X.; Liu, N.; Zhou, D.; Ju, Y. Characteristics of in situ stress and its influence on coal seam permeability in the Liupanshui Coalfield, Western Guizhou. *Energy Sci. Eng.* **2021**, *9*, 1773–1786. [CrossRef]
25. Teichmüller, M. The genesis of coal from the viewpoint of coal petrology. *Int. J. Coal Geol.* **1989**, *12*, 1–87. [CrossRef]
26. Yang, Q.; Ren, D.; Pan, Z. Preliminary investigation on the metamorphism of Chinese Coals. *Int. J. Coal Geol.* **1982**, *2*, 31–48. [CrossRef]
27. Chen, X. Distribution law and genesis of Permian coal in Guizhou. *Coal Geol. Explor.* **1995**, *23*, 21–24.
28. Xiong, M.; Qin, Y.; Yi, T. Sedimentary patterns and structural controls of late permian coal-bearing strata in Guizhou, China. *J. China Univ. Min. Technol.* **2006**, *35*, 778–782.
29. Zeng, R.; Zhao, J.; Zhuang, X. Quality of Late Permian coal and its controlling factors in Shuicheng mining district of Liupanshui area, Guizhou. *Acta Petrol. Sin.* **1998**, *14*, 549–558.
30. Li, M.; Jiang, B.; Liu, J.; Zhu, P.; Cheng, G. Geological models and structural controls of tectonically deformed coal in Tucheng syncline, western Guizhou Province. *J. China Coal Soc.* **2018**, *43*, 1565–1571.
31. Zhou, P.; Gao, W.; Deng, L.; Fu, W. Regional distribution and geotectonic control of Late Permian tectonically deformed coal in Zhina coalfield. *Coal Geol. Explor.* **2020**, *48*, 29–34.
32. Jia, T.; Wang, W.; Yan, J.; Tang, C. The rule of tectonic control and the zoning division of coal mine gas occurrence in Guizhou Province. *Earth Sci. Front.* **2014**, *21*, 281–288.
33. Song, Y.; Liu, S.; Ma, X.; Ju, Y.; Hong, F.; Jiang, L. Accumulation models for CBM in medium- to high-rank coals: Examples from the southern Qinshui Basin and southeastern Ordos Basin. *Aust. J. Earth Sci.* **2018**, *65*, 575–590. [CrossRef]

34. Hildenbrand, A.; Krooss, B.-M.; Busch, A.; Gaschnitz, R. Evolution of methane sorption capacity of coal seams as a function of burial history-a case study from the Campine Basin, NE Belgium. *Int. J. Coal Geol.* **2006**, *66*, 179–203. [CrossRef]
35. Han, S.; Sang, S.; Zhou, P.; Liang, J. The evolutionary history of methane adsorption capacity with reference to deep Carboniferous-Permian coal seams in the Jiyang Sub-basin: Combined implementation of basin modeling and adsorption isotherm experiments. *J. Pet. Sci. Eng.* **2017**, *158*, 634–646. [CrossRef]
36. Liu, D.; Yao, Y.; Cai, Y.; Zhang, B.; Zhang, K.; Li, J. Characteristics of porosity and permeability and their geological control of Permo-Carboniferous coals in the north China. *Geoscience* **2010**, *24*, 1198–1203.
37. Fan, Q.; Cai, Y.; Bei, J.; Wang, W.; Zhang, X. Pore and Fracture Structure of Coal Reservoir Constrained by Coal Metamorphism. *Geoscience* **2020**, *34*, 273–280.
38. Sun, F.; Li, W.; Sun, Q.; Sun, B.; Tian, W.; Chen, Y.; Chen, Z. Low-rank CBM exploration in Jiergalangtu sag, Erlian basin. *Acta Pet. Sin.* **2017**, *38*, 485–492.
39. Wang, B.; Yao, H.; Wang, H.; Zhao, Y.; Li, M.; Hu, Q.; Fan, M.; Yang, C. Favorable and major geological controlling factors for CBM accumulation and high production in the Chengzhuang Block, Qinshui Baisn. *Oil Gas Geol.* **2018**, *39*, 366–372.
40. Wu, C.; Liu, X.; Zhang, S. Construction of index system of "Hierarchical progressive" geological selection of CBM in multiple seam area of eastern Yunnan and western Guizhou. *J. China Coal Soc.* **2018**, *43*, 1647–1653.
41. Shao, L.; Wang, X.; Zhang, J.; Hou, H.; Tang, Y.; Wang, J.; Lin, Y. CBM accumulation characteristics and exploration target selection in northeastern Yunnan, China. *Nat. Gas Ind.* **2018**, *38*, 17–27.
42. Shi, J.; Durucan, S.; Sinka, I.C. Key parameters controlling CBM cavity well performance. *Int. J. Coal Geol.* **2002**, *49*, 19–31. [CrossRef]
43. Meng, Z.; Li, G.; Yang, Y.; Li, C.; Qiao, Y. Study on key technology for surface extraction of CBM in coal mine goaf from Sihe Wells Area, Jincheng. *Coal Sci. Technol.* **2021**, *49*, 240–247.
44. Sang, S.; Zhou, X.; Liu, S.; Wang, H.; Cao, L.; Liu, H.; Li, Z.; Zhu, S.; Liu, C.; Huang, H.; et al. Research advances in theory and technology of the stress release applied extraction of CBM from tectonically deformed coals. *J. China Coal Soc.* **2020**, *45*, 2531–2543.
45. Wu, H. Study and practice on technology of three-zones linkage 3D CBM drainage in coal mining area. *J. China Coal Soc.* **2011**, *36*, 1312–1316.
46. Li, G.; He, H.; Liu, G.; Jiao, H.; Liu, X. Three region linkage three-dimensional gas drainage theory and mode of coal bed methane in coal mining area. *Coal Sci. Technol.* **2012**, *40*, 7–11.
47. He, T.; Wang, B.; Tian, Y. Development and issues with coal and coal-bed methane simultaneous exploitation in Jincheng mining area. *J. China Coal Soc.* **2014**, *39*, 1779–1785.

energies

Article

Biogenic Methane Accumulation and Production in the Jurassic Low-Rank Coal, Southwestern Ordos Basin

Chao Zheng [1,2], Dongmin Ma [1,3,*], Yue Chen [1], Yucheng Xia [1], Zheng Gao [4], Guofu Li [3] and Weibo Li [5]

1 School of Geology and Environment, Xi'an University of Science and Technology, Xi'an 710054, China; zhengc0624@126.com (C.Z.); cyxust@126.com (Y.C.); xiayc823@163.com (Y.X.)
2 School of Energy Engineering, Long Dong University, Qingyang 745000, China
3 State Key Laboratory of Co-Mining Coal and Coalbed Methane Technology, Jincheng 048000, China; liguofu114@126.com
4 School of Energy and Power Engineering, Xi'an Jiaotong University, Xi'an 710049, China; gaozheng@stu.xjtu.edu.cn
5 Shaanxi Geological Science and Technology Center, Xi'an 710054, China; liweibo022@126.com
* Correspondence: mdm6757@126.com

Abstract: Geological conditions are the key for coalbed methane (CBM) accumulation and production. However, the geological feature of CBM accumulation and production in the Jurassic of Ordos Basin lacks systematic and detailed evaluation, resulting in poor CBM production in this area. This study has determined the genetic types of gas according to geochemistry characteristics of the gas, the geological factors to control CBM accumulation and production performance were revealed, and a comprehensive method was established to evaluate favorable areas based on 32 sets of CBM well production data from Jurassic Yan'an Formation. The results show the coal macerals are rich in inertinite (41.13~91.12%), and the maximum reflectance of vitrinite ($R_{o,max}$) in coal is 0.56~0.65%. According to gas compositions and carbon isotopes analysis, the $\delta^{13}C(CH_4)$ is less than −55‰, and the content of heavy hydrocarbon is less than 0.05%. The value of $C_1/(C_2 + C_3)$ is 6800~98,000, that is, the CBM is a typical biogenic gas of low-rank coal. The CBM accumulation model is the secondary biogenic on the gentle slope of the basin margin, in which gas content is closely related to buried depth and hydrodynamic environment, i.e., the high gas content areas are mainly located in the groundwater weak runoff zone at the burial depth of 450 m~650 m, especially in the syncline. Meanwhile, gas production mainly depends on the location of the structure. The high gas production areas of vertical wells were distributed on the gentle slope with high gas content between anticline and syncline, and the horizontal wells with good performance were located near the core of the syncline. According to the above analysis combined with the random forest model, the study area was divided into different production favorable areas, which will provide a scientific basis for the CBM production wells.

Keywords: low-rank coal; biogenic methane; geological factor; accumulation; production

Citation: Zheng, C.; Ma, D.; Chen, Y.; Xia, Y.; Gao, Z.; Li, G.; Li, W. Biogenic Methane Accumulation and Production in the Jurassic Low-Rank Coal, Southwestern Ordos Basin. *Energies* **2022**, 3, 3255. https://doi.org/10.3390/en15093255

Academic Editors: Mofazzal Hossain, Shu Tao, Dengfeng Zhang, Huazhou Huang, Shuoliang Wang and Yanjun Meng

Received: 16 March 2022
Accepted: 26 April 2022
Published: 29 April 2022

1. Introduction

China is a large country of coal production and energy consumption, the large-scale and high-efficiency development of CBM can not only reduce mine gas accidents but also supplement the shortage of conventional natural gas [1,2]. Coalbed methane (CBM) is an important environment-friendly energy resource [3,4]. CBM includes biogenic gas and thermogenic gas, of which biogenic gas is generated dominantly by anaerobic bacteria and methanogens via coal biodegradation at low temperatures (<5 °C) [5–7]. Influenced by generation time and geological evolution, the biogas nowadays in coal reservoirs is secondary biogenic gas, which mainly is the reduction in carbon dioxide according to the gas component [8]. The resources of CBM are abundant in China, and commercial development for the middle-high rank CBM has achieved technological breakthroughs [9–13]. Although the CBM resources of low-rank are rich [14,15], the development of low-rank CBM in western China is still not satisfactory [16–20].

Both scientific research and production practice show that it is a key step to making clear the geological characteristics of CBM accumulation and production. It was considered that CBM generation, accumulation and production are controlled by sedimentation, coalification, tectonism, hydrodynamics, as well as other geological factors [14,16,21–23]. The coalbed methane well productivity performance was a result of the strong interaction of cumulative thickness, burial depth, gas content, permeability, and reservoir pressure of fractured coal seam [24], in which the gas content is positively correlated with the coal burial depth, coal thickness [25]. The geologic structural setting and hydrogeological influenced the spatial distribution of gas content and permeability and so on [26–28]. Hydrodynamic confinement results in relatively high gas content of coal reservoir in the groundwater stagnation zone, while gas content is comparatively low in the recharge and runoff zones [11,29,30]. Characteristics of in situ stress indirectly control coalbed methane development by affecting permeability [31–33]. In a word, if the main controlling factors, such as structure, coal-forming environment, and hydrologic geology, are matched well, the enriched coalbed methane zone with high production would be formed [15].

Although much literature has been published on the geologic feature of CBM reservoirs [33–36], previous investigations on the CBM in the study area were mainly focused on geological background and CBM enrichment conditions [37–39]. Furthermore, the usual methods for the prediction of favorable areas are empirical methods, such as expert scoring method, fuzzy comprehensive evaluation method, etc., which have subjective human factors to a large extent [40,41]. Few studies on the model of coalbed methane accumulation and production based on the genetic type of coalbed methane have been conducted in this area. In particular, the detailed evaluation of low-rank coal biogas in western China. In this study, the random forest data training method was used to train and classify the sample data in order to comprehensively analyze the accumulation and production of CBM taking the Dafosi minefield, Binchang mining area as the study area. Firstly, we determine the gas compositions and the genetic types; secondly, we analyze the key factors affecting the CBM accumulation and production and then reveal the relationship between the gas content, gas production and geological conditions. Finally, the rank of factors is determined with grey correlation, and the favorable area for CBM development was predicted in the study area by using the random forest method.

2. Methodology

2.1. Data Acquisition and Processing

The coal samples in this paper were all collected from the No. 4 coal seam of the Middle Jurassic Yan'an Formation in the Dafosi minefield, Binchang mining area. The 10 coal samples were selected for the determination of maceral composition and the reflectance of vitrinite in coal according to the Chinese national standard GB/T 16773-2008, GB/T 8899-2013 and GB/T 6948-2008. The coal samples were ground to a particle size less than 1.0 mm air-dried base sample, made into pulverized coal slices, and then the samples were polished and carried out maceral composition determination. Meanwhile, the samples were repolished and dried in a drying oven at 30 °C~40 °C for 4 h before carrying out the reflectance of vitrinite in coal determination.

Carbon isotopic characteristics of CH_4 are important to the gas performance of coal reservoirs [42]. The gas samples came from the surface CBM wells, including multi-branch horizontal wells DFS-C02, DFS-05, DFS-06, DFS-09, and vertical wells DFS-C01, U-shaped well DFS-C03 and V-shaped well DFS-04. Wasson-7890A gas chromatograph was used for gas composition determination, and thermo MAT253 gas isotope mass spectrometer was employed in doing carbon isotope. Taking a certain volume of CBM sample with an airtight syringe and injecting it into the Trace GC ultra; the hydrocarbon components in the sample were separated by the meteorological chromatograph Trace GC ultra and converted into carbon dioxide in the oxidation unit GC combustion, and then the gas flow entered the isotope mass spectrometer MAT253 IRMS for the determination of carbon isotope components. Gas compositions and carbon isotopes are based on GB/T 13610-2003 and SY/T 5238-2008. The basic information of the test sample is shown in Table 1.

Table 1. Basic information of test sample.

Gas Compositions Sample			Coal Maceral Compositions		
Sample ID	**Burial Depth**	**Sample Size**	**Sample ID**	**Burial Depth**	**Sample Size**
DFS-C01	435 m	10~20 mL	DFS-C01	435 m	0.18~0.25 mm
DFS-C02	610 m	10~20 mL	DFS-D48	554 m	0.18~0.25 mm
DFS-C03	540 m	10~20 mL	DFS-D62	568 m	0.18~0.25 mm
DFS-C04	510 m	10~20 mL	DFS-M85	615 m	0.18~0.25 mm
DFS-05	590 m	10~20 mL	DFS-05	590 m	0.18~0.25 mm
DFS-06	480 m	10~20 mL	DFS-U53	435 m	0.18~0.25 mm
DFS-09	535 m	10~20 mL	DFS-09	535 m	0.18~0.25 mm
/	/	/	DFS-128	500 m	0.18~0.25 mm
/	/	/	DFS-134	448 m	0.18~0.25 mm
/	/	/	DFS-122	585 m	0.18~0.25 mm

2.2. Data Interpretation

In order to analyze the main controlling factors affecting CBM productivity, the grey correlation analysis method is introduced to quantitatively determine the correlation degree between each influencing factor and productivity, so as to objectively evaluate the influence degree of each geological factor on coalbed methane well productivity [39,43]. Grey correlation calculation is mainly divided into two processes, (1) correlation coefficient calculation, (2) calculation of correlation degree [44]. Random forest was used for favorable area evaluation and prediction. Random forest takes decision trees as the basic model to generate a series of differentiated decision tree models by building different training data sets and different feature spaces [45–48]. The random forest was constructed based on classification and regression trees in this paper. In the process of establishing each classification and regression tree, the splitting process of each node is completed by calculating the "purity" of the sample after the split. The classification and regression tree employ the *gini* coefficient to measure this so-called "purity", i.e., random forest uses the *gini* index to split the tree to complete the decision. The smaller the *gini* coefficient, the higher the purity of the sample and the better the effect of tree division. Assuming that the sample set T contains k categories, the Gini coefficient of the sample set can be expressed as [49]:

$$gini(T) = 1 - \sum_{i=1}^{k} p_i^2 \tag{1}$$

where, p_i is the probability that T contains class i. If T is divided into two subsets T_1 and T_2, the divided *gini* coefficient can be expressed as:

$$gini(T_1, T_2) = \frac{|T_1|}{|T|} gini(T_1) + \frac{|T_2|}{|T|} gini(T_2) \tag{2}$$

where, $|*|$ represents the number of elements in the current sample set.

The selection of CBM development prospect evaluation indicators vary in different areas. For low-rank coal development areas, the gas generation potential, storage performance and preservation conditions of the coal reservoir should be chosen. Through the analysis of the drilling and experimental data of the CBM wells, the main factors influencing the development potential and productivity of CBM, including resource conditions (gas content, ash, net coal seam thickness) and occurrence conditions (structural location, roof thickness), development conditions (permeability, reservoir pressure, burial depth) were selected as evaluation indicators. There are few faults in this area, mainly folds. In order to quantitatively study the influence of folds on the development of CBM, qualitative indicators (structural locations) are assigned and quantified separately. The flank is 0.8, the syncline is 1, and the anticline is 0.6.

In order to prevent the strong correlation between the factors, resulting in a decrease in the running rate of the evaluation model and over-fitting of the running results, the correlation between the factors needs to be tested. This paper adopts the Pearson correlation coefficient (*PCC*) to analyze the degree of correlation between the factors. The larger the value, the stronger the correlation of the factors. The calculation formula of the correlation coefficient between the evaluation factors is [50].

$$PCC = \frac{\sum_{i=1}^{n} (x_i - \overline{x}) \sum_{j=1}^{n} (y_i - \overline{y})}{\sqrt{\sum_{i=1}^{n} (x_i - \overline{x})^2 \sum_{j=1}^{n} (y_i - \overline{y})^2}} \tag{3}$$

where, x_i, y_i is the variable values of X_i and Y_i, respectively, $\overline{x}$ and $\overline{y}$ are the average values of X_i and Y_i, respectively. When $0 \leq |PCC| \leq 0.4$, it indicates that the factor is irrelevant or weakly correlated. When $|PCC| > 0.6$, it indicates that the factor is strongly correlated.

3. Geology Setting and Analysis

3.1. Geology Condition

The Binchang mining area is located in the Miaobin depression area on the northern margin of the Weibei flexure belt in the southern Ordos Basin, the surface is covered by loess formation, and the occurrence of Cretaceous strata exposed in the valley is relatively gentle [51]. The deep Jurassic hidden structure is generally a monoclinic structure with N 60°~70° E in strike and NW~NNW in the dip. A set of broad and gentle folds are developed on it, from south to north, Binxian anticline, Lujia–Xiaolingtai anticline, Mengcun syncline, Qilipu–Xipo anticline in order. According to the borehole, no fault structure is found in the area. The overall structure is a monoclinic structure inclined from southeast to northwest inherited from the Triassic basement, the stratum dip is gentle, on average 3~5°. The stratigraphic division belongs to the Ordos areas, from bottom to top, it is the Hujiacun Formation of the Upper Triassic System (T_{3h}), the Fuxian Formation of the Lower Jurassic System (J_{1f}), the Middle Jurassic Yan'an Formation (J_{2y}), Zhiluo Formation (J_{2z}), Anding Formation (J_{2a}), Lower Cretaceous Yijun Formation (K_{1y}), Luohe Formation (K_{1l}), Huachi Formation (K_{1h}), Neogene and Quaternary strata, as shown in Figure 1.

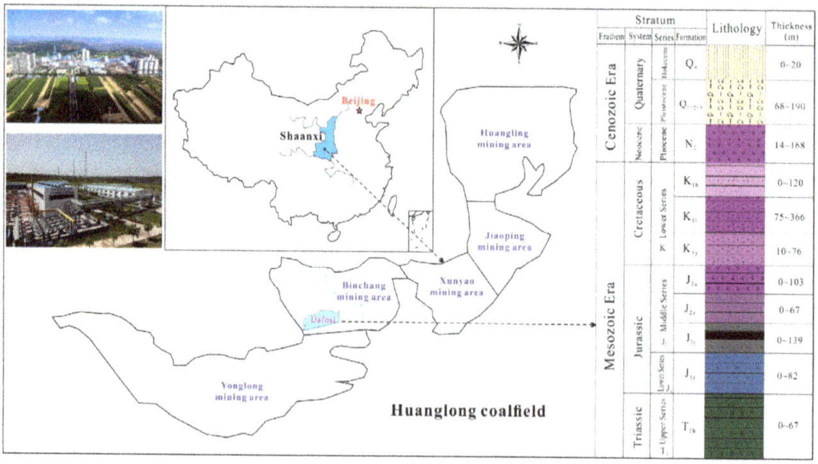

Figure 1. The location and stratum of Binchang mining area.

The overall structure of the study area is simple. As shown in Figure 2, the main structure is located on the southern flank of the Anhua syncline area to the north of the Binxian anticline. Due to the influence of synsedimentary structure, there are Anhua syncline, Qijia anticline, and Shijiadian syncline in order from north to south. The coal-bearing stratum in the study area is the Jurassic Middle Yan'an Formation, with a total of six coal-bearing seams, of which No. 4 coal is a relatively stable and main mineable coal seam in the minefield, distributed throughout the minefield. The stratigraphic occurrence and structural morphology show that the strata above Jurassic in this area have not been subjected to strong tectonic extrusion and deformation. The structure is closely related to the coal system, the deposition distribution of coal seams and their thickness changes. The coal seam is thicker at the syncline axis and thinner at the anticline axis. This pattern is mainly related to synsedimentary tectonics, and also reflects the inheritance of later tectonics.

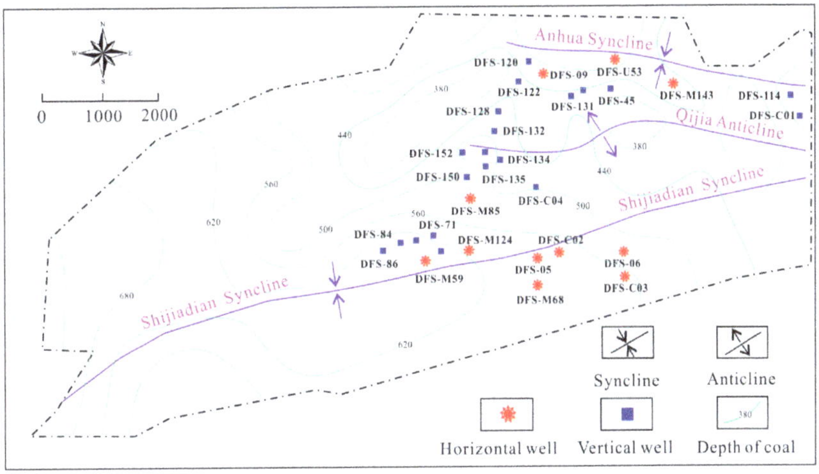

Figure 2. Structure characteristic of the study area.

CBM proved resources/reserves of the study area are 35.89 × 10^8 m^3, which are 25.5 × 10^8 m^3 in the No. 4 coal reservoir, and the average resource abundance is 1.44 × 10^8 m^3/km^2. CBM gas production capacity is 5.2 × 10^4 m^3/d, the average value of a single well can reach 1625 m^3/d. Consequently, the development potential of CBM is significant. The basic data used in this study, such as gas content and gas production, coal burial depth, thickness, lithology of overburden, etc., were collected from 32 CBM wells as shown in Table 2.

Table 2. Geological information of CBM well.

Well No.	Gas Production (m^3/d)	Gas Content (m^3/t)	Burial Depth (m)	Coal Thickness (m)	Permeability (mD)	Reservoir Pressure (MPa)	Ash of Coal (%)	Roof Thickness (m)
DFS-C02	11,270.97	4.30	610	11.00	7.70	2.80	14.00	2.20
DFS-05	10,600.96	2.80	590	16.80	10.00	3.30	11.00	2.10
DFS-M68	9625.14	3.60	589	14.00	9.20	3.10	12.00	2.00
DFS-M85	4025.73	8.10	615	11.80	6.60	3.20	15.50	2.10
DFS-09	3998.08	1.80	535	8.30	7.30	3.80	14.90	1.20
DFS-M143	3791.89	6.80	465	12.60	15.00	1.20	15.20	6.00
DFS-M124	3759.98	8.00	490	13.60	7.80	1.40	13.90	0.70
DFS-128	3256.75	3.30	500	10.30	3.80	2.90	10.50	0.70

Table 2. *Cont.*

Well No.	Gas Production (m^3/d)	Gas Content (m^3/t)	Burial Depth (m)	Coal Thickness (m)	Permeability (mD)	Reservoir Pressure (MPa)	Ash of Coal (%)	Roof Thickness (m)
DFS-C04	3145.28	8.30	510	12.20	11.00	5.50	15.90	3.00
DFS-148	2638.34	3.80	505	17.60	7.80	0.80	12.00	1.20
DFS-133	2590.39	8.00	490	10.20	6.80	4.30	17.00	0.60
DFS-CO1	2032.45	10.00	435	13.90	12.30	3.60	14.40	1.80
DFS-U53	1978.87	2.20	435	14.30	11.80	2.40	15.20	1.80
DFS-105	1936.58	7.70	500	8.30	7.40	5.50	15.80	1.40
DFS-45	1936.16	3.20	540	11.80	7.30	1.30	11.90	1.40
DFS-131	1876.32	2.80	570	10.20	4.80	2.50	19.00	1.50
DFS-134	1696.16	9.10	448	10.80	12.50	4.60	15.20	3.60
DFS-06	1339.94	8.30	480	11.00	18.80	5.60	15.10	6.80
DFS-C03	1248.76	3.50	540	12.30	6.00	6.00	16.00	0.80
DFS-132	836.56	8.20	505	8.30	7.40	6.00	15.20	1.40
DFS-122	683.15	8.60	585	20.30	6.50	3.30	17.00	2.60
DFS-114	579.37	9.30	630	13.50	8.00	3.80	16.80	3.00
DFS-152	554.33	8.10	510	12.30	8.10	4.80	16.30	1.00
DFS-M59	460.25	7.30	510	10.30	7.90	4.60	14.50	1.40
DFS-73	458.52	10.00	440	14.20	11.10	3.60	14.30	0.90
DFS-135	450.94	8.50	628	14.00	7.00	3.40	16.20	2.60
DFS-150	434.58	8.60	600	11.00	8.10	3.70	16.20	3.00
DFS-69	344.81	9.30	445	12.00	11.10	2.60	15.90	3.80
DFS-71	340.54	8.20	630	14.70	7.50	4.00	17.20	2.60
DFS-84	256.95	6.50	565	17.80	14.00	3.00	13.00	3.60
DFS-86	229.48	8.10	650	13.00	6.80	3.90	17.80	2.30
DFS-120	212.47	8.10	500	12.20	11.00	2.50	15.80	3.70

3.2. Geological Factors of Gas Content

3.2.1. Thickness of Coal Seam

Generally speaking, under the same conditions, the greater the thickness of the coal seam, the greater the amount of gas generated. The escape of gas is dominated by diffusion, and the concentration difference between two points in space is the main driving force for its diffusion [52]. The thickness of the No. 4 coal seam is 0.3 m~20.5 m in the study area, statistical analysis has indicated that the gas content is positively correlated with the thickness of the coal seam as shown in Figure 3a. The thicker the coal seam, the longer the gas diffusion path between the roof and floor, and the greater the diffusion resistance, which is more conducive to gas accumulation.

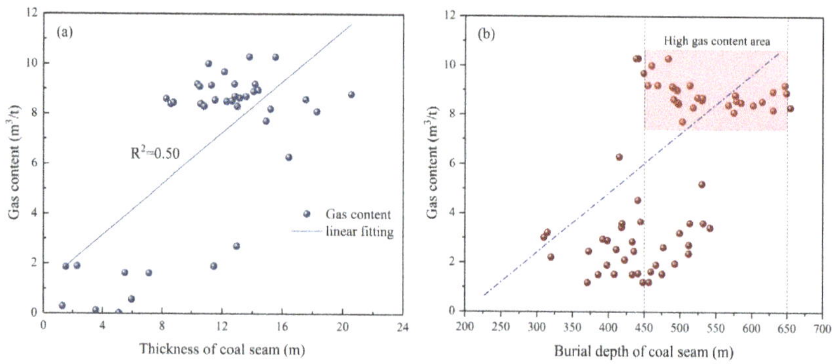

Figure 3. The relationship between gas content and thickness of coal seam, burial depth: (**a**) thickness of coal seam; (**b**) burial depth of coal seam.

3.2.2. Burial Depth and Structure

Burial depth is positively correlated with the reservoir pressure, and the larger reservoir pressure is conducive to the adsorption of gas. The migration distance of CBM to the surface with burial depth increases correspondingly, which is beneficial to the preservation of CBM. Therefore, the gas content generally increases with the increase in the burial depth of the coal seam. The burial depth of the No. 4 coal seam is 270 m~780 m from northeast to southwest. According to the analysis of the relationship between the burial depth and gas content, the area with gas content greater than 2 m^3/t is mainly distributed in 400 m~650 m as shown in Figure 3b, the gas content is low when the burial depth is less than 400 m. The main reason is that the shallow burial depth is in the groundwater recharge zone, and the overburden is thin, so the CBM is easy to escape. The groundwater stagnant zone with a buried depth exceeding 650 m is not conducive to the reproduction of methanogens, gas generation capacity itself is low. At present, the burial depth of coal seam with high gas content is mainly 450~650 m.

The structure of the study area is dominated by broad and gentle folds. Folds mainly affect the gas content by controlling the thickness of coal and the flow of groundwater, i.e., "syncline rich gas, anticline poor gas". There is a significant difference at different locations on the anticline and syncline. Above the neutral surface of the anticline axis, the coal seam and roof develop tensile fissures, and the gas is easy to escape. The groundwater flows from the axis of the anticline to the flanks, taking away part of the gas, which is not conducive to the accumulation of CBM. However, the coal rock above the neutral plane of the syncline is subjected to compressive stress, which is conducive to the preservation of gas with high gas content. It can be seen from Figure 4 that the content of the gas is more obviously controlled by structure style. The area with the highest gas content is the Shijiadian syncline axis area; the gas content of the Qijia anticline axis is relatively low.

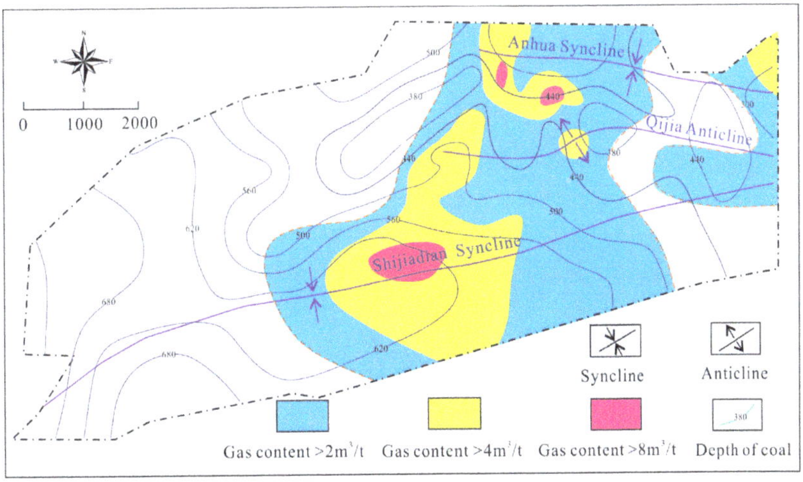

Figure 4. Geological structure and gas content distribution.

3.2.3. Hydrogeological Conditions

Hydrogeological conditions are an important factor affecting the occurrence of CBM [53,54], a positive relationship was found between gas content and basin hydrodynamics. As illustrated in Figure 5, it was found that the ions in the water of the Jurassic aquifer (Yan'an Formation and Zhiluo Formation) are mainly Cl^- and Na^+, the hydrochemical type is Cl^--Na^+, and the mineralization range is 13,781 mg/L~16,490 mg/L (Figure 6), which represents the water-bearing conditions of deepwater circulation and poor runoff, also provided favorable conditions for CBM accumulation. The main anions of the Cretaceous aquifers (Yijun Formation and Luohe Formation) are $SO_4{}^{2-}$ and $HCO_3{}^-$, the main cations are Na^+ and Mg^{2+}, and the hydrochemica types are $SO_4{}^{2-}$•Cl^--Na^+, $SO_4{}^{2-}$•(HCO^{3-})-Na^+•Mg^{2+}, the mineralization range are 859 mg/L~5790 mg/L. The hydrochemica type of the Xiaozhanggou Formation is HCO^{3-}-Na^+•Mg^{2+}•Ca^{2+}, and the mineralization is 274 mg/L~461 mg/L. The ions in the loose layer of the Quaternary system are HCO^{3-}, Na^+, Mg^{2+} mainly, the hydrochemical type is HCO^{3-}-Na^+•Mg^{2+}•Ca^{2+}, the water quality is good, and the mineralization is between 271 mg/L~317 mg/L (Figure 6). Low-rank coal located in the area of weak groundwater runoff, and low mineralization is conducive to secondary biogenic gas generation and hydrodynamic pressure-bearing sealing is conducive to the preservation of CBM.

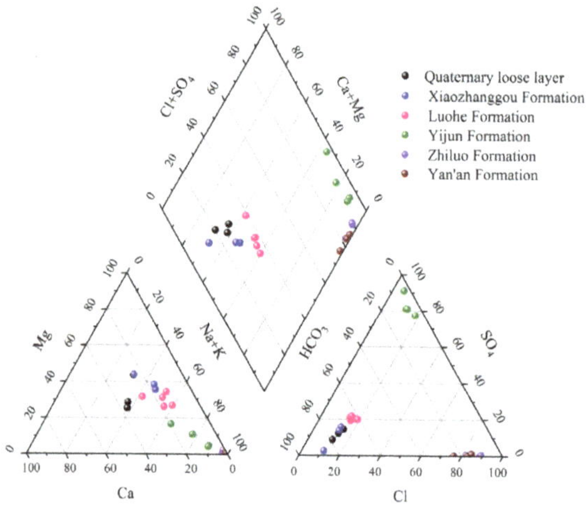

Figure 5. Hydrochemical type of groundwater.

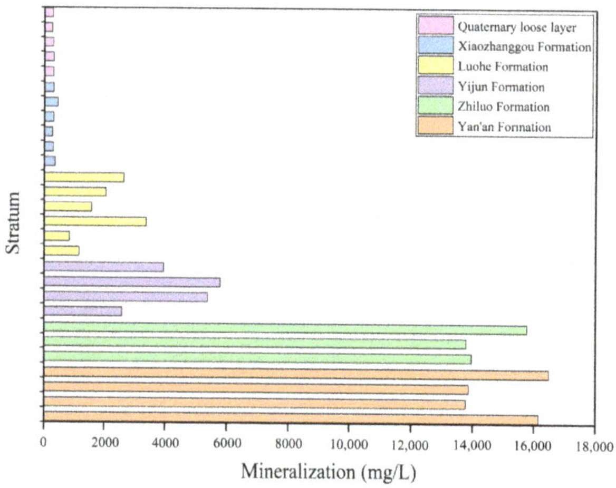

Figure 6. Mineralization degree of groundwater.

3.2.4. Lithology of Overburden Stratum

The lithology, thickness, development of joints and fissures of the roof above coal reservoirs are closely related to the accumulation of CBM. There are three mechanisms for the sealing effect of the roof on CBM, pressure and capillary can prevent gas migration, and hydrocarbon concentration mainly restrains gas diffusion [55]. The proper thickness and good sealing of the roof make the coal seam have the potential to preserve higher gas content. The tighter the roof lithology is, the more beneficial it is for the accumulation of CBM. So, the greater the proportion of mudstone thickness in the Yan'an formation on the roof, the higher the gas content (Figure 7a). It is particularly emphasized that the thickness of mudstone within 10 m of the roof plays a more key role in the migration of CBM. As shown in Figure 7b, although the correlation between mudstone thickness within 10 m of the roof and gas content is not very significant if the immediate roof of the coal seam is mudstone, which is conducive to the accumulation of CBM.

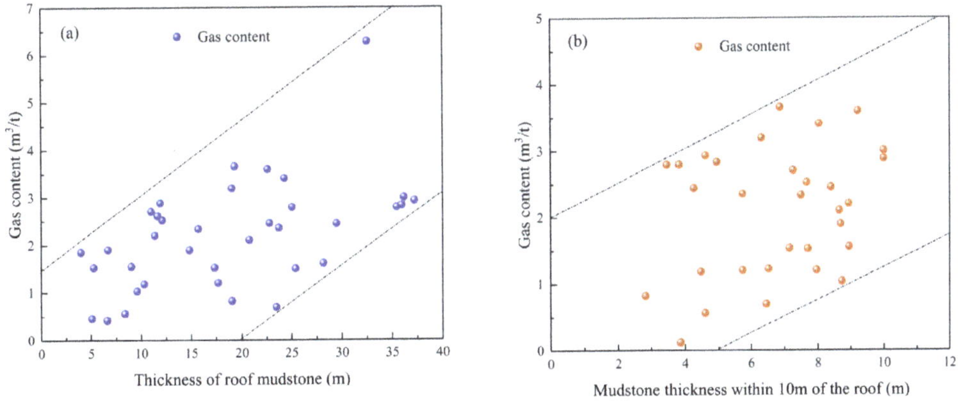

Figure 7. Relationship between gas content and roof: (**a**) mudstone thickness; (**b**) mudstone thickness within 10 m of the roof.

3.3. Geological Factors of Gas Production

3.3.1. Gas-Bearing Properties of Coal Reservoir

Gas content is a basic indicator of CBM storage performance and sufficient condition in the evaluation of CBM development prospects. Areas with high gas content are often favorable areas for CBM development. The gas content is mainly concentrated in 2 m^3/t~10 m^3/t, the gas production fluctuates in a wide range during the stable production stage (Figure 8).

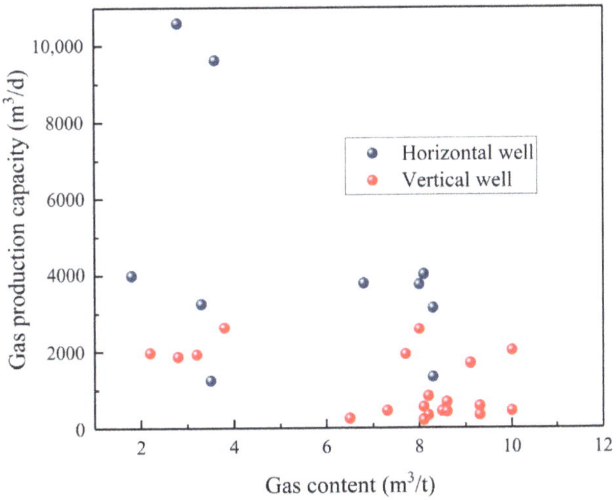

Figure 8. The relationship between gas production and gas content.

3.3.2. Burial Depth and Thickness of Coal Seam

The burial depth of coal seams directly affects coal seam permeability, in situ stress, and reservoir pressure. The burial depth of coal seams is between 400 m~650 m. It can be seen from Figure 9a that when the depth is greater than 550 m, the tendency of gas production to decrease with the increase in burial depth becomes more obvious. The thickness reflects the abundance of resources to a certain extent. The greater the thickness, the higher the resource abundance, and the gas production also increases. There is a positive correlation between the gas production and the thickness of the coal seam, especially with horizontal wells as shown in Figure 9b.

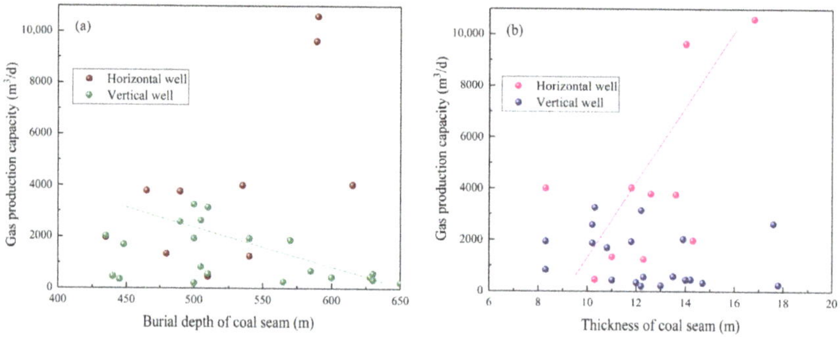

Figure 9. The relationship between gas production and burial depth, thickness of coal seam: (**a**) burial depth of coal seam; (**b**) thickness of coal seam.

3.3.3. Permeability and Reservoir Pressure

Permeability and reservoir pressure play an important role in the development of CBM and restrict the desorption and migration of CBM. According to field data, the permeability of the No. 4 coal reservoir is 2 mD~20 mD, the reservoir pressure is 2 MPa~8 MPa, and the gas production has no obvious correlation with permeability and reservoir pressure (Figure 10).

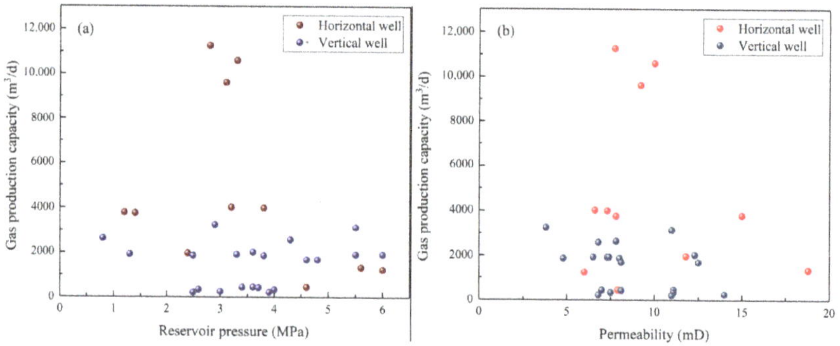

Figure 10. The relationship between gas production and reservoir pressure, permeability: (**a**) reservoir pressure; (**b**) permeability.

3.3.4. Structure Conditions

It was found that the fractures in the fold flanks are moderately developed, and the permeability of coal reservoirs as well. In the process of gas production, although it receives recharge from the anticline, underground water flows to the syncline core to make it easier for depressurization and gas migration. Therefore, the gas content of the fold flank and the permeability of the coal reservoir reach a suitable match, and the stable gas production of the CBM well is generally better. The Anhua syncline, Qijia anticline and Shijiadian syncline are developed in the study area. As shown in Figure 11, the gas production is closely related to the structure position, but the difference between different well types is obvious. The vertical wells DFS-128 and DFS-C04 are located in the gentle slope between the fold flanks, which average gas production were 3256.75 m^3/d and 3145.28 m^3/d, respectively, in the stable production stage. However, the gas production of vertical wells was poor near the syncline axis. Horizontal wells DFS-C02, DFS-M124, and DFS-05 are located near the Shijiadian syncline axis. The average gas production in the stable production stage is about 11,270.97 m^3/d, 3759.98 m^3/d, and 10,600.96 m^3/d. Therefore, it can be summarized that the high production area of horizontal wells is between the syncline core and the vertical wells are between the fold flanks.

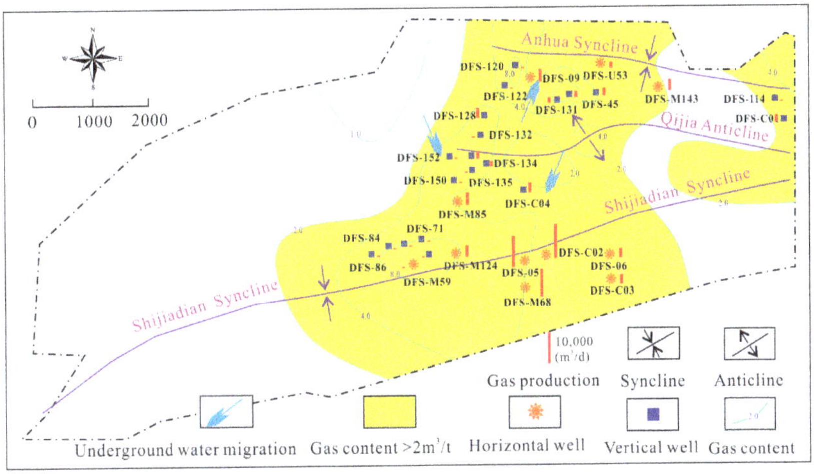

Figure 11. The relationship between gas production and structure.

4. Results and Discussion

4.1. Coal Sample Characteristics

The maximum vitrinite reflectance ($R_{o,max}$) of the coals is 0.56~0.65%. The coal macerals are mainly inertinite (41.13~91.12%), followed by vitrinite (6.93~49.3%), and the content of liptinite is the lowest (<9.52%) as shown in Table 3. Statistical analysis of the data found that the vitrinite is positively correlated with the gas content, and the fitting result R^2 value is 0.73, while the inertinite is negatively correlated (Figure 12), which is also consistent with the results of previous studies [51].

Table 3. Maceral of coal sample.

Sample ID	$R_{o,max}$/%	Maceral Compositions/%		
		Vitrinite	Inertinite	Liptinite
DFS-D48	0.58	6.93	92.12	0.95
DFS-D62	0.56	16.65	82.32	1.03
DFS-M85	0.65	18.94	77.84	3.22
DFS-09	0.59	20.17	74.55	5.28
DFS-U53	0.61	20.83	76.41	2.76
DFS-05	0.63	22.14	74.33	3.53
DFS-128	0.58	25.29	72.54	2.17
DFS-134	0.62	27.46	70.89	1.65
DFS-122	0.57	29.09	68.15	2.76
DFS-C01	0.63	49.30	41.18	9.52

Note: $R_{o,max}$ is maximum vitrinite reflectance of coal.

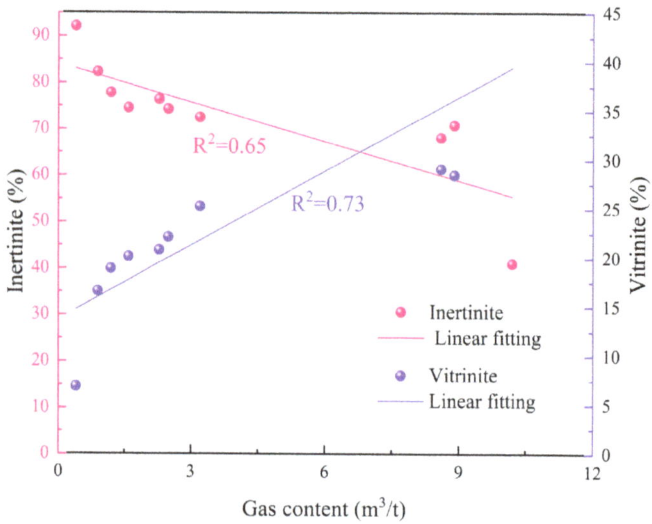

Figure 12. Correlation between gas content and macerals.

4.2. Geochemistry Characteristic of the Gas

The gas compositions (Table 4) show that the gas compositions include CH_4, CO_2, N_2, a small amount of O_2 and a trace of C_2H_6. The content of CH_4 is 68.35~97.95%, and C_2H_6 is 0.001~0.028%. The geochemical characteristics of carbon isotope determined that the $C_1/(C_2 + C_3)$ are 6800~98,000. The dry and wet index $C_1/C_{1\text{-}5}$ of the hydrocarbon composition is 0.99991. The non-hydrocarbon compositions are mainly N_2, the content is 1.256~29.75%, and the content of CO_2 is 0.24~2.375%, with an average of 1.16%. The results of the isotope test showed that the $\delta^{13}C(CH_4)$ was −87.2~−68.9‰, with an average of −78.75‰, and the $\delta^{13}C(CO_2)$ was between −41.693‰ and −7.0650‰, with an average of −20.43‰.

Table 4. Gas compositions and carbon isotopes analysis.

Well No.	Initial Compositions						Stable Isotopes ‰	
	CH_4%	C_2H_6%	O_2%	N_2%	CO_2%	$C_1/C_2 + C_3$	$\delta^{13}CH_4$	$\delta^{13}CO_2$
DFS-C01	97.95	0.001	0.052	1.256	0.736	97,954.8	−80.52	−24.427
DFS-C02	89.25	0.009	0.267	8.785	1.684	9917.26	−77.25	−12.623
DFS-C03	88.84	0.01	0.273	9.684	1.198	7724.68	−78.88	−7.524
DFS-C04	91.58	0.009	0.211	5.825	2.375	10,175.57	−78.06	−41.693
DFS-05	78.7	0.028	0.4	19.47	1.32	7870	−76.50	−14.8
DFS-06	88.01	0.007	0.331	10.73	0.918	12,573.24	−75.21	−7.065
DFS-09	76.33	0.001	0.22	23.03	0.24	92,940.6	−73.78	−11.419
DFS-45	70.33	-	0.25	28.68	0.44	7033	−76.1	-
DFS-M68	77.88	-	0.24	21.12	0.65	7788	−68.9	−13.2
DFS-69	78.36	-	0.23	20.53	0.78	7836	−82.3	−20.3
DFS-128	84.32	-	0.12	13.49	2.01	8432	−83.7	−22.7
DFS-131	77.34	-	0.18	21.97	0.31	7734	−77.6	-
DFS-133	85.35	-	0.16	12.09	2.34	8535	−87.2	−36.6
DFS-148	68.35	-	0.39	29.75	1.3	6835	−86.5	−32.9

Note: DFS-45, DFS-M68, DFS-69, DFS-128, DFS-131, DF-133, DFS-148 [56].

The $\delta^{13}C_1$ is less than −55‰, and the content of heavy hydrocarbon is less than 0.05%. By comparing $\delta^{13}CH_4$ with $C_1/(C_2 + C_3)$, and the methane-genesis identification chart [57], it can be determined that the gas is a biogenic gas (Figure 13). The N_2 is 1.256~29.75%, with an average of 16.17%. There is an obvious linear negative correlation between the content of N_2 and CH_4 (Figure 14), indicating that the coal seam was exposed to the surface or connected with surface water, N_2 in the atmosphere seeped into the coal seam with the water flow, resulting in the high N_2 content. The CO_2 content is 0.24~2.375% and the value of $\delta^{13}C$ (CO_2) is between −36.693‰ and −7.0650‰, it belongs to CO_2 of organic origin.

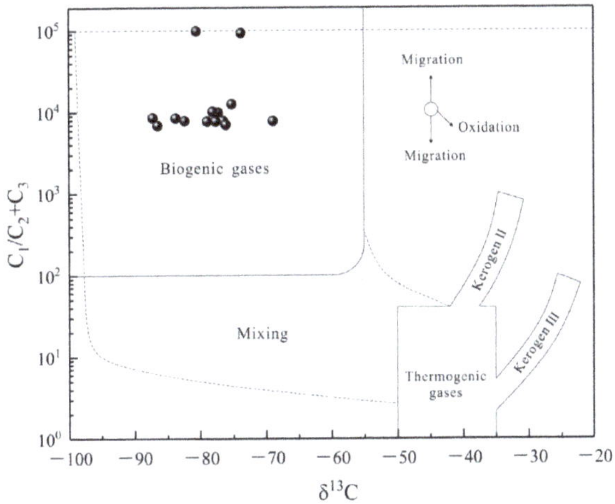

Figure 13. Identification chart of the genetic of CBM.

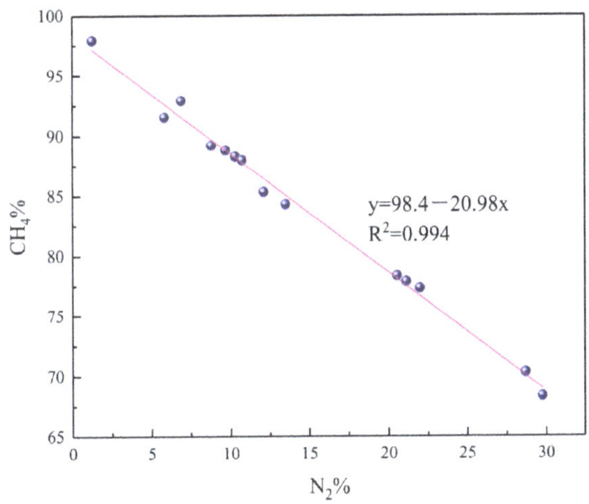

Figure 14. Correlation between N_2 and CH_4 in gas compositions.

The maximum vitrinite reflectance ($R_{o,max}$) has indicated that the coal of the study area was a low metamorphism degree. The coal seam was at the end of the primary biogenic gas and has not yet entered the thermal genetic stage of massive methane generation, which provided favorable material conditions for the generation of secondary biogenic gas. Based on the above comprehensive analysis, CBM was the typical secondary biogenic gas.

4.3. Geological Model of CBM Accumulation

In the geological evolution process of coal reservoirs, tectonic evolution plays a key role in gas generation through stratum subsidence and thermal evolution. Wang et al. [58] proposed the feature of the later tectonic of the Ordos coal-forming basin is the symbiosis of the western thrust, the southern uplift and the fault depression. Near south-north thrust nappe structures developed in the western margin of the basin, leading to exposure and large-scale denudation of the Jurassic coal-bearing rock series. The study area is a northeast inclined monoclinic structure, the coal seam belongs to the Jurassic coal seam with shallow burial depth and low metamorphism degree ($R_{o,max}$ < 0.65%). When the coal seam was raised to the shallow area, the stratum fissure developed and atmospheric precipitation and surface runoff directly recharged the coal reservoir, which provided suitable geological environmental conditions for the reproduction of bacteria. Under the action of the methane bacteria, the organic matter turned into methane under physical and chemical. It was found that the relatively enriched area of CBM was mainly concentrated in the weak runoff zone of the basin margin slope with a burial depth of 450 m~650 m. The weak runoff zone has weak groundwater activity, which is conducive to the preservation of CBM reservoirs. The deep stagnant zone (>650 m) is insufficient to recharge the coal seams and the water mineralization is high, which is not conducive to the reproduction of methane bacteria and gas generation. According to the analysis results of 4.1 and 4.2, CBM was the secondary biogenic gas of low-rank coal. Based on the above comprehensive analysis, the CBM accumulation model is the secondary biogenic gas accumulation on the gentle slope of the basin margin (Figure 15).

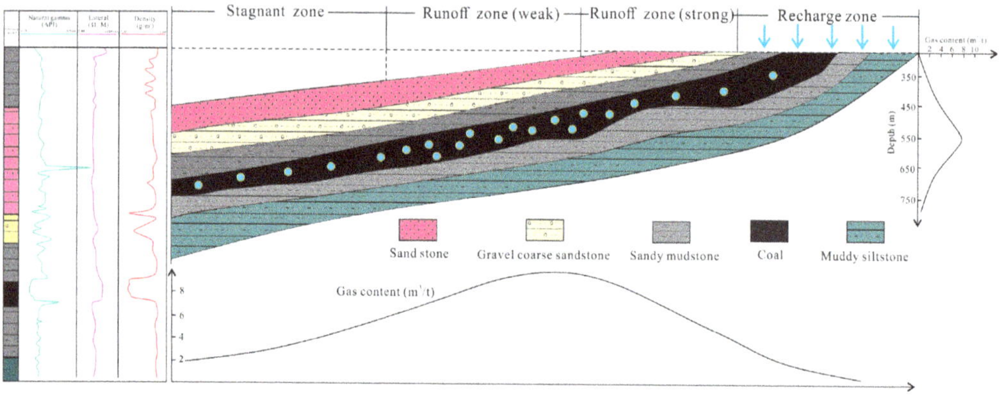

Figure 15. Geological model of CBM accumulation and enrichment.

4.4. CBM Production Analysis

4.4.1. Geological Factors Evaluation of CBM Production

As mentioned before, the main controlling factors of CBM production capacity include gas content, burial depth, coal thickness, structural, hydrological conditions, permeability, and reservoir pressure. Daily gas production is a key indicator reflecting production capacity, the average daily gas production at the stable production stage is considered as the dependent variable and other related factors as independent variables. The main control factors of gas production capacity were analyzed by the gray correlation. The calculation process of correlation degree refers to the literature [44,59,60]. The lithology of the No. 4 coal seam roof is all sandy mudstone or carbonaceous mudstone, and the difference in lithology is small. Therefore, the factors that can affect the CBM production capacity (gas production) include coal thickness, permeability, gas content, reservoir pressure, ash, roof thickness and burial depth. According to the results of the correlation degree, it can be seen that the main control factors of the productivity between vertical wells and horizontal wells are quite different, as shown in Table 5.

Table 5. Grey correlation quantitative evaluation of controlling factors of CBM production.

Vertical Well			Horizontal Well		
Factors	**Correlation**	**Rank**	**Factors**	**Correlation**	**Rank**
Coal thickness	0.662	1	Gas content	0.656	1
Burial depth	0.572	2	Coal thickness	0.622	2
Permeability	0.541	3	Reservoir pressure	0.579	3
Ash of coal	0.493	4	Ash of coal	0.579	4
Roof thickness	0.415	5	Permeability	0.555	5
Gas content	0.404	6	Roof thickness	0.427	6
Reservoir pressure	0.196	7	Burial depth	0.394	7

4.4.2. High Production Model of CBM

The study area as a whole is a monoclinic structure dipping from northeast to southwest. Except for DFS114 and DFS-C01 located in the eastern part of the minefield, the rests were distributed in the central area of the minefield. In terms of gas production, there are 19 production wells above 1000 m^3/d, in which horizontal wells DFS-C02 and DFS-05 have gas production rates of more than 10,000 m^3/d. High production vertical wells (>3000 m^3/d) include DFS-128 and DFS -C04.

Through the connected well profile of the high production well in the Dafosi minefield, DFS-128(V)—DFS-C04(V)—DFS-C02/-05(H)—DFS-M168(H) showed that the vertical wells DFS-128 and DFS-C04 were mainly distributed on the flanks of the anticline, while horizontal wells DFS-C02, DFS-05 and DFS-M168 were located near the core of the Shijiadian syncline, and the gas content of the high production wells are basically more than 4 m^3/t (Figure 16). The well positions of different well types have obvious differences, the high production area of vertical wells is located at the folding flank, and horizontal wells are located at the syncline axis. The roof mudstone and weak underground water sealing effect are conducive to forming the high gas content in the fold flank. Under the action of tensile stress, the fractures above the neutral of the fold are more developed and the permeability is good, which is conducive to rapid depressurization of the reservoir and gas production by drainage, e.g., the gas production of well DFS-128 during the stable production period is more than 3200 m^3/d. The thickness (12 m~20 m) and burial depth (500 m~600 m) of the coal seam at the syncline axis of Shijiadian are large. The hydrodynamic environment is a weak runoff zone. Due to the long horizontal section of the horizontal well, the influencing range in the coal seam is relatively wide. Good drainage and depressurization effect, high gas production, e.g., the gas production of DFS-C02 well during steady production period is about 11,000 m^3/d. Of course, apart from geological factors, drilling and production technologies, fracturing technologies and drainage systems all have a significant impact on gas production, but we will not discuss them here.

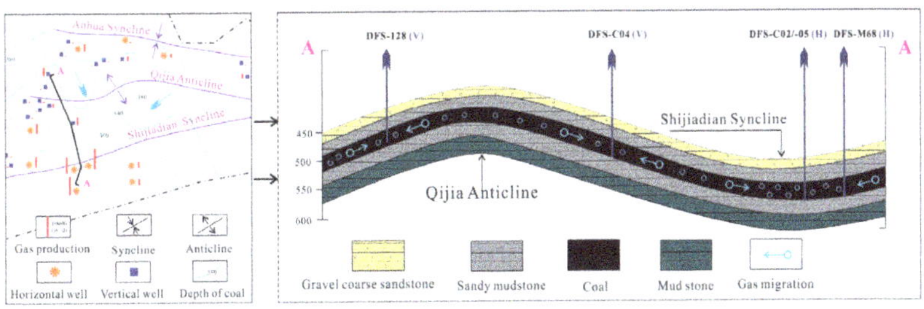

Figure 16. High-production model of CBM.

4.4.3. The Favorable Areas for CBM High Production

It can be seen in Table 6, that the absolute value of the correlation coefficient between all factors was less than 0.4, and the correlation coefficient of the structure position was also close to 0.4. Therefore, the factors can be considered independent of each other. The gas production of CBM wells is quite different, the classification standard of high production and low production is the daily gas production in the stable production stage. If it is higher than 1500 m^3/d, it is regarded as high production, less than 1500 m^3/d is the low production. The attribute data of 12 high production wells and the same number of non-high production wells are randomly selected as the training sample set, and the remaining five high production wells and the rest of the non-high production well points form a test sample set. Based on MATLAB, the random forest model training sample set is used for learning, and the learned model is used in the test sample. The prediction rate of the model is 70%. Therefore, the model is considered reliable and can be used to evaluate the whole area in the next step.

Table 6. Correlation coefficient of main controlling factor.

Controling Factors	Gas Content	Burial Depth	Coal Thickness	Permeability	Reservoir Pressure	Structural Location	Ash of Coal	Roof Thickness
Gas content	1							
Burial depth	−0.130	1						
Coal thickness	0.044	0.186	1					
Permeability	0.284	−0.475	0.166	1				
Reservoir pressure	0.312	0.011	−0.396	0.021	1			
Structure location	0.417	0.199	−0.198	−0.122	0.350	1		
Ash of coal	0.194	−0.127	0.141	0.218	0.231	0.075	1	
Roof thickness	0.318	−0.047	0.106	0.768	−0.012	0.146	0.271	1

Subsequently, the attribute database of the study area was brought into the trained model, and the susceptibility index of high production CBM wells based on the random forest model was obtained, and the interval is [0, 0.948]. The natural discontinuity method was used to classify its susceptibility levels into three categories, which are the unfavorable area [0.03, 0.386], generally the favorable area [0.386, 0.51], and the extremely favorable area [0.51, 0.948], the predicted result of the generated favorable area of CBM is shown in Figure 17.

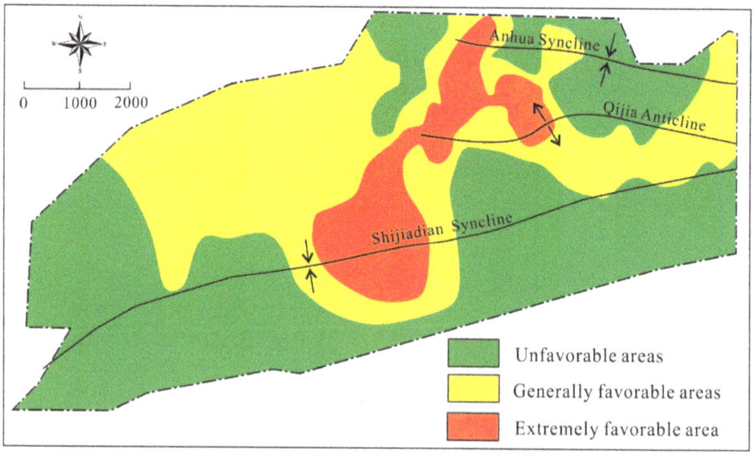

Figure 17. Prediction results of favorable areas in the study area.

5. Conclusions

(1) The maximum reflectance of vitrinite ($R_{o,max}$) in coal are 0.56~0.65%. The result of gas compositions and carbon isotopes analysis shows that CBM is typical biogenic gas of low-rank coal.

(2) The gas content is closely related to buried depth and hydrodynamic environment, i.e., the high gas content areas are mainly located in the groundwater weak runoff zone at the burial depth of 450 m~650 m, especially in the syncline. The CBM accumulation model is the secondary biogenic on the gentle slope of the basin margin.

(3) The productivity of CBM wells is closely related to the structural location, but vertical wells gas production performance is different from horizontal wells. The high gas production area of vertical wells is distributed on the gentle slope with high gas content between anticline and syncline, the horizontal wells with good performance are located near the core of the syncline.

(4) According to the random forest model, the study area is divided into three areas, which are the unfavorable area, the generally favorable area, and the extremely favorable area. The predicted results will provide the scientific basis for the CBM production wells.

(5) The CBM development of low-rank coal in Western China is still in the early stage. The study area is relatively small, and the number of samples is few. It is suggested that future research should be based on more production data. In addition, mining technology is an indispensable aspect of the research on CBM development, and the evaluation of geological factors should be combined with mining technology.

Author Contributions: Conceptualization, C.Z. and D.M.; methodology, Y.X.; formal analysis, Z.G.; investigation, C.Z.; data curation, W.L.; supervision, Y.X.; project administration, G.L.; funding acquisition, D.M. and Y.C.; writing—original draft preparation, C.Z.; writing—review and editing, D.M. All authors have read and agreed to the published version of the manuscript.

Funding: This work is supported by National Natural Science Foundation of China (Grant No. 41902175), the Shanxi Province Science and Technology Major Special Funding Project (Grant No. 20201101002).

Conflicts of Interest: The authors declare no conflict of interest.

References

1. Li, Y.; Wang, Z.; Tang, S.; Elsworth, D. Re-evaluating adsorbed and free methane content in coal and its ad- and desorption processes analysis. *Chem. Eng. J.* **2022**, *428*, 131946. [CrossRef]
2. Li, Y.; Yang, J.; Pan, Z.; Meng, S.; Wang, K.; Niu, X. Unconventional Natural Gas Accumulations in Stacked Deposits: A Discussion of Upper Paleozoic Coal-Bearing Strata in the East Margin of the Ordos Basin, China. *Acta Geol. Sin.-Engl. Ed.* **2019**, *93*, 111–129. [CrossRef]
3. Moore, T.A. Coalbed methane: A review. *Int. J. Coal Geol.* **2012**, *101*, 36–81. [CrossRef]
4. Mastalerz, M.; Drobniak, A. Coalbed Methane: Reserves, Production, and Future Outlook. In *Future Energy*; Elsevier: Amsterdam, The Netherlands, 2020; pp. 97–109. [CrossRef]
5. Guo, H.; Zhang, J.; Han, Q.; Huang, Z.; Urynowicz, M.A.; Wang, F. Important Role of Fungi in the Production of Secondary Biogenic Coalbed Methane in China's Southern Qinshui Basin. *Energy Fuels* **2017**, *31*, 7197–7207. [CrossRef]
6. Stolper, D.A.; Lawson, M.; Davis, C.L.; Ferreira, A.A.; Santos Neto, E.V.; Ellis, G.S.; Lewan, M.D.; Martini, A.M.; Tang, Y.; Schoell, M.; et al. Gas formation. Formation temperatures of thermogenic and biogenic methane. *Science* **2014**, *344*, 1500–1503. [CrossRef] [PubMed]
7. Li, Y.; Tang, D.; Fang, Y.; Xu, H.; Meng, Y. Distribution of stable carbon isotope in coalbed methane from the east margin of Ordos Basin. *Sci. China Earth Sci.* **2014**, *57*, 1741–1748. [CrossRef]
8. Whitiar, M.J.; Faber, E.; Schole, M. Biogenic methane formation in marine and freshwater environments: CO_2 reduction VS. acetate fermentation-Isotope evidence. *Gmchrm. Et Cosmwhim. Acta* **1986**, *50*, 693–709. [CrossRef]
9. Cai, Y.; Liu, D.; Yao, Y.; Li, J.; Qiu, Y. Geological controls on prediction of coalbed methane of No. 3 coal seam in Southern Qinshui Basin, North China. *Int. J. Coal Geol.* **2011**, *88*, 101–112. [CrossRef]
10. Chen, Y.; Tang, D.; Xu, H.; Li, Y.; Meng, Y. Structural controls on coalbed methane accumulation and high production models in the eastern margin of Ordos Basin, China. *J. Nat. Gas Sci. Eng.* **2015**, *23*, 524–537. [CrossRef]
11. Li, Y.; Tang, D.; Xu, H.; Elsworth, D.; Meng, Y. Geological and hydrological controls on water coproduced with coalbed methane in Liulin, eastern Ordos basin, China. *AAPG Bull.* **2015**, *99*, 207–229. [CrossRef]
12. Su, X.; Lin, X.; Liu, S.; Zhao, M.; Song, Y. Geology of coalbed methane reservoirs in the Southeast Qinshui Basin of China. *Int. J. Coal Geol.* **2005**, *62*, 197–210. [CrossRef]
13. Yao, Y.; Liu, D.; Tang, D.; Tang, S.; Che, Y.; Huang, W. Preliminary evaluation of the coalbed methane production potential and its geological controls in the Weibei Coalfield, Southeastern Ordos Basin, China. *Int. J. Coal Geol.* **2009**, *78*, 1–15. [CrossRef]

14. Fu, H.; Tang, D.; Xu, H.; Xu, T.; Chen, B.; Hu, P.; Yin, Z.; Wu, P.; He, G. Geological characteristics and CBM exploration potential evaluation A case study in the middle of the southern Junggar Basin, NW China. *J. Nat. Gas Sci. Eng.* **2016**, *30*, 57–570. [CrossRef]
15. Wang, B.; Li, J.; Zhang, Y.; Wang, H.; Liu, H.; Li, G.; Ma, J. Geological characteristics of low rank coalbed methane, China. *Pet. Explor. Dev.* **2009**, *36*, 30–34. [CrossRef]
16. Qin, Y.; Moore, T.A.; Shen, J.; Yang, Z.; Shen, Y.; Wang, G. Resources and geology of coalbed methane in China: A review. *Int. Geol. Rev.* **2017**, *60*, 777–812. [CrossRef]
17. Chen, Y.; Ma, Z.; Ma, D.; Zhang, Z.; Li, W.; Yang, F.; Ji, Y.; Peng, T.; Meng, Y. Characteristics of the Coal Fines Produced from Low-Rank Coal Reservoirs and Their Wettability and Settleability in the Binchang Area, South Ordos Basin, China. *Geofluids* **2021**, *2021*, 5560634. [CrossRef]
18. Gao, Z.; Ma, D.; Chen, Y.; Zheng, C.; Teng, J. Study for the Effect of Temperature on Methane Desorption Based on Thermodynamics and Kinetics. *ACS Omega* **2021**, *6*, 702–714. [CrossRef]
19. Zheng, C.; Ma, D.; Chen, Y.; Gao, Z.; Teng, J. Pore structure of different macroscopically distinguished components within low-rank coals and its methane desorption characteristics. *Fuel* **2021**, *293*, 120465. [CrossRef]
20. Yun, J.; Xu, F.; Liu, L.; Zhong, N.; Wu, X. New progress and future prospects of CBM exploration and development in China. *Int. J. Min. Sci. Technol.* **2012**, *22*, 363–369. [CrossRef]
21. Wang, X.; Cheng, Y.; Zhang, D.; Liu, Z.; Wang, Z.; Jiang, Z. Influence of tectonic evolution on pore structure and fractal characteristics of coal by low pressure gas adsorption. *J. Nat. Gas Sci. Eng.* **2021**, *87*, 103788. [CrossRef]
22. Yan, T.; Yao, Y.; Liu, D. Critical tectonic events and their geological controls on gas generation, migration, and accumulation in the Weibei coalbed methane field, southeast Ordos basin. *J. Nat. Gas Sci. Eng.* **2015**, *27*, 1367–1380. [CrossRef]
23. Li, Y.; Zhang, C.; Tang, D.; Gan, Q.; Niu, X.; Wang, K.; Shen, R. Coal pore size distributions controlled by the coalification process: An experimental study of coals from the Junggar, Ordos and Qinshui basins in China. *Fuel* **2017**, *206*, 352–363. [CrossRef]
24. Tang, S.; Tang, D.; Tang, J.; Tao, S.; Xu, H.; Geng, Y. Controlling factors of coalbed methane well productivity of multiple superposed coalbed methane systems: A case study on the Songhe mine field, Guizhou, China. *Energy Explor. Exploit.* **2017**, *35*, 665–684. [CrossRef]
25. Zhu, H.; Liu, P.; Chen, P.; Kang, J. Analysis of coalbed methane occurrence in Shuicheng Coalfield, southwestern China. *J. Nat. Gas Sci. Eng.* **2017**, *47*, 140–153. [CrossRef]
26. Lv, Y.; Tang, D.; Xu, H.; Luo, H. Production characteristics and the key factors in high-rank coalbed methane fields: A case study on the Fanzhuang Block, Southern Qinshui Basin, China. *Int. J. Coal Geol.* **2012**, *96–97*, 93–108. [CrossRef]
27. Xing, L.R.; Yao, Y.B.; Liu, D.M.; Liu, J.G.; Zhou, L.L.; Li, H.P. Geological Characteristics of Coalbed Methane Reservoir in Southern Shizhuang Block, Southeastern Qinshui Basin. *Appl. Mech. Mater.* **2013**, *295–298*, 3209–3212. [CrossRef]
28. Wang, H.; Yao, Y.; Liu, D.; Pan, Z.; Yang, Y.; Cai, Y. Fault-sealing capability and its impact on coalbed methane distribution in the Zhengzhuang field, southern Qinshui Basin, North China. *J. Nat. Gas Sci. Eng.* **2016**, *28*, 613–625. [CrossRef]
29. Du, Z.; Zhang, X.; Huang, Q.; Zhang, S.; Wang, C. The gas content distribution of coal reservoir at the Changzhi block, south-central Qinshui Basin, North China: Influences of geologic structure and hydrogeology. *Energy Explor. Exploit.* **2018**, *37*, 144–165. [CrossRef]
30. Wang, B.; Sun, F.; Tang, D.; Zhao, Y.; Song, Z.; Tao, Y. Hydrological control rule on coalbed methane enrichment and high yield in FZ Block of Qinshui Basin. *Fuel* **2015**, *140*, 568–577. [CrossRef]
31. Cao, L.; Yao, Y.; Cui, C.; Sun, Q. Characteristics of in-situ stress and its controls on coalbed methane development in the southeastern Qinshui Basin, North China. *Energy Geosci.* **2020**, *1*, 69–80. [CrossRef]
32. Lin, Y.; Qin, Y.; Ma, D.; Zhao, J. Experimental Research on Dynamic Variation of Permeability and Porosity of Low-Rank Inert-Rich Coal Under Stresses. *ACS Omega* **2020**, *5*, 28124–28135. [CrossRef] [PubMed]
33. Zhao, J.; Tang, D.; Xu, H.; Li, Y.; Tao, S.; Lin, W.; Liu, Z. Characteristic of In Situ Stress and Its Control on the Coalbed Methane Reservoir Permeability in the Eastern Margin of the Ordos Basin, China. *Rock Mech. Rock Eng.* **2016**, *49*, 3307–3322. [CrossRef]
34. Tao, S.; Tang, D.; Xu, H.; Gao, L.; Fang, Y. Factors controlling high-yield coalbed methane vertical wells in the Fanzhuang Block, Southern Qinshui Basin. *Int. J. Coal Geol.* **2014**, *134–135*, 38–45. [CrossRef]
35. Xu, H.; Tang, D.; Zhao, J.; Tao, S.; Li, S.; Fang, Y. Geologic controls of the production of coalbed methane in the Hancheng area, southeastern Ordos Basin. *J. Nat. Gas Sci. Eng.* **2015**, *26*, 156–162. [CrossRef]
36. Guo, C.; Qin, Y.; Wu, C.; Lu, L. Hydrogeological control and productivity modes of coalbed methane commingled production in multi-seam areas: A case study of the Bide–Santang Basin, western Guizhou, South China. *J. Pet. Sci. Eng.* **2020**, *189*, 107039. [CrossRef]
37. Ayers, W.B., Jr. Coalbed gas systems, resources, and production and a review of contrasting cases from the San Juan and Powder River Basins. *AAPG Bull.* **2002**, *86*, 1853–1890.
38. Bustin, R.M.; Clarkson, C.R. Geological controls on coalbed methane reservoir capacity and gas content. *Int. J. Coal Geol.* **1998**, *38*, 3–26. [CrossRef]
39. Zhao, J.; Tang, D.; Xu, H.; Lv, Y.; Tao, S. High production indexes and the key factors in coalbed methane production: A case in the Hancheng block, southeastern Ordos Basin, China. *J. Pet. Sci. Eng.* **2015**, *130*, 55–67. [CrossRef]
40. Wang, G.; Qin, Y.; Xie, Y.; Shen, J.; Zhao, L.; Huang, B.; Zhao, W. Coalbed methane system potential evaluation and favourable area prediction of Gujiao blocks, Xishan coalfield, based on multi-level fuzzy mathematical analysis. *J. Pet. Sci. Eng.* **2018**, *160*, 136–151. [CrossRef]

41. Wei, Q.; Li, X.; Hu, B.; Zhang, X.; Zhang, J.; He, Y.; Zhang, Y.; Zhu, W. Reservoir characteristics and coalbed methane resource evaluation of deep-buried coals: A case study of the No.13–1 coal seam from the Panji Deep Area in Huainan Coalfield, Southern North China. *J. Pet. Sci. Eng.* **2019**, *179*, 867–884. [CrossRef]
42. Zhang, J.; Liu, D.; Cai, Y.; Yao, Y.; Ge, X. Carbon isotopic characteristics of CH4 and its significance to the gas performance of coal reservoirs in the Zhengzhuang area, Southern Qinshui Basin, North China. *J. Nat. Gas Sci. Eng.* **2018**, *58*, 135–151. [CrossRef]
43. Li, Y.; Liu, D.M.; Yao, Y.B.; Wang, X.H.; Yin, Z.W. Evaluation of Favorable Area in the Weibei Coalbed Methane Field: A Study by Using Grey Correlation Analysis. *Appl. Mech. Mater.* **2013**, *295–298*, 3342–3345. [CrossRef]
44. Wang, W.; He, M.; Wang, X.; Jiang, P.; Peng, L.; Du, Y. Analysis on main controlling factors and comprehensive evaluation of coalbed methane production capacity of Junlian Block. *Coal Sci. Technol.* **2017**, *45*, 194–200. [CrossRef]
45. Han, J. Geochemical characteristics of South China Sea based on random forest algorithm and Wushu teaching action simulation. *Arab. J. Geosci.* **2021**, *14*, 1802. [CrossRef]
46. Zhu, Q.; Hu, Q.; Du, H.; Fan, B.; Zhu, J.; Zhang, B.; Zhao, Y.; Liu, B.; Tang, J. A gas production model of vertical coalbed methane well based on random forest algorithm. *J. China Coal Soc.* **2020**, *45*, 2846–2855. [CrossRef]
47. Dong, Y.; Du, B.; Zhang, L. Target Detection Based on Random Forest Metric Learning. *IEEE J. Sel. Top. Appl. Earth Obs. Remote Sens.* **2015**, *8*, 1830–1838. [CrossRef]
48. Wang, Q.; Chen, H.; Patnaik, S. Optimization of parallel random forest algorithm based on distance weight. *J. Intell. Fuzzy Syst.* **2020**, *39*, 1951–1963. [CrossRef]
49. Daho, M.E.H.; Chikh, M.A. Combining Bootstrapping Samples, Random Subspaces and Random Forests to Build Classifiers. *J. Med. Imaging Health Inform.* **2015**, *5*, 539–544. [CrossRef]
50. Abdi, J.; Hadipoor, M.; Esmaeili-Faraj, S.H.; Vaferi, B. A modeling approach for estimating hydrogen sulfide solubility in fifteen different imidazole-based ionic liquids. *Sci. Rep.* **2022**, *12*, 4415. [CrossRef]
51. Xu, H.; Tang, D.Z.; Liu, D.M.; Tang, S.H.; Yang, F.; Chen, X.Z.; He, W.; Deng, C.M. Study on coalbed methane accumulation characteristics and favorable areas in the Binchang area, southwestern Ordos Basin, China. *Int. J. Coal Geol.* **2012**, *95*, 1–11. [CrossRef]
52. Meng, Z.; Zhang, G.; Li, G.; Liu, J. Analysis of diffusion properties of methane in low rank coal. *Coal Geol. Explor.* **2019**, *47*, 84–89. [CrossRef]
53. Chen, S.; Tao, S.; Tian, W.; Tang, D.; Zhang, B.; Liu, P. Hydrogeological control on the accumulation and production of coalbed methane in the Anze Block, southern Qinshui Basin, China. *J. Pet. Sci. Eng.* **2021**, *198*, 108138. [CrossRef]
54. Zhang, Z.; Yan, D.; Zhuang, X.; Yang, S.; Wang, G.; Li, G.; Wang, X. Hydrogeochemistry signatures of produced waters associated with coalbed methane production in the Southern Junggar Basin, NW China. *Environ. Sci. Pollut. Res. Int.* **2019**, *26*, 31956–31980. [CrossRef] [PubMed]
55. Sang, S.; Fan, B.; Qin, Y. Conditions of Sealing and Accumulation in Coal-Bed Gas. *Oil Gas Geol.* **1999**, *20*, 104–107.
56. Bao, Y.; Wang, W.; Ma, D.; Shi, Q.; Ali, A.; Lv, D.; Zhang, C. Gas Origin and Constraint of δ13C(CH_4) Distribution in the Dafosi Mine Field in the Southern Margin of the Ordos Basin, China. *Energy Fuels* **2020**, *34*, 14065–14073. [CrossRef]
57. Whiticar, M.J. Carbon and hydrogen isotope systematics of bacterial formation and oxidation of methane. *Chem. Geol.* **1999**, *161*, 291–314. [CrossRef]
58. Wang, S. Ordos basin tectonic evolution and structural control of coal. *Geol. Bull. China* **2011**, *30*, 544–552.
59. Guo, G. Comprehensive evaluation study on favorable area of coalbed methane reservoir in southern Shizhuang Block. *Coal Sci. Technol.* **2019**, *49*, 200–206. [CrossRef]
60. Zou, Z.; Liu, D.; Cai, Y.; Wang, Y.; Li, J. Geological Factors and Reservoir Properties Affecting the Gas Content of Coal Seams in the Gujiao Area, Northwest Qinshui Basin, China. *Energies* **2018**, *11*, 1044. [CrossRef]

energies

MDPI

Article

Pressure Relief Mechanism and Gas Extraction Method during the Mining of the Steep and Extra-Thick Coal Seam: A Case Study in the Yaojie No. 3 Coal Mine

Hao Zhang [1], Lehua Xu [2], Mengmeng Yang [1], Cunbao Deng [1] and Yuanping Cheng [3,4,5,*]

1 College of Safety & Emergency Management Engineering, Taiyuan University of Technology, Taiyuan 030024, China; haozhang@cumt.edu.cn (H.Z.); yangmengmeng@tyut.edu.cn (M.Y.); dengcunbao@tyut.edu.cn (C.D.)
2 Key Laboratory of In-Situ Property-Improving Mining of Ministry of Education, Taiyuan University of Technology, Taiyuan 030024, China; xulehua@tyut.edu.cn
3 Key Laboratory of Coal Methane and Fire Control, Ministry of Education, China University of Mining and Technology, Xuzhou 221116, China
4 National Engineering Research Center for Coal and Gas Control, China University of Mining and Technology, Xuzhou 221116, China
5 School of Safety Engineering, China University of Mining and Technology, Xuzhou 221116, China
* Correspondence: ypcheng@cumt.edu.cn

Abstract: Gas disasters, such as coal and gas outburst and gas overflow, always occur during the mining of the steep and extra-thick coal seam in the horizontal, fully mechanized, top coal slice caving (HFMTCSC) method. To solve these issues and guarantee the safe and efficient mining in the Yaojie No. 3 coal mine, 3DEC software was used in this work to investigate the overburden movement and collapse law as well as the stress redistribution and coal-seam deformation characteristics below the goaf. The results show that a pressure arch structure and a hinge structure are formed in succession in the overburden rock, which induces stress redistribution in the coal below the goaf. During the mining of the upper slice, more than 75% of the coal in the lower slice is located at the effective pressure relief zone; therefore, the steep and extra-thick coal seam can then be protected slice by slice. Meanwhile, with the increase of mining depth, the efficient pressure relief range expands. Based on this pressure relief mechanism, crossing boreholes and bedding boreholes were reasonably designed to efficiently extract the pressure relief gas during the mining of the steep and extra-thick coal seam in the Yaojie No. 3 coal mine.

Keywords: gas extraction; overburden movement; expansion deformation; effective pressure relief range

Citation: Zhang, H.; Xu, L.; Yang, M.; Deng, C.; Cheng, Y. Pressure Relief Mechanism and Gas Extraction Method during the Mining of the Steep and Extra-Thick Coal Seam: A Case Study in the Yaojie No. 3 Coal Mine. *Energies* **2022**, *15*, 3792. https://doi.org/10.3390/en15103792

Academic Editors: Shu Tao, Dengfeng Zhang, Huazhou Huang, Shuoliang Wang and Yanjun Meng

Received: 7 February 2022
Accepted: 14 April 2022
Published: 21 May 2022

1. Introduction

As is known, the in situ stress decreases, and the coal permeability increases in the protected layer during the mining of the protective layer, which is beneficial to its extraction [1–7]. Therefore, protective layer mining, an effective measure for outburst prevention, has been widely adopted in China under different geological conditions, including different interlayer distances [8–10], different dip angels [11,12], different mining thicknesses [13,14] and so on. However, with the deepening of mining depth in China, it is always difficult to find a coal seam to be mined first as the protective layer because almost all the deep coal seams are at outburst risk [15–18].

In the past, traditional mining methods were always adopted during the mining of the steep and extra-thick coal seam with a dip angle greater than 45° and a thickness over decade meters. However, the mining efficiency was rather low due to the difficulty in the installation of the shearer under this special coal seam condition. With the development of mining technology, the horizontal, fully mechanized, top coal slice caving (HFMTCSC) method has been widely used in northwest China [19–22]. After adopting this new mining method, the mining efficiency improved significantly. Unfortunately, the gas control is rather difficult in steep and extra-thick coal seams because its gas resources are very rich. If there is no suitable coal seam to be mined as the protective layer, a large number of gas extraction boreholes should be constructed, and the gas extraction period will be very long. Therefore, gas disasters, such as coal and gas outburst and gas overflow, always occur during the mining process. Although various measures (such as symmetrical caving and pre-splitting blasting) have been adopted by many researchers to prevent the accumulation of the noxious gas caused by the large area dynamic collapse of the top coal [22–24], pre-mining gas extraction is the basic measure to solve this issue [25].

The main minable seam, a steep and extra-thick one, is of great outburst risk in the Yaojie No. 3 coal mine [26]. There is no suitable adjacent coal seam to be mined as the protective layer; thus, its gas extraction is rather difficult, and several serious coal and gas outburst accidents have been reported. The HFMTCSC method was first used in the Yaojie No. 3 coal mine in 1986. During the mining process in this method, the stress on the hydraulic support is rather low, and a large amount of gas originating from the coal below the goaf can desorb and flow freely into the working face. These phenomena suggest that pressure relief is maybe achieved in the coal below the goaf. To better understand this pressure relief mechanism, the overburden movement and collapse law as well as the stress and deformation characteristics of the coal seam below the goaf were analyzed by adopting the 3DEC software [27] according to the engineering geological conditions in the Yaojie No. 3 coal mine. Based on the pressure relief mechanism, crossing boreholes and bedding boreholes were reasonably designed to extract the pressure relief gas.

2. Geological Setting and Mining Conditions

Steep and extra-thick coal seams are widely developed in northwest China, especially in coal fields such as at Yaojie, Huating and Urumqi. The location of the Yaojie coal field is shown in Figure 1a. Yaojie No. 3 coal mine, a main recovery one, is sited at the north center of the coal field. The mine produced more than 2 million tons of coal in 2014. The mine generally shows a monoclonal structure with a northeast trend and a southwest dip, and several faults are also developed, as shown in Figure 1b.

In the Yaojie No. 3 coal mine, the #2 coal seam is the only one with commercial value. Its thickness is approximately 24.91 m, and its dip angle is approximately 55°, i.e., the #2 coal seam is a typical steep and extra-thick one. Meanwhile, its average gas content is approximately 10 m^3/t. The schematic cross section of the #2 coal seam is shown in Figure 1c. Moreover, several serious spontaneous combustions occurred in the shallow coal near the earth surface, which destroyed the coal's commercial value there.

The HFMTCSC method adopted in the mining of the #2 coal seam is shown in Figure 2a. During the mining process, the coal seam was artificially divided into different horizontal slices. In each slice, its total thickness was approximately 11.2 m, and the fully mechanized, top coal caving mining method was adopted. The mining height was approximately 2.8 m, while the caving height was approximately 8.4 m. The mining condition in each slice is shown in Figure 2b, and the coal roadway support condition is shown in Figure 2c. Moreover, the generalized stratigraphic column of the Yaojie No. 3 coal mine is shown in Figure 2d.

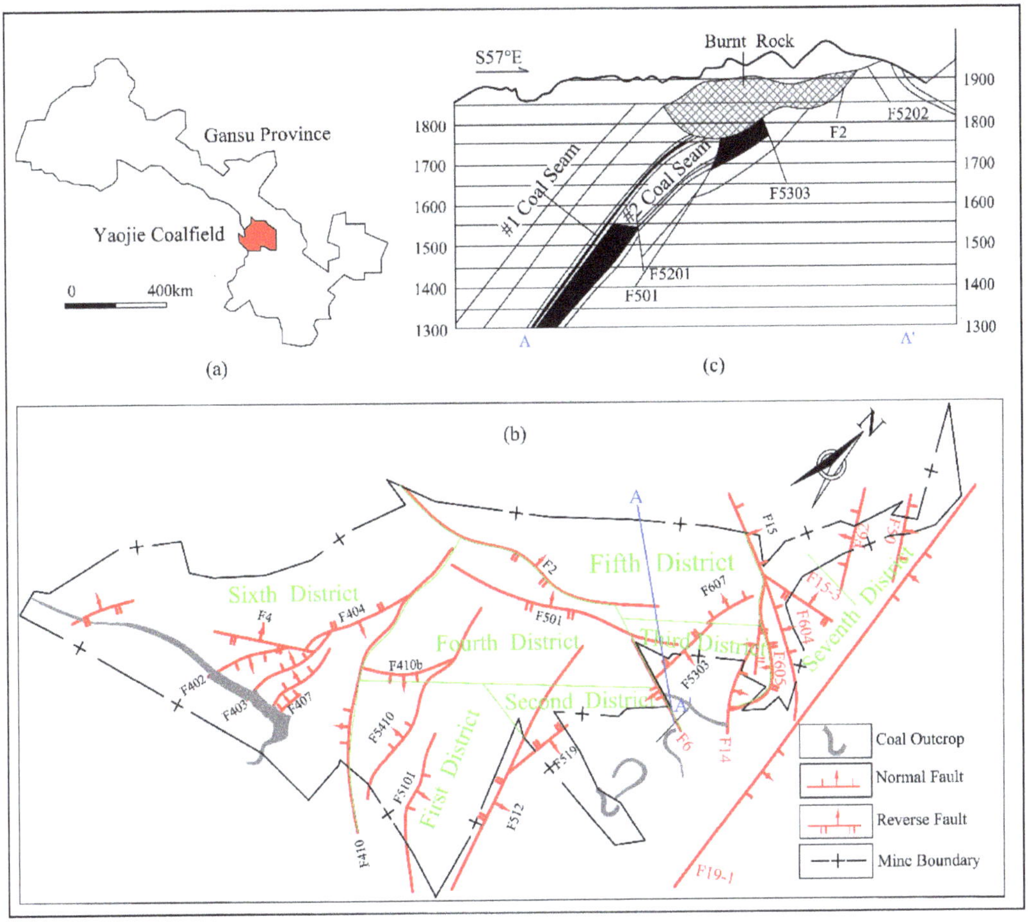

Figure 1. The location and the geological setting of the Yaojie No.3 coal mine: (**a**) location of the Yaojie coalfield; (**b**) geological setting of the Yaojie No.3 coal mine and (**c**) schematic cross section along the line A-A′ in (**b**).

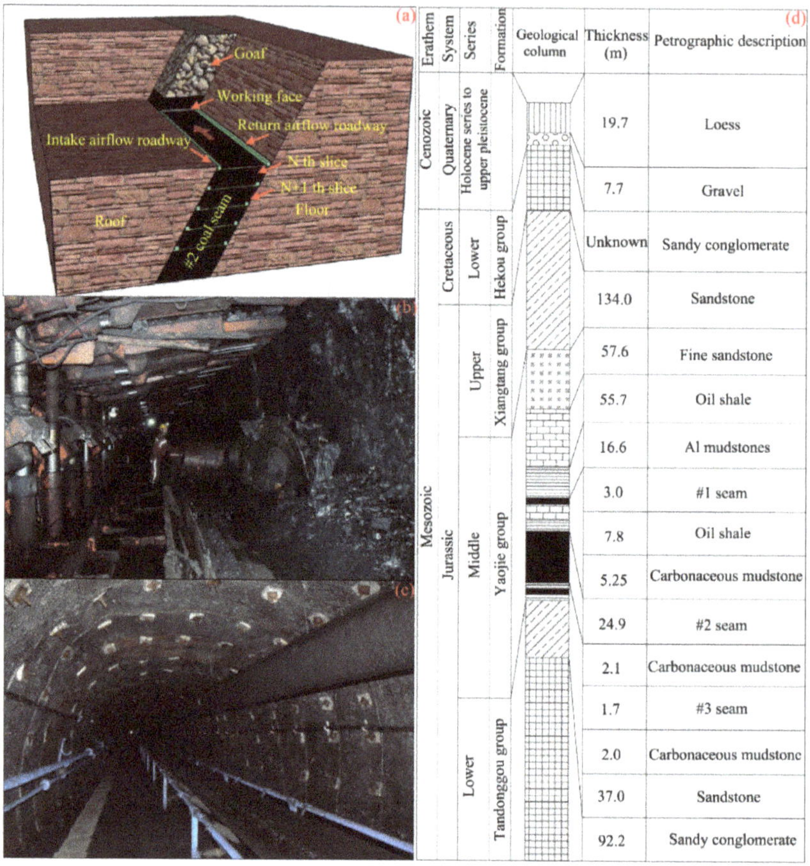

Figure 2. Mining method and generalized stratigraphic column in the Yaojie No. 3 coal mine: (**a**) schematic view of the HFMTCSC mining method; (**b**) mining condition; (**c**) coal roadway support condition and (**d**) the stratigraphic column.

3. Research Method

Numerical simulations, which can overcome many complicated problems in the analytical method, have been used widely in engineering and theoretical analysis in recent years [27–30]. A software based on the discrete element method and developed by Itasca Consulting Group Inc. (Minneapolis, MN, USA) [31], 3DEC, is one of the most important numerical software tools in current rock mechanics calculations. Therefore, 3DEC software was selected to analyze the overburden movement and collapse law as well as the stress and deformation characteristics of the coal seam below the goaf during the mining of the steep and extra-thick coal seam in the HFMTCSC method.

For the sake of simplification, a two-dimensional geometric model was built according to the plane strain assumption, as shown in Figure 3. Considering that the overburden strata will move towards the goaf sharply during the mining process of the steep and extra-thick coal seam, from the shallow to the deep slice by slice, the length in the x-direction was set as 700 m while the height was set as 550 m. Meanwhile, 16 slices with an average thickness of 11 m for each were built in the geometric model. The 1st slice is located at 220 m below the top side. The 16th slice is located at 154 m above the bottom side. As for the boundary conditions, the top side was set as the stress boundary with a vertical stress of 0.6 MPa. At the same time, the horizontal displacement at the lateral and the vertical displacement at the base are constrained. The Mohr Coulomb block model and the joint area contact Coulomb slip model were used during the simulation process [27]. The parameters of the main rock stratum and coal seams employed in the model are shown in Tables 1 and 2 and were tested in the laboratory.

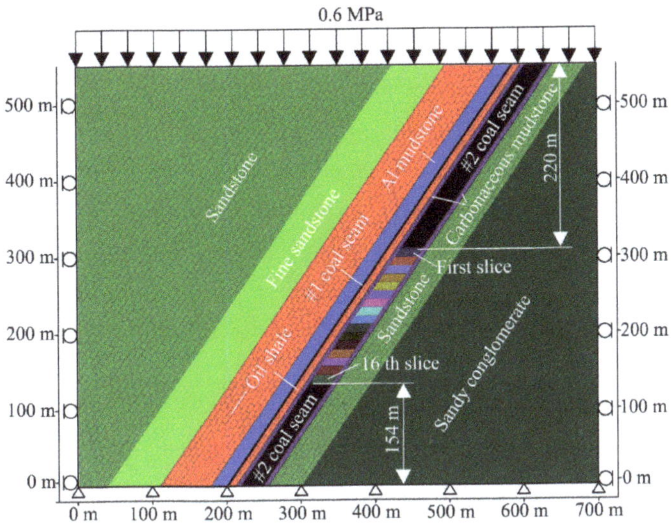

Figure 3. Geometric model.

Table 1. Physical and mechanics properties of the rocks and coal blocks.

Rock Strata	Density (kg/m^3)	Bulk Modulus (GPa)	Shear Modulus (GPa)	Cohesion (MPa)	Friction Angle (°)	Tensile Strength (MPa)
Sandstone	2600	8.7	8.3	6.0	35	9.0
Fine sandstone	2600	7.8	4.9	8.3	38	13.1
Oil shale	2100	3.3	3.4	2.3	39	3.5
Al mudstones	2500	3.0	3.0	1.6	27	4.5
Coal seam	1400	0.8	0.8	1.2	30	2.5
Carbonaceous mudstone	1250	2.6	1.3	2.3	42	3.2
Sandy conglomerate	2750	10.5	6.3	7.1	40	9.4

Table 2. Physical and mechanics properties of the rocks and coal joints.

Rock Strata	Normal Stiffness (GPa)	Shear Stiffness (GPa)	Cohesion (MPa)	Friction Angle (°)	Tensile Strength (MPa)
Sandstone	20.0	19.0	1.2	18	12.3
Fine sandstone	29.0	27.0	1.8	21	8.5
Oil shale	16.0	16.0	1.0	15	4.5
Al mudstones	35.0	34.5	0.6	18	4.2
Coal seam	24.0	23.5	0.5	12	1.5
Carbonaceous mudstone	22.0	20.0	0.8	22	3.5
Sandy conglomerate	19.0	18.0	1.5	20	10.2

4. Result and Discussion

4.1. Mining-Induced Overburden Movement and Collapse Law

The numerical simulation method provides us an opportunity to observe the overburden movement during the mining of the steep and extra-thick coal seam. The movements of the overburden rock during the mining process are shown in Figure 4.

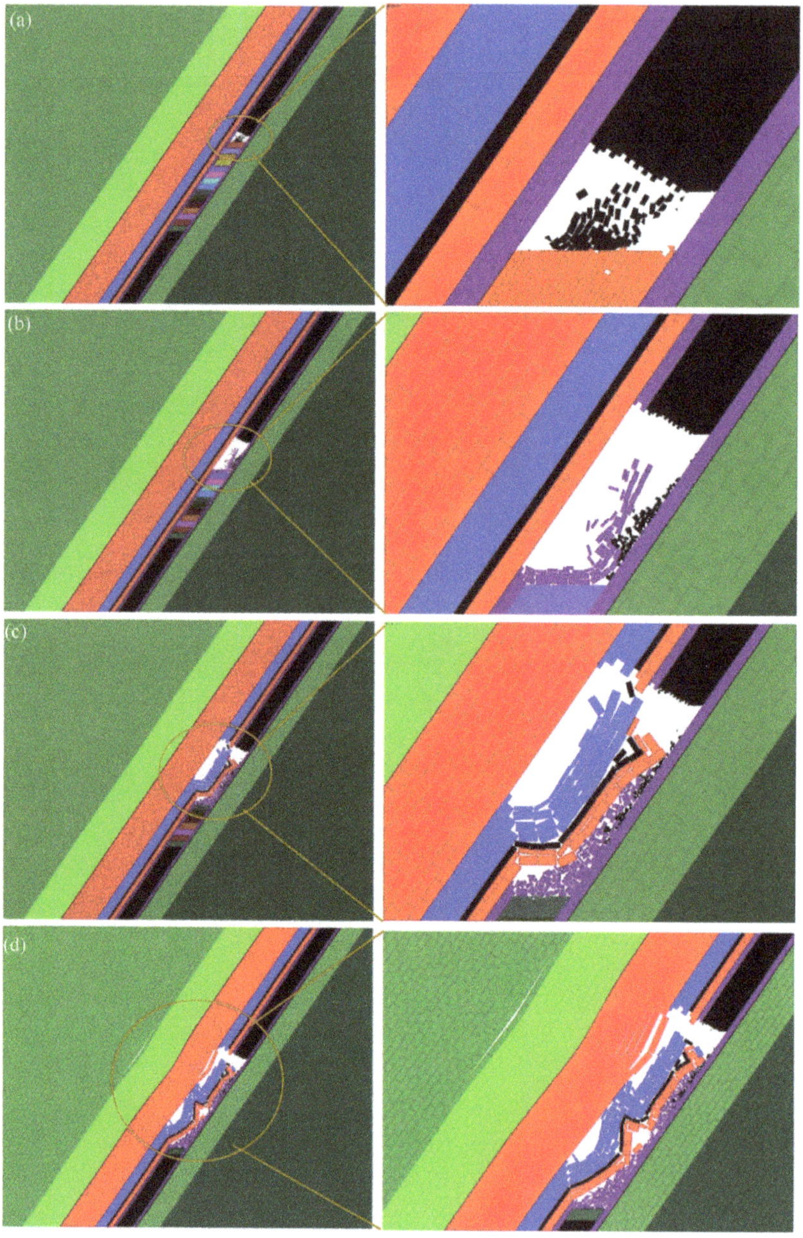

Figure 4. The motion states of the overburden rocks during the mining process: (**a**) the first slice; (**b**) the 5th slice; (**c**) the 8th slice and (**d**) the 15th slice.

During the mining of the first slice, the top coal and the immediate roof begin to collapse and bend towards the goaf, as shown in Figure 4a. The caving characteristic of the top coal is in good accordance with the simulation result in the Particle Flow Code programs by [23]. With the continual mining of the coal seam, the roof hanging distance becomes increasingly larger. During the mining of the 5th slice, the immediate roof collapses for the first time, as shown in Figure 4b. Subsequently, the immediate roof collapses, dilates, accumulates and fills the goaf endlessly. However, the caved rock cannot completely fill in the goaf. During the mining of the 8th and the 15th slices, the main roof collapses once and again, as shown in Figure 4c,d. Meanwhile, the goaf is gradually filled in with the periodical collapse of the main roof. Afterwards, it is difficult for the overburden rock to collapse under the support of the caved rocks filled in the goaf; thus, a new balance structure (namely a hinge structure) is formed along the dip in the overburden rock.

In a word, a pressure arch structure with an arch foot located at the coal above the goaf and the other at the coal below the goaf is formed in the overburden rock during the mining process from the 1st slice to the 14th slice. Meanwhile, a hinge structure is also formed in the overburden rock during the mining process from the 15th slice to the 16th slice.

4.2. The Pressure Relief Mechanism during the Mining of the Steep and Extra-Thick Coal Seam

As mentioned above, a pressure arch structure and a hinge structure are formed in succession during the mining of the steep and extra-thick coal seam. In the pressure arch structure, the vertical stress it bears could be transferred to the arch feet. Therefore, a vertical stress-unloading effect could be achieved in the coal below the pressure arch structure. However, in the coal around the arch feet, there is an obvious increasing tendency in the stress, which results in a stress concentration zone. In a word, the coal below the goaf can be divided into a pressure relief zone and a stress concentration zone, as shown in Figure 5a. A similar stress redistribution could also be achieved when the hinge structure is formed in the overburden rock (Figure 5b): the vertical stress the hinge structure bears could also be transferred to the caved rock filled in the goaf and the coal roof; as a result, the coal below the goaf could also be divided into a pressure relief zone and a stress concentration zone. According to the above analysis, the pressure relief effect in the coal below the goaf is mainly caused by the pressure relief structure and the hinge structure formed in the overburden rock during the mining process of the steep and extra-thick coal seam.

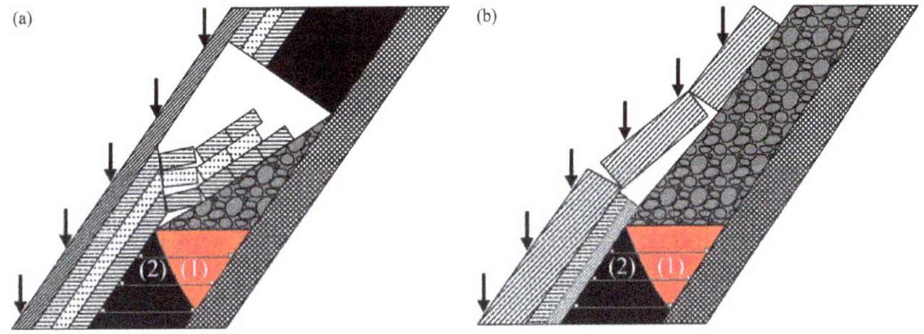

Figure 5. The stress redistribution mechanism of the steep and extra-thick coal seam: (**a**) the pressure arch structure stage and (**b**) the hinge structure stage. In both figures: (1) denotes the pressure relief zone, and (2) denotes the stress concentration zone.

The stress redistribution could result in a coal-seam deformation. With the unloading of the vertical stress, the mechanical energy stored in the coal could be released to a certain degree. Therefore, the coal in the pressure relief zone could expand and deform towards the goaf. At the same time, the horizontal stress may increase in the stress concentration zone. Under the squeezing effect of the horizontal stress, the coal's expansion deformation could be further enhanced. In summary, the coal in the pressure relief zone will expand and deform into the goaf under the effects of vertical pressure relief and horizontal squeezing during the mining process, as shown in Figure 6.

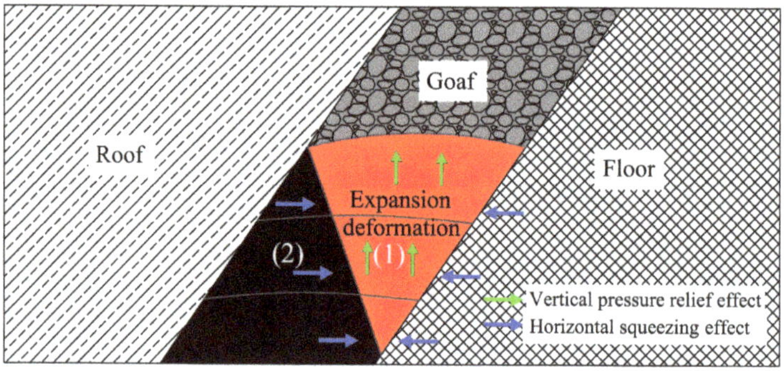

Figure 6. The expansion deformation mechanism: (1) denotes the pressure relief zone, and (2) denotes the stress concentration zone.

4.3. Stress Evolution Characteristics in the Coal Below the Goaf

To achieve a better understanding of the stress evolution characteristics in the coal below the goaf, the vertical stress in the coal at different distances below the goaf was monitored during the mining of the steep and extra-thick coal seam. Figure 7 illustrates the stress monitoring results during the mining of the 5th slice.

As shown in Figure 7, the vertical stress decreases in the pressure relief zone, whereas it increases in the concentration zone. Meanwhile, the pressure relief level decreases with the distance below the goaf increasing. As shown in Figure 7a, the vertical stress in the coal 2 m below the goaf decreases to approximately 0.5 MPa during the mining of the 5th slice. However, the vertical stress in the coal at 23 m below the goaf decreases to approximately 3 MPa, as shown in Figure 7h. At the same time, with the distance below the goaf increasing, the area of the pressure relief zone decreases while the area of the stress concentration zone increases. In China, the pressure relief angle has been widely adopted to describe the pressure relief range. According to the vertical stress monitoring results, the pressure relief angle could be obtained. As shown in Figure 8a, the pressure relief angle is approximately 72° during the mining of the 5th slice. During the mining process of the 10th and 15th slices, the vertical stress evolutions were also monitored in the coal below the goaf. According to the stress monitoring results, their pressure relief angles could also be obtained, as shown in Figure 8b,c. From these figures, we can see that the pressure relief angle is approximately 70° during the mining process of the 10th and 15th slices. Therefore, the pressure relief range almost remains the same during the mining process of the steep and extra-thick coal seam.

Moreover, the vertical stress is almost decreased to the same level during the mining of the different slices. The vertical stress at the same distance below the goaf was also monitored during the mining of different slices. Taking the 5th slice, the 10th slice and the 15th slice as examples, the vertical stress monitoring results at 5 m below the goaf are shown in Figure 9. From this figure, we can see that the vertical stresses in the pressure relief zones all decrease to approximately 0.5 MPa, although the initial vertical stress is approximately 3.0 MPa, 3.5 MPa and 4.0 MPa, respectively. In contrast, the vertical stress in the stress concentration zone increases to 4.0 MPa, 4.5 MPa and 6.5 MPa, respectively. In other words, the stress concentration level increases with the mining depth.

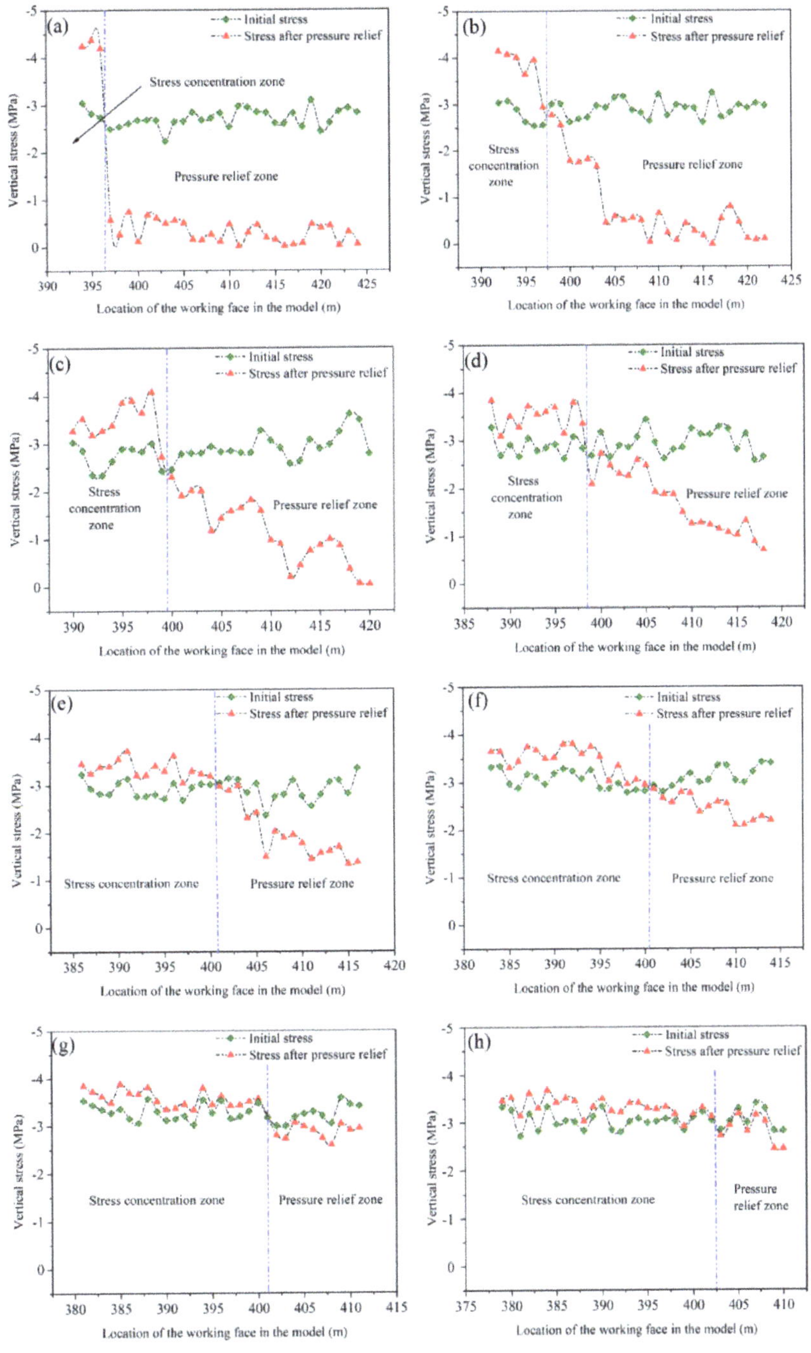

Figure 7. Vertical stress monitoring results during the mining of the 5th slice: (**a**) 2 m below the goaf; (**b**) 5 m below the goaf; (**c**) 8 m below the goaf; (**d**) 11 m below the goaf; (**e**) 14 m below the goaf; (**f**) 17 m below the goaf; (**g**) 20 m below the goaf and (**h**) 23 m below the goaf.

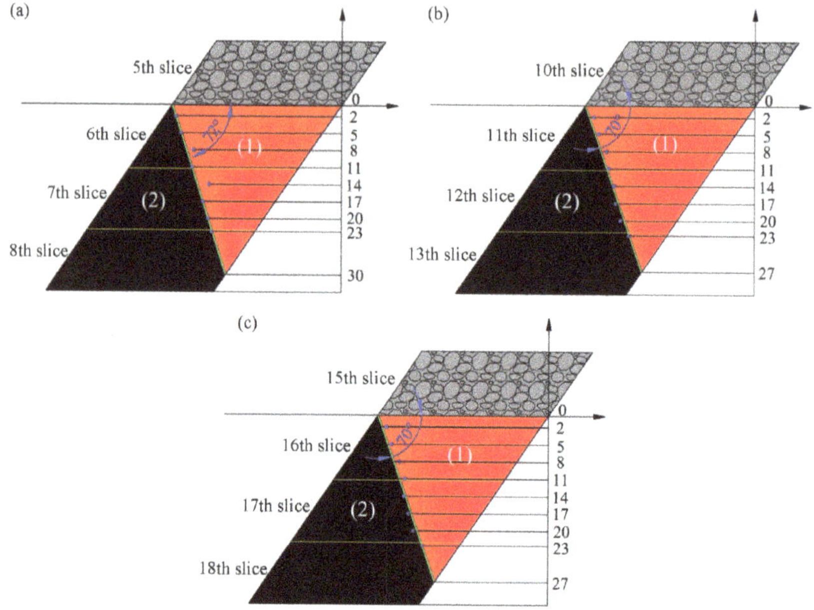

Figure 8. Pressure relief range in the coal below the goaf: (**a**) the 5th slice, (**b**) the 10th slice and (**c**) the 15th slice. In each figure, (1) denotes the pressure relief zone, and (2) denotes the stress concentration zone.

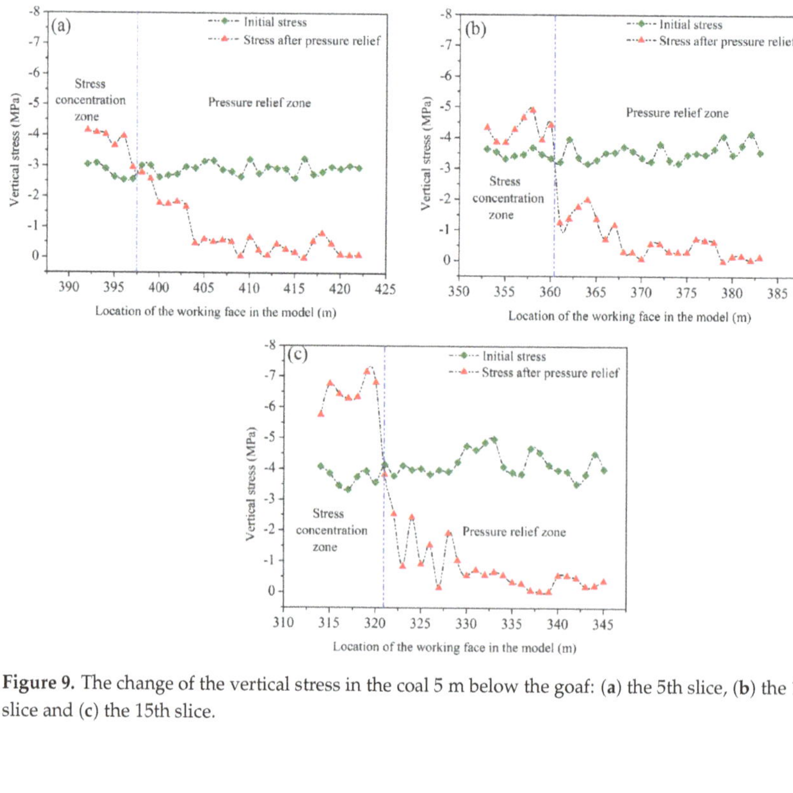

Figure 9. The change of the vertical stress in the coal 5 m below the goaf: (**a**) the 5th slice, (**b**) the 10th slice and (**c**) the 15th slice.

4.4. Expansion Deformation Characteristics in the Coal Below the Goaf

As we can find in Figure 6, the coal in the pressure relief zone will expand and deform towards the goaf under the effects of vertical pressure relief and horizontal squeezing. Figure 10 shows the expansion deformation cloud charts in the coal below the goaf during the mining process of the 5th, 10th and 15th slices. From this figure, we can find that the expansion deformation is increasingly large with the mining depth. The largest expansion deformation in the coal below the goaf is just 120 mm during the mining of the 5th slice, whereas those of the 10th and the 15th slices increase to 150 mm and 180 mm, respectively. The reason for this is because the vertical pressure relief effect is increasingly better in the deep coal.

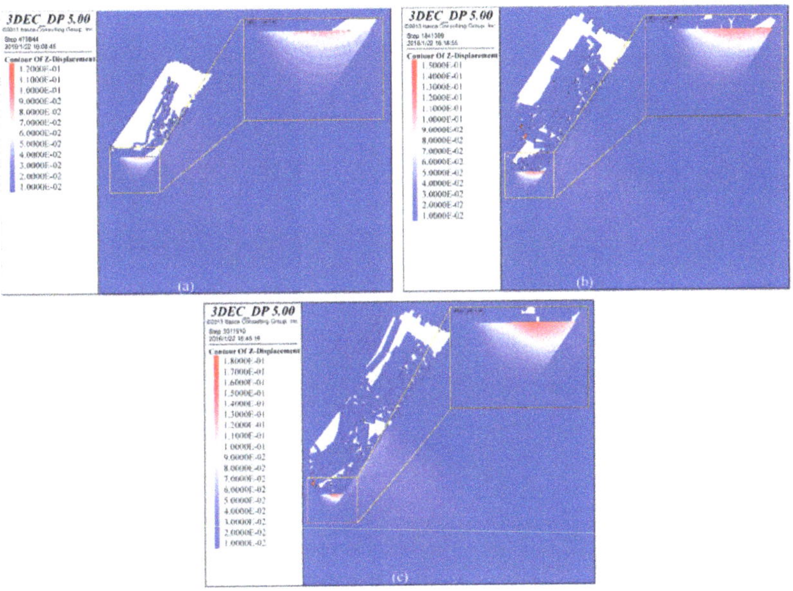

Figure 10. The expansion deformation cloud charts in the coal below the goaf: (**a**) the 5th slice; (**b**) the 10th slice and (**c**) the 15th slice.

4.5. The Effective Pressure Relief Range

The relative expansion deformation is one of the most important evaluation indices widely used to examine the pressure relief effect of the protective layer mining technology [11,32]. In China, the critical relative expansion deformation is 3‰ [17,18,33]. Once the relative expansion deformation in the protected layer is greater than 3‰, effective pressure relief could be achieved.

To obtain the relative expansion deformation in the coal at different distances below the goaf, we divide each slice into three sub-slices, namely, 0~3.7 m, 3.7~7.3 m and 7.3~11 m below the goaf. Hence, the relative expansion deformations of four sub-slices (0~3.7 m, 3.7~7.3 m, 7.3~11 m and 11~14.7 m below the goaf) were calculated during the mining of each slice. The expansion deformation calculation results during the mining of the 5th, 10th and 15th slices are shown in Figure 11.

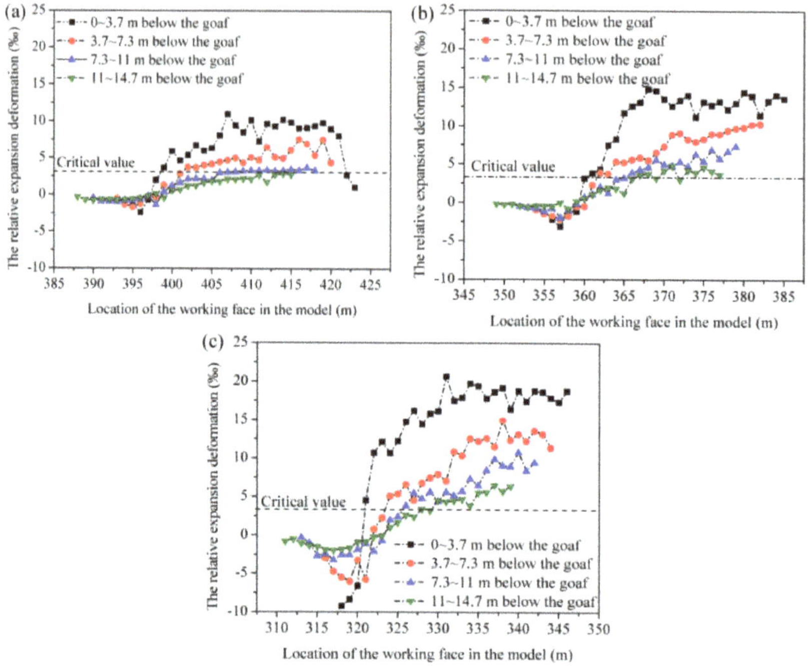

Figure 11. The relative expansion deformation calculation results: (**a**) the 5th slice, (**b**) the 10th slice and (**c**) the 15th slice.

As shown in Figure 11, with the distance below the goaf increasing, the relative expansion deformation in each sub-slice decreases gradually. For example, the relative expansion deformations in the sub-slices 0~3.7 m, 3.7~7.3 m, 7.3~11 m and 11~14.7 m below the goaf are approximately 10‰, 5‰, 3‰ and 1.5‰, respectively, as shown in Figure 11a. The same regulation is also found during the mining of the 10th and 15th slices, as shown in Figure 11b,c. Moreover, with the mining depth increasing, the relative expansion deformation tends to increase. Take the sub-slice 3.7~7.3 m below the goaf as an example; its relative expansion deformation is approximately 5‰ during the mining of the 5th slice, whereas it increases to approximately 9‰ and 13‰ during the mining of the 10th and 15th slices, respectively. This implies that the pressure relief effect increases with the mining depth. The effective pressure relief ranges obtained by the relative expansion deformation date are shown in Figure 12.

As we can see from Figure 12, the effective pressure relief range has an obvious increasing tendency with the mining depth. During the mining process of the fifth slice, the effective pressure relief angle is approximately 57°, and the effective pressure relief depth is approximately 11.0 m. Meanwhile, 75.6% of the coal in the sixth slice is located at the effective pressure relief zone. However, during the mining process of the 10th slice, the effective pressure relief angle and the effective pressure relief depth increase to 61° and 14.8 m, respectively. Moreover, 76.7% of the coal in the 11th slice is located at the effective pressure relief zone. During the mining process of the 15th slice, the effective pressure relief angle increases to 64°, and the effective pressure relief depth remains at 14.8 m, resulting in 78.2% of the coal in the 16th slice being located at the effective pressure relief zone.

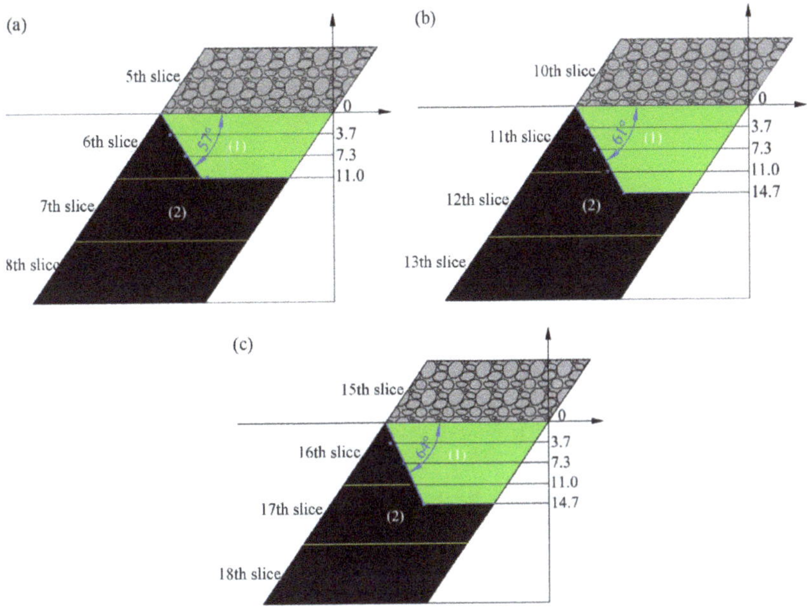

Figure 12. The effective pressure relief ranges in the coal below the goaf: (**a**) the 5th slice, (**b**) the 10th slice and (**c**) the 15th slice. In each figure: (1) denotes the effective pressure relief zone, and (2) denotes the stress concentration zone.

Therefore, over 75% of the coal in the lower slice can be protected effectively during the mining of the upper slice in the steep and extra-thick coal seam. In other words, the steep and extra-thick coal seam can be protected effectively slice by slice during the mining process in the HFMTCSC method.

5. Gas Extraction Method and Field Application

5.1. The Gas Extraction Method

The #2 coal seam in the Yaojie No. 3 coal mine is of great outburst risk, and several serious coal and gas outburst accidents have been reported in the past few decades. At the same time, the gas originating from the coal below the goaf, the mining slice and the remaining coal in the goaf all easily migrate towards the working face during the mining process; thus, gas overflow also usually occurs in the working face. The gas emission sources in the working face are shown in Figure 13. Therefore, pre-mining gas extraction measures should be adopted in the steep and extra-thick coal seam to eliminate the outburst hazard and decrease the gas emission in the working face.

In the Yaojie No. 3 coal mine, crossing boreholes and bedding boreholes were adopted to extract the gas in the coal below the goaf. Crossing boreholes are constructed from the rock roadway in the roof, as shown in Figure 14. Crossing boreholes are adopted to extract the gas in 4~5 slices below the goaf. Considering that the coal permeability values in the effective pressure relief zone and the stress concentration zone are rather different, the borehole spacing should also be differently set in different zones. During the actual application process, long boreholes used to extract the pressure relief gas are constructed first with an angle of 5° in the dip until all they are drilled through the coal floor. Next, enforced short boreholes are supplemented into the stress concentration zone. After the crossing boreholes are finished, the borehole spacing in the effective pressure zone is approximately 6~8 m, whereas that in the stress concentration zone is approximately 3~4 m. This arrangement of the crossing boreholes can satisfy the gas extraction demand both in the effective pressure relief zone and the stress concentration zone. Figure 14b shows the arrangement of bedding boreholes, from which we can see that bedding boreholes with a spacing of 3 m are constructed from the roadway of the working face in each slice.

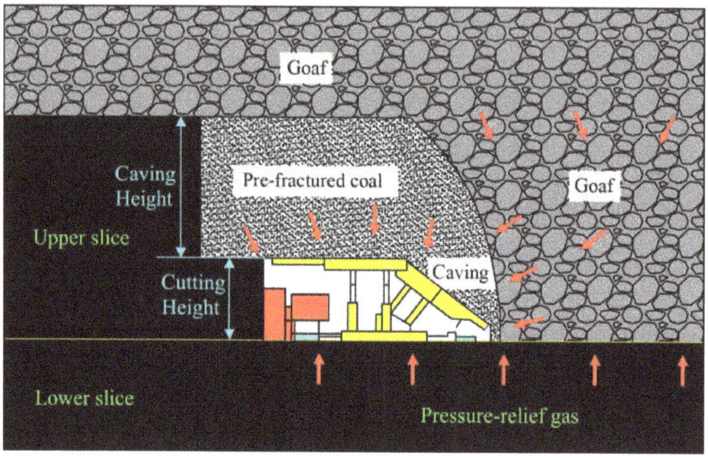

Figure 13. The gas emission sources in the working face.

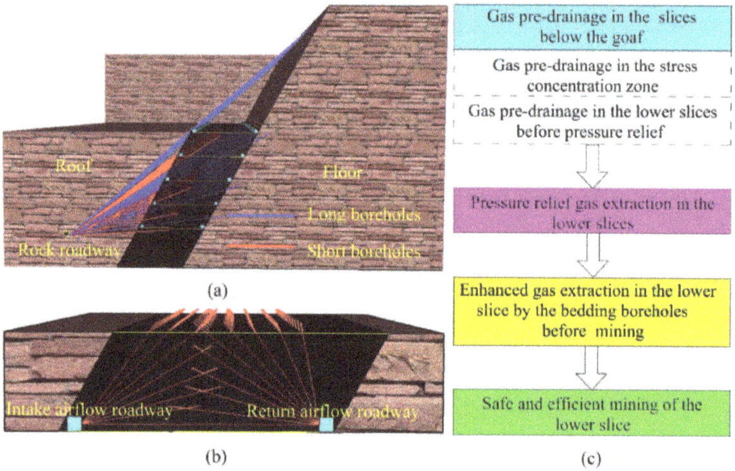

Figure 14. Schematic view of the gas extraction in the steep and extra-thick coal seam: (**a**) crossing boreholes from the rock roadway in the roof; (**b**) bedding boreholes from the roadway of the working face; and (**c**) the gas extraction flow chart.

During the mining process of each slice, crossing boreholes are set to extract the gas in 4~5 slices below the goaf simultaneously. Moreover, due to the fact that the effective pressure relief depth of each slice is just approximately 11 m to 14.8 m, the stress in the 3rd~5th slices below the goaf cannot be released during the mining process. During this period, the crossing boreholes can also be adopted to extract the gas in these slices, although the permeability is relatively low. Afterwards, when the upper coal seam is exploited, the stress in these slices will be released, and the permeability will increase gradually. Then, the crossing boreholes are used to extract the pressure relief gas in these slices. Therefore, the crossing boreholes in each roof roadway are used twice during the gas extraction process, with the first time during the pre-drainage before the pressure relief and the second time during the extraction of the pressure relief gas.

The outburst risk in one to two slices below the goaf will be eliminated by the gas extraction in the crossing boreholes during the mining of each slice. Thus, the roadway of the next slice below the goaf can be excavated, and the bedding boreholes can be constructed. By the enhanced gas extraction in the bedding boreholes, the gas content in the slice to be mined next will decrease furthermore, thereby guaranteeing the safe and efficient mining in the steep and extra-thick coal seam. The flow chart of the gas extraction in the steep and extra-thick coal seam is shown in Figure 14c.

5.2. Field Application

Since this gas extraction method was adopted in 1996, the gas extraction condition in the Yaojie No. 3 coal seam has improved significantly. The gas extraction percentage increased from 25% to 70%, and the daily coal production rose from 521 t to 3300 t, as shown in Figure 15. The only coal and gas outburst accident occurred near a major fault during the mining process in 2003 [34]. The occurrence of the fault destroyed the continuity of the coal seam; thus, the stress in coal below the goaf cannot be released effectively during the mining process. Therefore, more attention should be paid during the mining process near faults.

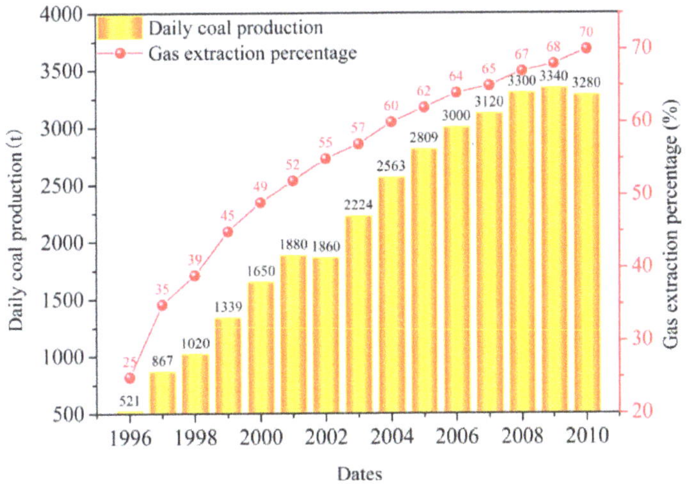

Figure 15. The gas extraction percentage and daily coal production in recent years.

6. Conclusions

The following conclusions can be drawn in this work:

(1) During the mining of the steep and extra-thick coal seam, a pressure arch structure and a hinge structure are formed in succession in the overburden rock, inducing stress redistribution in the coal below the goaf. Therefore, the coal below the goaf can be divided into a pressure relief zone and a stress concentration zone.

(2) The pressure relief range with a pressure relief angle of approximately 70° in the coal below the goaf is almost the same, regardless of whether it is a pressure arch structure or a hinge structure in the overburden rock. Moreover, the vertical stress at a same distance below the goaf is decreased to the same level during the mining of different slices, whereas the stress in the stress concentration zone increases gradually with the mining depth.

(3) As the mining depth increases, the pressure relief effect in the coal below the goaf is increasingly notable and the effective pressure relief range increases gradually. When the mining depth increases from the fifth slice, the effective pressure relief angel increases from 57° to 64°, and the effective pressure relief depth increases from 11.0 m to 14.7 m.

(4) The gas in the coal below the goaf both in the effective pressure relief zone and in the stress concentration zone can be well extracted by the reasonably designed crossing

boreholes and bedding boreholes. Since this gas extraction method was adopted in 1996, the gas extraction condition in the Yaojie No. 3 coal seam has improved significantly. The gas extraction percentage increased from 25% to 70%, and the daily coal production rose from 521 t to 3300 t.

Author Contributions: H.Z.: Conceptualization, methodology, writing-original draft. L.X.: Writing—review and editing. M.Y.: Writing—review and editing. C.D.: Writing—review and editing. Y.C.: Conceptualization, writing—review and editing. All authors have read and agreed to the published version of the manuscript.

Funding: This work was funded by the National Natural Science Foundation of China (No. 52104210), the Natural Science Foundation of Shanxi Province (No. 20210302124350 and No. 20210302123146) and Research Project Supported by Shanxi Scholarship Council of China (No. 2021-044).

Institutional Review Board Statement: Not applicable.

Informed Consent Statement: Not applicable.

Data Availability Statement: Data will be available from the author on request.

Acknowledgments: The authors are grateful to the editor and reviewers of this work for their insightful comments, criticisms, and suggestions.

Conflicts of Interest: The authors declare that they have no known competing financial interest or personal relationships that would have appeared to influence the work reported in this paper.

References

1. Noack, K. Control of gas emissions in underground coal mines. *Int. J. Coal Geol.* **1998**, *35*, 57–82. [CrossRef]
2. Aguado, M.B.D.; Nicieza, C.G. Control and prevention of gas outbursts in coal mines, Riosa–Olloniego coalfield, Spain. *Int. J. Coal Geol.* **2007**, *69*, 253–266. [CrossRef]
3. Yang, W.; Lin, B.; Qu, Y.; Li, Z.; Zhai, C.; Jia, L.; Zhao, W. Stress evolution with time and space during mining of a coal seam. *Int. J. Rock Mech. Min. Sci.* **2011**, *48*, 1145–1152. [CrossRef]
4. Moore, T.A. Coalbed methane: A review. *Int. J. Coal Geol.* **2012**, *101*, 36–81. [CrossRef]
5. Saghafi, A.; Pinetown, K.L. A new method to determine the depth of the de-stressed gas-emitting zone in the underburden of a longwall coal mine. *Int. J. Coal Geol.* **2015**, *152*, 156–164. [CrossRef]
6. Wang, L.; Lu, Z.; Chen, D.; Liu, Q.; Wen, Z. Safe strategy for coal and gas outburst prevention in deep-and-thick coal seams using a soft rock protective layer mining. *Saf. Sci.* **2020**, *129*, 104800. [CrossRef]
7. Xu, C.; Yang, G.; Wang, K.; Fu, Q. Uneven stress and permeability variation of mining-disturbed coal seam for targeted CBM drainage: A case study in Baode coal mine, eastern Ordos Basin, China. *Fuel* **2021**, *289*, 119911. [CrossRef]
8. Wang, H.; Cheng, Y.; Wu, D.; Liu, H. Gas emission and parameter optimization of gas extraction in mining face of short distance protective seam. *J. China Coal Soc.* **2010**, *4*, 590–594. (In Chinese)
9. Liu, H.; Cheng, Y.; Chen, H.; Mou, J.; Kong, S. Characteristics of mining gas channel expansion in the remote overlying strata and its control of gas flow. *Int. J. Min. Sci. Technol.* **2013**, *23*, 481–487. [CrossRef]
10. Zhang, R.; Cheng, Y.; Zhou, H.; Yuan, L.; Li, W.; Liu, Q.; Jin, K.; Tu, Q. New insights into the permeability-increasing area of overlying coal seams disturbed by the mining of coal. *J. Nat. Gas Sci. Eng.* **2018**, *48*, 352–364. [CrossRef]
11. Hu, G.; Wang, H.; Li, X.; Fan, X.; Yuan, Z. Numerical simulation of protection range in exploiting the upper protective layer with a bow pseudo-incline technique. *Min. Sci. Technol.* **2009**, *19*, 58–64. (In Chinese) [CrossRef]
12. Yang, W.; Lin, B.; Qu, Y.; Zhao, S.; Zhai, C.; Jia, L.; Zhao, W. Mechanism of strata deformation under protective seam and its application for relieved methane control. *Int. J. Coal Geol.* **2011**, *85*, 300–306. [CrossRef]
13. Li, D. Underground hydraulic mining of thin sub-layer as protective coal seam in coal mines. *Int. J. Rock Mech. Min. Sci.* **2014**, *67*, 145–154. [CrossRef]
14. Liu, H.; Liu, H.; Cheng, Y. The elimination of coal and gas outburst disasters by ultrathin protective seam drilling combined with stress-relief gas drainage in Xinggong coalfield. *J. Nat. Gas Sci. Eng.* **2014**, *21*, 837–844. [CrossRef]
15. Kong, S.; Cheng, Y.; Ren, T.; Liu, H. A sequential approach to control gas for the extraction of multi-gassy coal seams from traditional gas well drainage to mining-induced stress relief. *Appl. Energy* **2014**, *131*, 67–78. [CrossRef]
16. Yuan, Y.; Chen, Z.; Zhang, X.; Wang, Z. Intermediate coal pillar instability and permeability evolution in extremely thin protective seam by auger mining. *Arab. J. Geosci.* **2019**, *12*, 322. [CrossRef]
17. Xu, Y.; Lun, Z.; Pan, Z.; Wang, H.; Zhou, X.; Zhao, C.; Zhang, D. Occurrence space and state of shale oil: A review. *J. Pet. Sci. Eng.* **2022**, *211*, 110183. [CrossRef]

18. Zhang, H.; Cheng, Y.; Deng, C.; Shu, L.; Pan, Z.; Yuan, L.; Wang, L.; Liu, Q. A novel in-seam borehole discontinuous hydraulic flushing technology in the driving face of soft coal seams: Enhanced gas extraction mechanism and field application. *Rock Mech. Rock Eng.* **2022**, *55*, 885–907. [CrossRef]
19. Tu, S.; Yuan, Y.; Li, N.; Dou, F.; Wang, F. Hydraulic support stability control of fully mechanized top coal caving face with steep coal seams based on instable critical angle. *J. Coal Sci. Eng.* **2008**, *14*, 382–385. (In Chinese) [CrossRef]
20. Wang, N.; Li, L.; Lai, X.; Chai, X. Comprehensive analysis of safe mining to heavy and steep coal seam under complex geophysics environment. *J. Coal Sci. Eng.* **2008**, *14*, 378–381. (In Chinese) [CrossRef]
21. Zhang, J.; Zhao, Z.; Gao, Y. Research on top coal caving technique in steep and extra-thick coal seam. *Procedia Earth Planet. Sci.* **2011**, *2*, 145–149. [CrossRef]
22. Lai, X.; Shan, P.; Cao, J.; Sun, H.; Suo, Z.; Cui, F. Hybrid assessment of pre-blasting weakening to horizontal section top coal caving (HSTCC) in steep and thick seams. *Int. J. Min. Sci. Technol.* **2014**, *24*, 31–37. [CrossRef]
23. Miao, S.; Lai, X.; Cui, F. Top coal flows in an excavation disturbed zone of high section top coal caving of an extremely steep and thick seam. *Min. Sci. Technol.* **2011**, *21*, 99–105. (In Chinese)
24. Zhao, J.; Lai, X. Application of preblasting to high-section top coal caving for steep-thick coal seam. *J. Coal Sci. Eng.* **2011**, *17*, 113–118. (In Chinese) [CrossRef]
25. Karacan, C.Ö.; Ruiz, F.A.; Cotè, M.; Phipps, S. Coal mine methane: A review of capture and utilization practices with benefits to mining safety and to greenhouse gas reduction. *Int. J. Coal Geol.* **2011**, *86*, 121–156. [CrossRef]
26. Li, W.; Cheng, Y.; Wang, L. The origin and formation of CO_2 gas pools in the coal seam of the Yaojie coalfield in China. *Int. J. Coal Geol.* **2011**, *85*, 227–236. [CrossRef]
27. Wang, W.; Cheng, Y.; Wang, H.; Liu, H.; Wang, L.; Li, W.; Jiang, J. Fracture failure analysis of hard–thick sandstone roof and its controlling effect on gas emission in underground ultra-thick coal extraction. *Eng. Fail. Anal.* **2015**, *54*, 150–162. [CrossRef]
28. Tang, C.; Kaiser, P. Numerical simulation of cumulative damage and seismic energy release during brittle rock failure—part I: Fundamentals. *Int. J. Rock Mech. Min. Sci.* **1998**, *35*, 113–121. [CrossRef]
29. Nadimi, S.; Shahriar, K.; Sharifzadeh, M.; Moarefvand, P. Triaxial creep tests and back analysis of time-dependent behavior of Siah Bisheh cavern by 3-Dimensional Distinct Element Method. *Tunn. Undergr. Space Technol.* **2011**, *26*, 155–162. [CrossRef]
30. Coggan, J.; Gao, F.; Stead, D.; Elmo, D. Numerical modelling of the effects of weak immediate roof lithology on coal mine roadway stability. *Int. J. Coal Geol.* **2012**, *90*, 100–109. [CrossRef]
31. Itasca. *3DEC-3-Dimensional Distinct Element Code*; Version 4.10; Itasca Consulting Group: Minneapolis, MN, USA, 2013.
32. Wang, L.; Wang, Z.; Xu, S.; Zhou, W.; Wu, J. A field investigation of the deformation of protected coal and its application for CBM extraction in the Qinglong coalmine in China. *J. Nat. Gas Sci. Eng.* **2015**, *27*, 367–373. [CrossRef]
33. Liu, H.; Cheng, Y. The elimination of coal and gas outburst disasters by long distance lower protective seam mining combined with stress-relief gas extraction in the Huaibei coal mine area. *J. Nat. Gas Sci. Eng.* **2015**, *27*, 346–353. [CrossRef]
34. Cao, J.; Dou, L.; Zhu, G.; He, J.; Wang, S.; Zhou, K. Mechanism of rock burst in horizontal section mining of a steeply inclined extra-thick coal seam and prevention technology. *Energies* **2020**, *13*, 6043. [CrossRef]

energies

Article

A New Relative Permeability Characterization Method Considering High Waterflooding Pore Volume

Guangfeng Qi [1], Jingang Zhao [1], Hu He [1], Encheng Sun [1], Xin Yuan [1] and Shuoliang Wang [2,*]

1 Sinopec Shengli Oilfield Technology Testing Center, Dongying 257062, China; zzlcc1323@gmail.com (G.Q.); shuyao_yang@163.com (J.Z.); bdfvhj@126.com (H.H.); 13854662554@139.com (E.S.); 15666381577@163.com (X.Y.)
2 Faculty of Engineering, School of Energy, China University of Geosciences, Beijing 100083, China
* Correspondence: wangshuoliang@cugb.edu.cn; Tel.: +86-135-0101-7546

Abstract: In the process of waterflooding development, high waterflooding PVs will make the fluid percolation in the reservoir more complicated, resulting in lower efficiency of waterflooding. High waterflooding PVs will affect the relative permeability and change the seepage law of oil–water two-phase flow in a high water-cut period. In this study, we performed high waterflooding PVs relative permeability experiments using nine natural cores. The unsteady measurement method is used to test the relative permeability curve. The results show that: (1) the relative permeability is affected by the waterflooding PVs, the recovery efficiency of 2000 waterflooding PVs is 10.72% higher than that of 50 waterflooding PVs on the core scale; (2) it makes water mobility increase sharply, while oil phase flow capacity remains low and decreases at high water cut stage. A new relative permeability characterization method considering high waterflooding PVs is established, which is applied to the numerical simulator. It shows that the remaining oil saturation of the high-permeability belt is higher than the calculation results of the traditional numerical simulator. It means that the injected water does not diffuse much into the low-permeability zone of the formation. The modified simulator is validated with the actual China offshore oilfield model. The numerical saturation of the key section of the passing well is in good agreement with the actual logging interpretation results, and the water cut curve fits better in the whole area. The modified simulator could predict oil production accurately after high waterflooding PVs treatment.

Keywords: relative permeability; high waterflooding PVs; numerical simulation; physical simulation

Citation: Qi, G.; Zhao, J.; He, H.; Sun, E.; Yuan, X.; Wang, S. A New Relative Permeability Characterization Method Considering High Waterflooding Pore Volume. *Energies* **2022**, *15*, 3868. https://doi.org/10.3390/en15113868

Academic Editors: Riyaz Kharrat and Mofazzal Hossain

Received: 31 March 2022
Accepted: 19 May 2022
Published: 24 May 2022

1. Introduction

The development of oil and gas reservoirs generally goes through three stages. The first stage is the increased production period. The second is the constant production stage, and the third is characterized by declining production. As the reservoir enters its production decline period, there are several EOR methods available to help the field achieve maximum recovery [1,2]. Increasing the waterflooding PVs is an economical and straightforward technique, considering the economic cost. In the process of actual reservoir water flooding development, which is affected by reservoir heterogeneity, water is injected into the planes and longitudinal high-permeability bands, the water–oil ratio rises sharply, and the waterflooding PVs can reach hundreds or thousands of times in the mainline or near the well area [3,4]. The uneven distribution of waterflooding PVs in the reservoir leads to the invalid circulation of injected water. It has been found that the method of flow diagnosis that has been applied to profile control and water plugging measures, by defining the relationship of the connectivity volume, flux, distribution factors of injection flow, production wells in the reservoir, and Lorentz coefficient, can semi-quantitatively judge the waterflooding PVs in the reservoir and provide well reference for profile control and water plugging [5–9]. The traditional relative permeability test is usually performed under the conditions of 30–50 PV displacement multiple and 50–60% oil displacement efficiency. Laboratory tests in the Shengli oilfield show that the oil recovery efficiency can reach 70–80% by increasing waterflooding PVs and displacement pressure gradient. The oil recovery efficiency of the Daqing oilfield is nearly 100% when the waterflooding PVs are 26,331 PVs [10,11]. After long-term waterflooding, the pore throat structure of loose sandstone reservoirs will change [12]. It is suggested that the formation is formed in large water channels. In the stage of high water cut and high recovery of China's Bohai SZ oilfield, the coring data of new drilling wells shows that the microscopic pore throat characteristics of the reservoir went through significant changes after long-term waterflooding. The south China sea offshore oil field laboratory experimental study shows that the oil displacement efficiency can be increased from 60.67% in the case of waterflooding (100 PVs) to 71.27% in the case of extra high waterflooding (2000 PVs), and the residual oil saturation can be reduced from 29.56% to 21.72%, indicating that high water-cut oil fields still have great exploitation potential [13–15]. Similar characteristics are also found in the Daqing Lasaxing oilfield and some foreign oilfields in the North Sea. The remaining oil saturation of the strongly water washed oil layer is lower than the laboratory experiment results [16–18]. A large number of field practice and laboratory physical simulation experiments prove that increasing the waterflooding PVs can effectively improve the oilfield development effect.

However, there are several understandings of the mechanism of high waterflooding PVs in EOR. First, long-term waterflooding can cause a change in the reservoir; the essence of high waterflooding PVs is the long-term reconstruction of underground reservoir conditions. If the sandstone is poorly cemented, long-term waterflooding will cause the migration of particles and clay swelling, even sanding, which can cause physical parameters such as reservoir porosity, permeability, and reservoir microscopic pore structure changes. This affects fluid percolation characteristics in the reservoir [19–24]. Second, in the whole process of waterflooding development, the residual oil saturation is not a constant, and the oil displacement efficiency measured in the laboratory at the early stage of development does not represent the ultimate recovery efficiency. High waterflooding PVs will reduce the critical capillary number of the reservoir, thus improving the oil displacement efficiency [17,25]. Third, increasing waterflooding PVs can improve the wettability of reservoir and oil recovery [26]. Either of the above mechanisms or the combination of two or more mechanisms will result in significant changes in the relative permeability curves that describe the characteristics of oil–water two-phase seepage.

Fang Yue et al. [27] conducted water–oil displacement experiments with high waterflooding PVs. They found that under high waterflooding PVs, oil-phase permeability decreased slowly while water-phase permeability increased significantly, and the waterflooding characteristic curve showed an upward "inflection point". After the "inflection point", water consumption increased sharply. The higher the permeability, the higher the displacement efficiency, and increasing injection speed can improve oil displacement efficiency. Hong-min Yu [28] found that high water waterflooding PVs relative permeability curve has the characteristic of semi-logarithmic piecewise linear bounded by the turning point of water cut, the relative permeability of oil and water does not change before the turning point, the residual oil saturation decreases after the turning point, and the relative permeability of oil and water extends to the limit. Zhang Wei et al. [13] found that the isotonic point of the relative permeability curve shifts to the right after the high waterflooding PVs, indicating that the water wettability is enhanced, which is beneficial to water flooding. Microscopic remaining oil flow patterns are classified and studied by Chun lei Yu [29] through the micro-glass etching model experiment and computer image recognition processing technology. The results show that the remaining oil flow patterns can be divided into clusters of flow, porous flow, columnar, membrane, and flows in dropwise flow. In five classes considering oil-water and the contact relation between the pore throat, cluster flow accounts for the largest proportion. With the increase in water saturation, the cluster flow gradually transformed into porous flow, columnar flow, membrane flow, and droplet flow. At the same time, the reason for the nonlinear relative permeability curve and the mobility law of oil and water in the ultra-high water cut period is explained from the microscopic point of view.

From the above research results, it can be clearly seen that the physical simulation of waterflooding with high PVs is relatively mature. However, different types of reservoirs with different properties also have significant differences in their production systems [30–35]. Therefore, the performance of oil displacement efficiency and relative permeability is also quite different for the high waterflooding PVs treatment. Moreover, the application of the understanding of high waterflooding PVs' relative permeability in numerical simulation is still unclear. In other words, there are bottlenecks when the in-house experimental results are applied to the field.

The relative permeability in the traditional numerical simulation model can not change automatically according to the actual production strategy and reservoir properties changes. Therefore, we took the natural core of Q oilfield in the Bohai Sea as the research object and carried out high waterflooding PVs experiments. According to the experimental results, the relationship between relative permeability and water saturation (Sw) and waterflooding PVs is established. The knowledge is coupled to the numerical simulator. The effectiveness of the simulator is verified by the latest logging interpretation results of the passing well and the fitting of the water cut in the well area.

2. Experiments

2.1. Experimental Design

Bohai oilfield is a typical fluvial facies reservoir with high permeability and porosity. According to the property variation range of the main reservoir, the law of high waterflooding PVs' relative permeability of different properties (244~9380 mD) was studied. A total of 9 groups of waterflooding experiments were carried out using typical reservoir cores. The simulated oil viscosity is 28 mPa·s. Standard brine used in the experiments has a salinity of 10,000 mg/L. The test fluid was designed according to the actual reservoir oil and water properties of the Q32-6 oilfield. The waterflooding rate was 1.0 mL/min. The maximum waterflooding PVs is 2000 PVs.

2.2. Sample and Devices

In this study, natural cores with different permeability (244–4814.8 mD) were selected to carry out high waterflooding PVs displacement experiments. We describe the cores used in the experiment in Figure 1. All of these cores are moderately wet. Table 1 shows a summary of cores conditions.

Table 1. Summary of the experiment samples.

Core No.	Core Length (cm)	Core Diameter (cm)	Porosity (f)	Permeability (md)	Irreducible Water Saturation (%)	Residual Oil Saturation (%)
1	5.96	2.502	37.5	244.4	28.8	25.99
2	5.88	2.51	37	659.4	27.3	24.72
3	6.04	2.498	37.3	1536.6	25.6	23.21
4	6.33	2.5	38.9	1793.1	22.1	24.79
5	6.24	2.496	38.3	2531.1	20.9	21.99
6	5.97	2.492	38	2681.3	24.4	18.37
7	6.25	2.506	39.5	2811.5	24.2	18.87
8	5.76	2.504	41.1	4035.3	19.3	20.26
9	6.14	2.497	38.7	4814.8	19.6	18.89

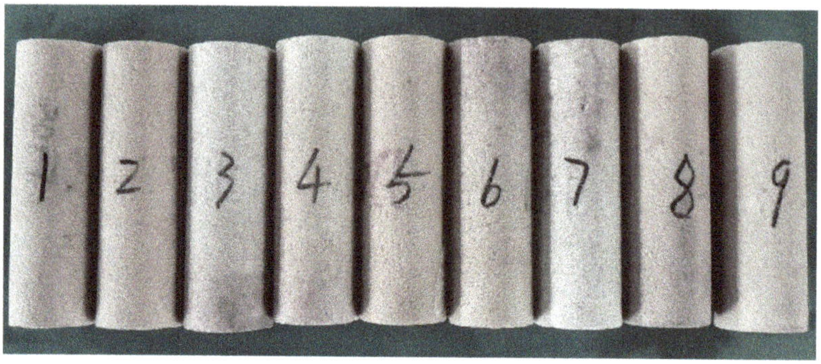

Figure 1. Cores used in the experiments.

The experimental process consists of five parts: pressure control system, temperature control system, core holder, metering, and data acquisition system (Figure 2).

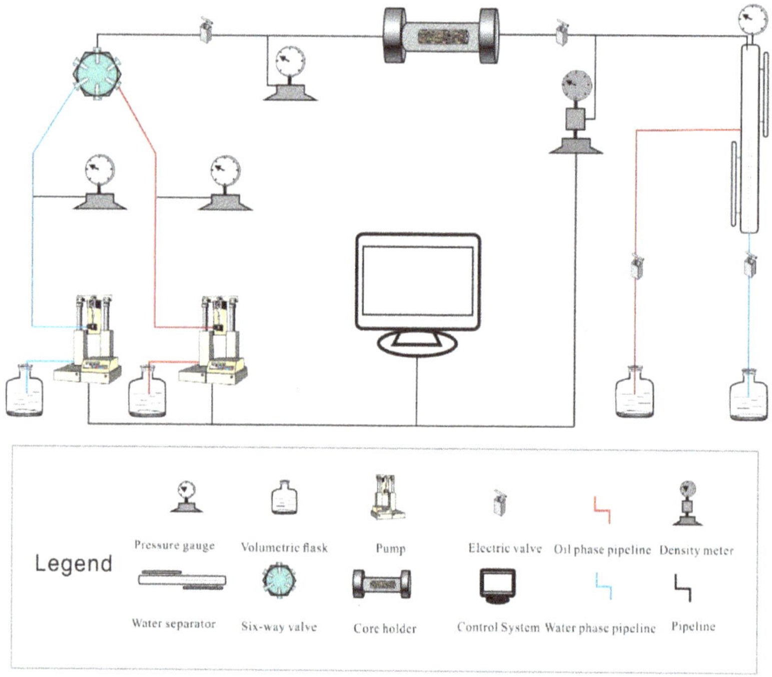

Figure 2. Schematic representation of the experimental apparatus of the relative permeability measurement.

The relative permeability of different cores at high waterflooding PVs was measured.

2.3. *Experimental Procedure*

In this experiment, the core is vacuumed and then saturated with standard saltwater, followed by oil flooding to create irreducible water conditions, and finally, continuous water flooding at a specified temperature. The variation of oil and water at the outlet end with time and the displacement pressure difference is recorded. The specific steps are as follows:

A. The simulated formation water is prepared and stands for 1 day, then filtered and loaded into an intermediate container.

B. The cores treated by washing oil and salt are dried, weighed, vacuumed, and saturated with simulated formation water.

C. The core is loaded into the core holder, the irreducible water is established by simulated oil flooding, and the oil phase permeability under the condition of irreducible water is measured.

D. Simulated formation water is used to displace the core with constant speed. The data of breakthrough time, cumulative oil production, cumulative liquid production, displacement velocity, and the displacement pressure difference between the two ends of the core are accurately recorded.

E. At the beginning, encrypt the record and gradually lengthen the time interval of the record with the continuous decrease in oil production. The experiment was finished after a water injection of 2000 PVs.

3. Experimental Result

3.1. High Waterflooding PVs Relative Permeability

Nine groups of high waterflooding PVs' relative permeability curves show similar characteristics. Here, we take the relative permeability curve of Core #4 as an example (Figure 3, other high waterflooding PVs relative permeability curves see Appendix A). It can be seen that with the increase in water saturation, the relative permeability of the oil phase decreases slowly, while that of the water phase increases sharply, and the increase in the water phase is much larger than the decrease in the oil phase. As waterflooding PVs increase from 50 PVs to 2000 PVs, water saturation increases from 60.8 to 75.2. If water waterflooding PVs stop at 50 PVs, oil-phase relative permeability is considered 0, and the water saturation is 60.8. Water-phase relative permeability is 13.4. If the waterflooding PVs increase to 2000 PVs, the relative permeability of the water phase increases from 13.4 to 66.4, and the relative permeability of the oil phase still decreases slowly with the water saturation increases from 60.8 to 66.4. During the increase in waterflooding PVs, the mobility of the oil phase is very low, accompanied by a continuous decrease, and the mobility of the water phase increases sharply.

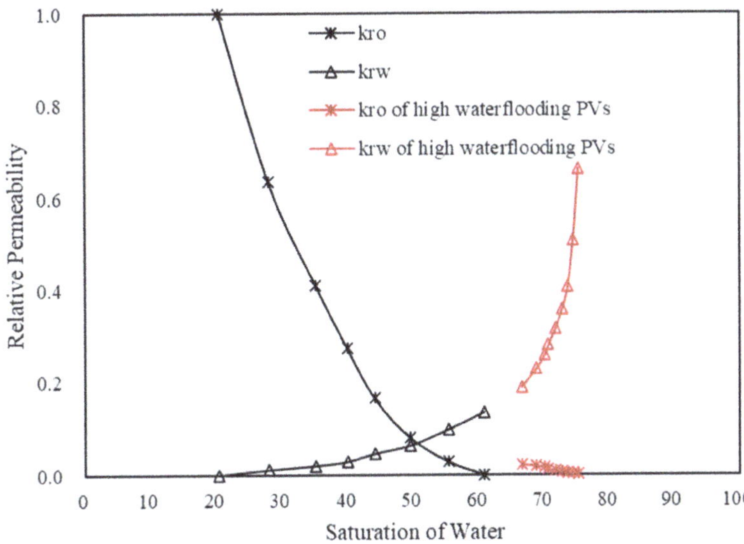

Figure 3. Relative permeability curve of #4 (K = 1793.1 mD).

The waterflooding characteristic curve of no. 4 natural core at high waterflooding PVs (2000 PV) was obtained (Figure 4). The waterflooding characteristic curve upturns after the water cut reaches 98%. Before the "inflection point", the waterflooding PVs is 67.26 PV, the water consumption is 3.36%, the cumulative time is 941.44 min, and the oil displacement efficiency is 55.74%. From the "inflection point" to the end of the experiment, another 1933 PV water was injected, which accounted for 96.64% of the total water consumption. The cumulative time was 27,052 min, and the oil displacement efficiency increased by 13%. It is suggested that after the waterflooding characteristic curve is upturned, the water consumption increases sharply, but the oil displacement efficiency increases slowly, and the development effect gradually deteriorates.

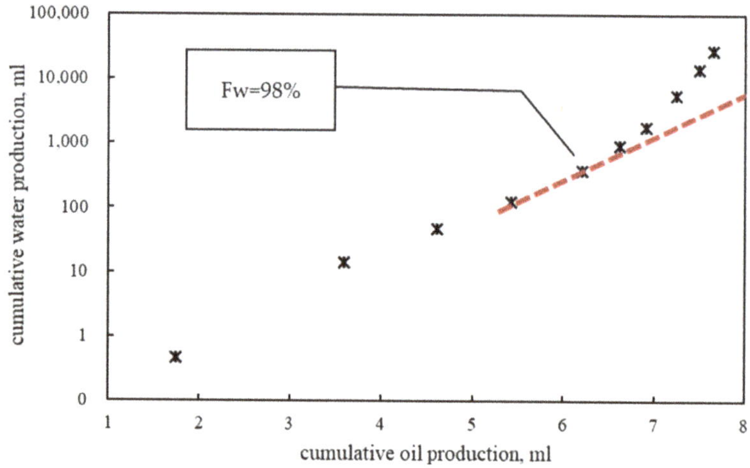

Figure 4. Type A waterflood characteristic curve of #4.

The experimental results of relative permeability at high waterflooding PVs show that increasing waterflooding PVs can improve oil displacement efficiency. However, the mobility of the water phase increases sharply at high waterflooding PVs, especially after the "inflection point" of a waterflooding characteristic curve appears; the same production of oil needs to consume more injection water. Therefore, before the "inflection point" appears in the waterflooding characteristic curve, it is the key to increasing waterflooding recovery efficiency economically and effectively. After the "inflection point", appropriate measures can be taken to control the profile and plug water in order to improve the oil displacement contribution of injected water. Injection–production correlation analysis and flow diagnosis [5–9] can effectively help to identify high water consumption brands and provide a basis for better waterflooding development of oilfields.

3.2. *Relationship between Residual Oil Saturation and Waterflooding PVs*

The amplitude of oil displacement efficiency improvement in different waterflooding PVs was counted (Figure 5). When the waterflooding PVs were increased from 50 PV to 500 PV, the average oil displacement efficiency was increased by 10.5%; when the displacement was further increased from 500 PV to 1000 PV, the oil displacement efficiency was increased by 1.5%; when the displacement was further increased from 1000 PV to 2000 PV, the oil displacement efficiency was only increased by 0.9%. The ultimate oil displacement efficiency is closely related to permeability. With the increase in permeability, the final displacement efficiency and the improved range of displacement efficiency increase. In the actual oilfield development process, the oil displacement efficiency under 2000 PV can be regarded as the ultimate displacement efficiency under the waterflooding state.

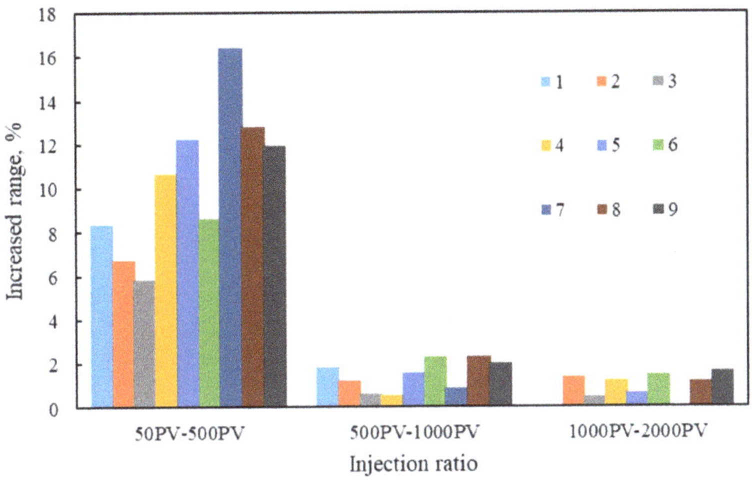

Figure 5. The range of oil displacement efficiency improvement in different water waterflooding PVs.

3.3. *Mechanism Analysis*

The reason for high waterflooding PVs' relative permeability of high water cut stage shows these characteristics: (1) With the increase in water saturation in the process of waterflooding, the oil phase is continuously segmented, and the degree of discontinuity is continuously intensified, which is characterized by large quantity, small volume, and strong dispersion. (2) The increase in discontinuity is due to the increase in discontinuous flow patterns, such as cluster flow and porous flow, and the oil flow becomes more and more dispersed. (3) With the increase in water saturation, the proportion of discontinuous flow morphology (droplet flow, membrane flow, columnar flow, and porous flow) continues to increase, while the proportion of continuous flow morphology (cluster flow) gradually decreases. (4) In the continuous phase, the cluster flow has strong mobility and relatively high relative permeability, while the remaining oil of the other four forms has weak mobility and low relative permeability in the discontinuous phase, while the inflection point of the relative permeability curve occurs when the discontinuous phase with low relative permeability begins to develop. In other words, inhibiting the transformation and development of the continuous phase and discontinuous phase and maintaining the proportion of the continuous phase at a certain level can delay the occurrence of the inflection point of the phase permeability curve to some extent, which is helpful in improving the recovery factor [36–39].

4. Numerical Simulation

According to the experimental results, there is a significant difference between the relative permeability of high waterflooding PVs and the traditional relative permeability, especially in the high water cut stage. Therefore, based on MRST, we introduce the high waterflooding PVs' relative permeability into the traditional numerical simulator (the new numerical simulator), update the relative permeability curve by automatically converting the injected water multiple, and compare the results with the traditional numerical simulator (a constant phase permeability curve is used in the whole simulation process). We set up a numerical simulation model: a heterogeneous model with high permeability bands along the diagonal distribution. The numerical simulation, calculating results obtained from the modified simulator, is compared with that from a traditional numerical simulator. The basic parameters of the numerical simulation model are shown in Table 2, and the relative permeability curve used in the numerical simulation model is shown in Figure 3.

Table 2. Simulation model parameters.

Grid node	40 × 40 × 1	D_x (m)	10
D_y (m)	10	D_z (m)	10
Top deep (m)	1000	Initial water saturation (f)	0.205
Porosity (f)	0.35	Permeability (10^{-3} μm^2)	300/3000 (high permeability band)
Water viscosity (mPa·s)	1	Oil viscosity (mPa·s)	28
Exploit scheme	Water flooding	Well pattern	One injection well and one production well

4.1. High Water Waterflooding PVs Relative Permeability Characterization

MRST (MATLAB Reservoir Simulation Toolbox) is free, open-source software for reservoir modeling and simulation, developed primarily by the Computational Geosciences group in the Department of Mathematics and Cybernetics at SINTEF Digital. It provides a friendly environment for the realization of applications of new understanding or mechanisms on the influence of two-phase flow in the reservoir. It was used to write the empirical formula of residual oil saturation, respectively, into the simulator, which calculates the dynamic residual oil saturation of each grid at each time step and ensures that the value is less than the limit value of residual oil saturation.

The relationship between oil displacement efficiency and residual oil saturation is as follows:

$$E_d = \frac{1 - S_{\text{or}} - S_{\text{wirr}}}{1 - S_{\text{wirr}}} \times 100\% \tag{1}$$

In the numerical simulation, residual oil saturation is mainly used to calibrate key parameters such as the relative permeability curve. When the waterflooding PVs are less than 50, the results of conventional waterflooding experiments and relative permeability can be used, and when the water waterflooding PVs are greater than 50, it is necessary to dynamically characterize the residual oil saturation. After processing the experimental data of each group, the relationship between residual oil saturation and waterflooding PVs is established, and the power function is used as the time-varying expression of residual oil saturation (Figure 6).

$$S_{\text{or}} = A \cdot PV^B \tag{2}$$

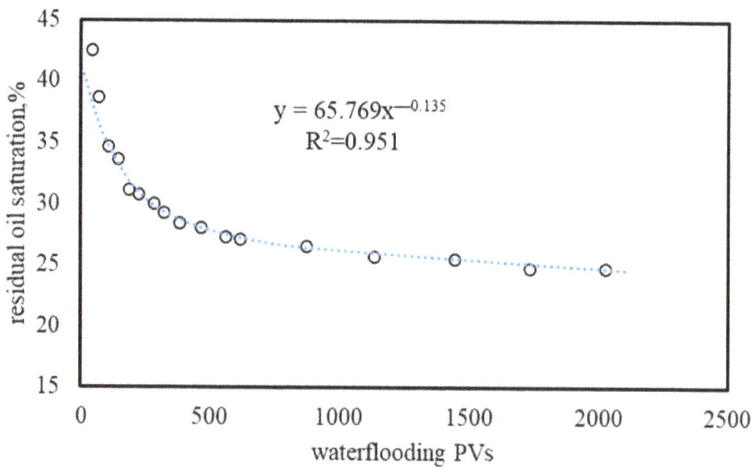

Figure 6. The relationship between the waterflooding PVs and the residual oil saturation.

After fitting 9 groups of experimental data, the values of coefficient A and coefficient B under different permeability are obtained. The expressions of parameter A and parameter B with physical properties and mobility parameters are established (Figures 7 and 8).

$$A = 12.027\, ln\left(\frac{k}{\varphi}\right) + 0.1634 \quad (3)$$

$$B = 0.0122 \times ln\left(\frac{k}{\mu}\right) + 0.0937 \quad (4)$$

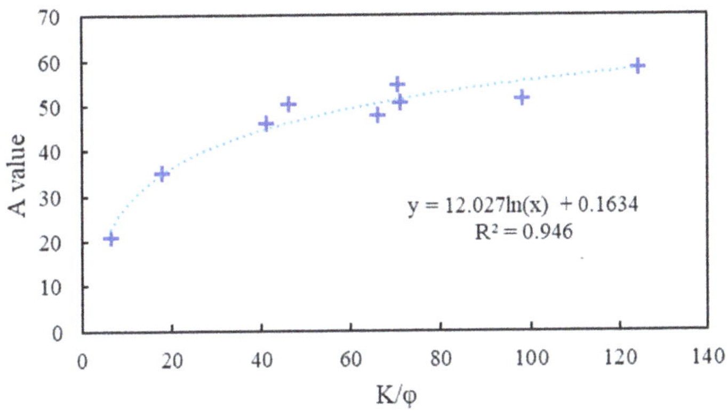

Figure 7. The relationship between the value of coefficient A and the ratio of k/φ.

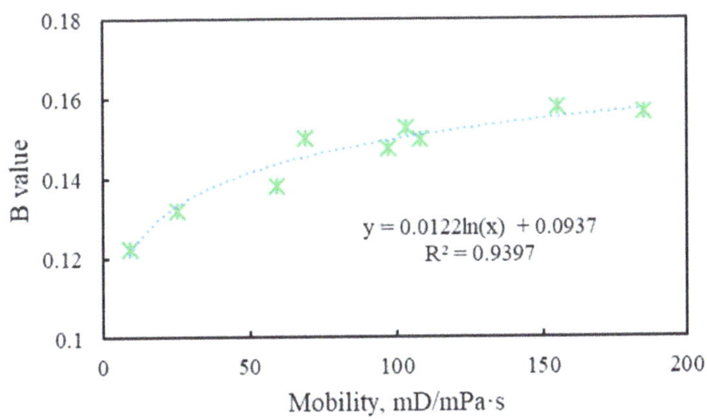

Figure 8. The relationship between the value of coefficient B and mobility.

The empirical formula of residual oil saturation in the oil field is obtained by substituting Equations (3) and (4) into (2).

The actual core experiment and field practice show that there is an extreme value that does not change when the PV number is maximum. Therefore, it is necessary to restrict the extreme value of residual oil saturation. The residual oil saturation under 2000 PV is used as the limit residual oil saturation value.

There is a correlation between irreducible water saturation and core permeability (Figure 9). According to the regression relationship of relevant experimental data in the target oil field, the calculation formula of irreducible water saturation is obtained.

$$S_{wirr} = -3.116 \times lg\,(k) + 46.868 \quad (5)$$

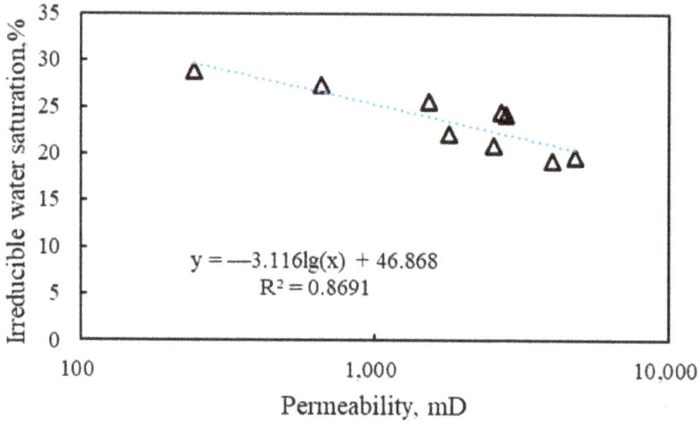

Figure 9. The relationship between the irreducible water saturation and permeability.

The expression of ultimate residual oil saturation in the target oil field can be deduced by simultaneous Equations (1), (2), and (5).

$$S_{\text{orjx}} = \left(1.266 + 7.626\, lg\ K - 0.829 ln\frac{K}{\mu} - 0.486\, lg\ K\ ln\frac{K}{\mu}\right)/100 \quad (6)$$

The model was simulated with MRST, which automatically converted the water multiple, corrected the residual oil saturation endpoint, and calculated the phase permeability iteratively at each step. However, the initially written equation often cannot achieve the best fitting state, and the dynamic residual oil saturation formula needs to be modified iteratively according to the historical fitting effect.

4.2. Numerical Simulation with High Waterflooding PVs Relative Permeability

In this section, we study the effect of high waterflooding PVs' relative permeability on remaining oil saturation. It is indicated that waterflooding is uniform, and the remaining oil is symmetrically distributed along the injection-production wells' diagonal line in the traditional simulator. The calculation results of the numerical simulator considering the relative permeability of high waterflooding PVs show that the overall oil saturation distribution is more uneven, the water saturation of the high-permeability belt is higher, and the remaining oil saturation of the relatively low-permeability region far from the high-permeability belt is higher than the calculation results of the traditional numerical simulator (Figure 10). This means that the injected water does not flood much into the low-permeability zone of the formation but flows along the high-permeability strip to the production well. In heterogeneous reservoirs, with the continuous enhancement of waterflooding, the high permeability area receives more water and is more likely to be washed out, and the water phase mobility is stronger. The low permeability area does not suffer more erosion due to the increase in water injection, which is consistent with the results of parallel flooding experiments and large 3D physical simulations in the laboratory [31,37,40–42]. Thus, the injected water is more inefficient in the process of increasing waterflooding PVs and further enhances reservoir heterogeneity. Because of the above reasons, the actual reservoir sweep area is more complex, and the remaining oil distribution is far more complex than the traditional numerical simulation results.

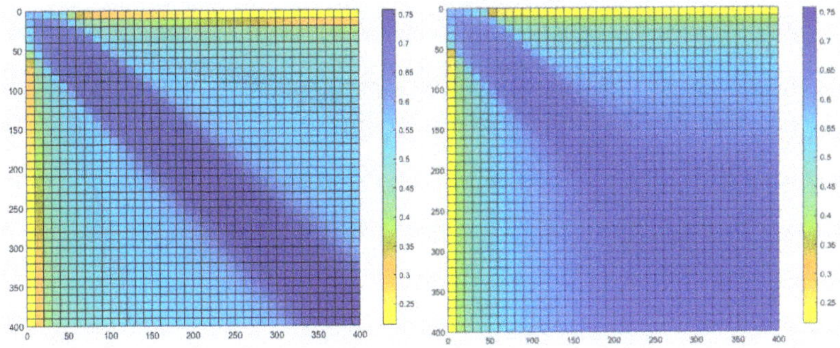

Figure 10. The water saturation distribution at the 50th time step (The left is the numerical simulation results considering the relative permeability of high waterflooding PVs, and the right is the traditional numerical simulation results).

The pressure distribution calculated by the two numerical simulation methods shows that the pressure spreads uniformly from the water injection well to the production well. However, the pressure distribution of the numerical simulation considering the high waterflooding PVs' relative permeability is more uniform, and the pressure difference between the injection well and the production well is smaller than that of the traditional numerical simulation result, indicating that the injected water can flow well to the oil production well along the high permeability channel, so the pressure propagation is more smooth. There is an obvious high-pressure area near the injection well, and an obvious low-pressure area can also be seen near the production well in the traditional numerical simulation result, indicating that the pressure propagation speed is slow (Figure 11). This is also the reason for the difference in saturation distribution.

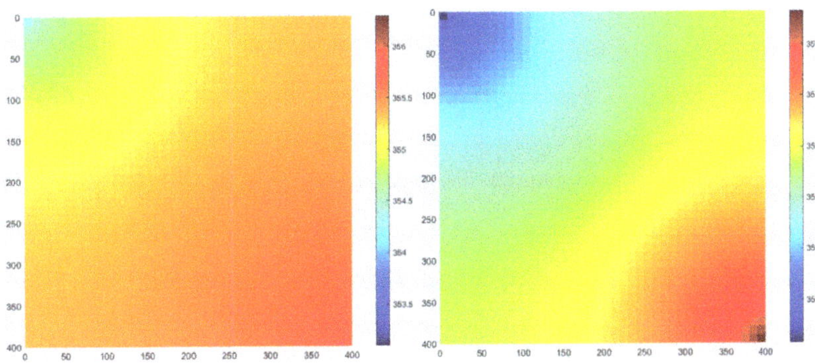

Figure 11. The water saturation distribution at the 50th time step (the left is the numerical simulation results considering the relative permeability of high waterflooding PVs, and the right is the traditional numerical simulation results).

As we have seen, in the initial oil saturation condition, whether considering the high waterflooding PVs' relative permeability or traditional relative permeability, the calculated results of both simulators are consistent. There is anhydrous oil recovery period, and the water cut rises rapidly until point A. When breakthrough occurs in the production well, the water erosion degree of the high permeability channel will be higher and higher, and far more than the low permeability zone. Therefore, after point A, due to lower residual oil saturation of high waterflooding PVs' relative permeability, it is true that more injected water flows to production wells along the high permeability belt, and the sweeping speed of the surrounding low permeability area is slow. When considering the relative permeability of high waterflooding PVs, the water cut of the simulation result is lower than that of the traditional numerical simulation (AB section in Figure 12). As the injection multiple continues to increase, the oil displacement efficiency gradually approaches the limit oil displacement efficiency, and the water cut increases gradually, exceeding the water cut calculated by the traditional numerical simulation.

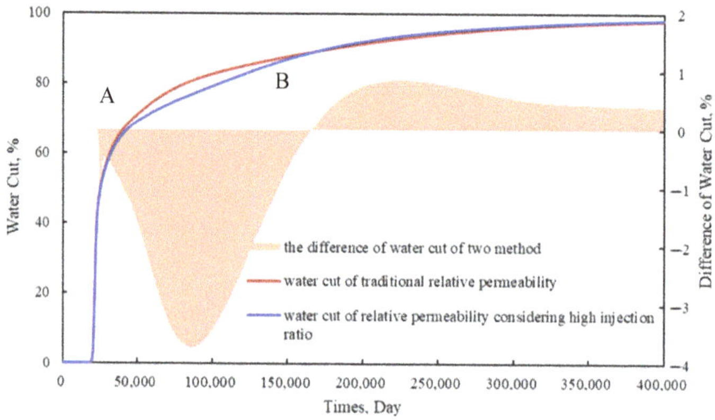

Figure 12. The water cut curve of numerical simulation of traditional simulator and simulator that considering high water waterflooding PVs.

5. Simulation Validations

In order to verify the accuracy of the calculation results of the new numerical simulator, we compared the logging interpretation saturation of the target layer of the newly drilled passing well with that of the new numerical simulator and the traditional simulator. Taking the reservoir model of a certain area of the Q32-6 oilfield as an example, a fine numerical simulation considering high waterflooding PVs' relative permeability was carried out under the condition of small adjustments of local permeability, conductivity, interlayer plugging performance, and other parameters. The coincidence between the oil saturation of the target layer calculated by the new numerical simulator and the logging interpretation of oil saturation is obviously higher than that calculated by the traditional numerical simulator (Figure 13). The water cut of the whole area fit well (Figure 14). Therefore, the modified simulator could provide a more accurate remaining oil saturation prediction for further developing the potential of the reservoir.

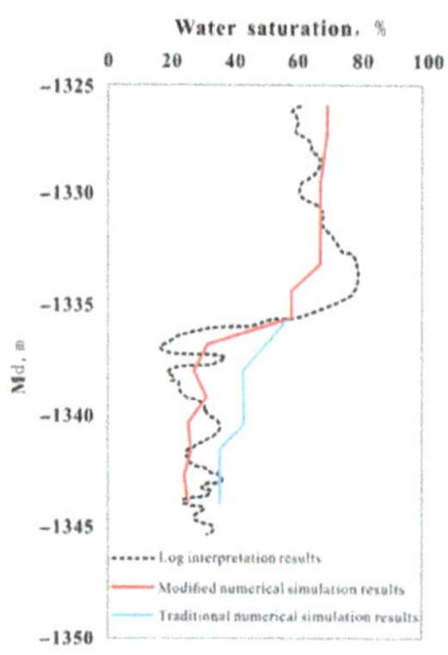

Figure 13. The comparison between the actual logging interpretation oil saturation curve and the model oil saturation curve of a passing well.

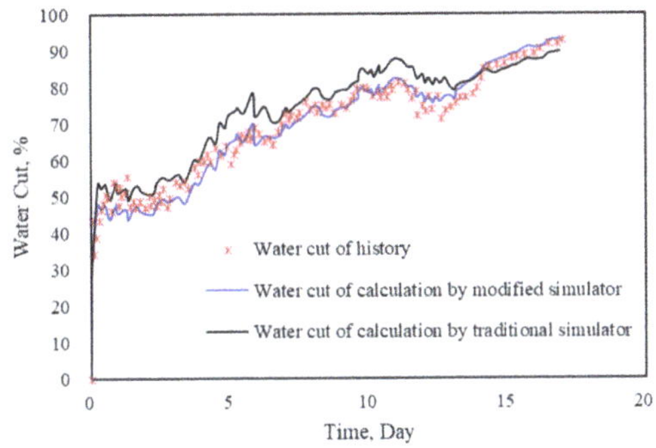

Figure 14. Fitting results of water cut in the whole well area.

6. Conclusions

A total of nine groups of typical cores with different permeability were selected to carry out high waterflooding PVs experiments. The results show that oil displacement efficiency increases 12.9% on average after increasing water waterflooding PVs to 2000 PV, and the potential of the high waterflooding PVs stage cannot be ignored.

The expression of high waterflooding PVs' relative permeability and dynamic displacement efficiency is obtained. The new understanding of high waterflooding PVs' relative permeability is introduced into the numerical simulator. It is considered that in heterogeneous reservoirs, the waterflooding PVs in high permeability areas is much larger than that in low permeability areas. As a result, the oil recovery in the high permeability area and the residual saturation in the low permeability area are underestimated.

Taking the saturation of the key section of the passing well as the verification, the saturation of the passing well at the corresponding time point is in good agreement with the actual logging interpretation results, and the water cut curve fits well in the whole area.

Author Contributions: Conceptualization, S.W.; Formal analysis, J.Z.; Investigation, J.Z.; Methodology, H.H. and S.W.; Software, H.H.; Validation, E.S.; Visualization, X.Y.; Writing—original draft, G.Q.; Writing—review & editing, S.W. All authors have read and agreed to the published version of the manuscript.

Funding: This research received no external funding.

Institutional Review Board Statement: Not applicable.

Informed Consent Statement: Not applicable.

Data Availability Statement: Not applicable.

Conflicts of Interest: The authors declare no conflict of interest.

Nomenclature

E_d	oil displacement efficiency, %
S_{or}	residual oil saturation, %
S_{wirr}	irreducible water saturation, %
PV	pore volume, mL
A	coefficient
B	coefficient
K	permeability, mD
φ	porosity, %
μ	viscosity, mpa·s
k_{ro}	oil relative permeability, decimal
k_{rw}	water relative permeability, decimal
S_{orjx}	ultimate residual oil saturation, %

Appendix A Relative Permeability Curves of No. 1–No. 3 and No. 5–No. 9 Cores

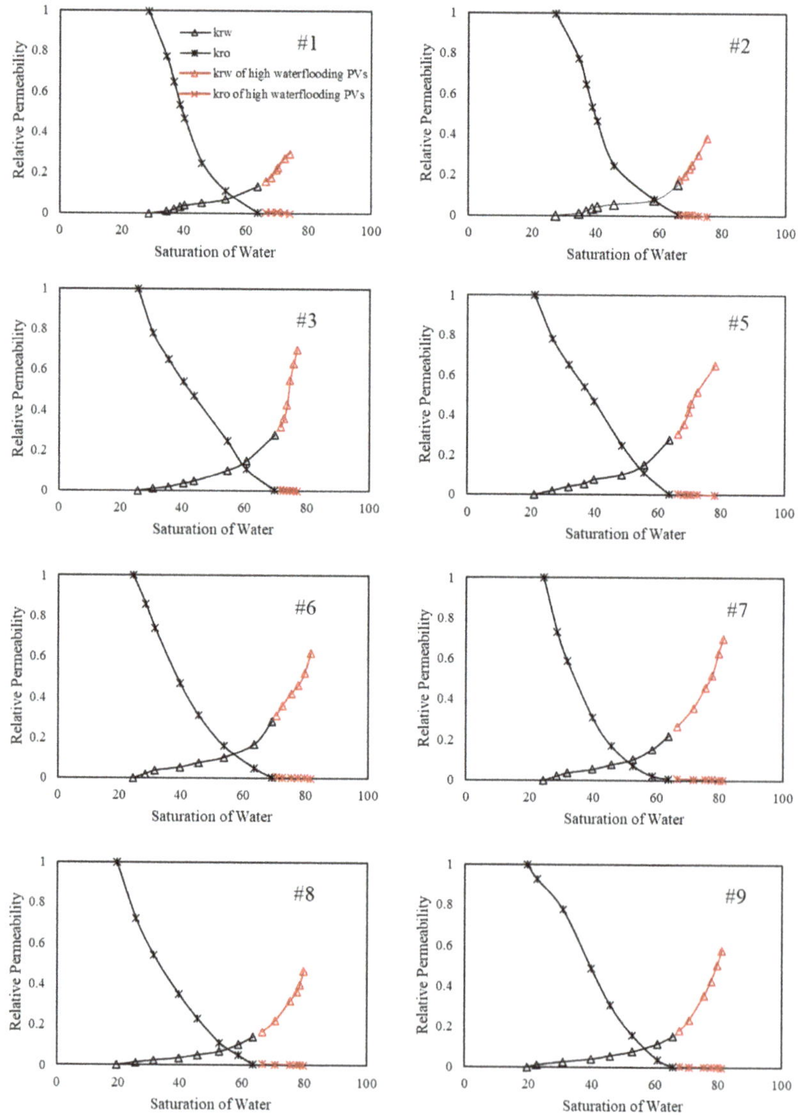

Figure A1. High waterflooding PVs' relative permeability curves of other cores.

References

1. You, Q.; Wang, H.; Zhang, Y.; Liu, Y.; Fang, J.; Dai, C. Experimental study on spontaneous imbibition of recycled fracturing flow-back fluid to enhance oil recovery in low permeability sandstone reservoirs. *J. Pet. Sci. Eng.* **2018**, *166*, 375–380. [CrossRef]
2. You, Q.; Wen, Q.; Fang, J.; Guo, M.; Zhang, Q.; Dai, C. Experimental study on lateral flooding for enhanced oil recovery in bottom-water reservoir with high water cut. *J. Pet. Sci. Eng.* **2019**, *174*, 747–756. [CrossRef]
3. Hou, X.; Wang, Y.; Yang, Q. A new calculation method for unstable state oil/water relative permeability curve. *Pet. Geol. Oilfield Dev. Daqing* **2008**, *27*, 54–56.
4. Li, L.; Song, K.; Gao, L.; Wang, P. Water Flooding Behavior of High Water-Cut Oilfield. *Pet. Drill. Tech.* **2009**, *37*, 91–94.

5. Shahvali, M.; Mallison, B.; Wei, K.; Gross, H. An alternative to streamlines for flow diagnostics on structured and unstructured grids. *SPE J.* **2012**, *17*, 768–778. [CrossRef]
6. Chen, R. Streamline Tracing and Time of Flight Diagnostics for Waterflooding Optimization: Theory and Application. 2015. Available online: https://hdl.handle.net/1969.1/156392 (accessed on 13 October 2021).
7. Møyner, O.; Krogstad, S.; Lie, K.-A. The application of flow diagnostics for reservoir management. *SPE J.* **2015**, *20*, 306–323. [CrossRef]
8. Lie, K.A.; Møyner, O.; Krogstad, S. Application of flow diagnostics and multiscale methods for reservoir management. In Proceedings of the SPE Reservoir Simulation Symposium OnePetro, Houston, TX, USA, 23–25 February 2015.
9. Zhang, Z.; Geiger, S.; Rood, M.; Jacquemyn, C.; Jackson, M.; Hampson, G.; De Carvalho, F.M.; Marques Machado Silva, C.C.; Machado Silva, J.D.; Costa Sousa, M. A tracing algorithm for flow diagnostics on fully unstructured grids with multipoint flux approximation. *SPE J.* **2017**, *22*, 1946–1962. [CrossRef]
10. Zhou, F.; Liu, Z.; Zhang, C. A practical method for determining the relative permeability of oil and water by steady-state method. *Pet. Geol. Eng.* **2009**, *23*, 105–106.
11. Bing, S. *Study on Water Drive Development Characteristics Based on the Oil-Water Two Phase Flow of Ultra-High Water Cut Stage*; Southwest Petroleum University: Chengdu, China, 2013.
12. Wu, Z.; Cui, C.; Hao, Y.; Sun, Y.; Lv, G.; Sun, D.; Zhang, Z. Relative permeability model taking the roughness and actual fluid distributions into consideration for water flooding reservoirs. *Arab. J. Sci. Eng.* **2019**, *44*, 10513–10523. [CrossRef]
13. Zhang, W.; Cao, R.; Luo, D.; Sun, C.; Li, W. Displacement characteristics of high-multiple water drive in marine sandstone reservoirs in the Pearl River Mouth Basin, South China Sea. *Pet. Geol. Recovery Effic.* **2018**, *25*, 64–71.
14. Hu, Z.; Ma, K.; Liu, Z.; Huang, K.; Zhang, J. Research and Application of the Properties Variation Regularity of Water Injection Reservoirs in S Offshore Oilfield. *Sci. Technol. Eng.* **2014**, *14*, 164–168.
15. Wang, X. Influence of Water Injection Multiple on Displacement Efficiency. *Henan Sci.* **2017**, *35*, 139–143.
16. Deng, S.; Wang, N.; Meng, L.; Wu, Z.; Chen, Z. Establishment and Application of The New Two-type Water-Flooding Characteristic Curves at High Water Cut Stage. *Pet. Geol. Oilfield Dev. Daqing* **2017**, *36*, 58–63. [CrossRef]
17. Ji, S.; Tian, C.; Shi, C.; Ye, J.; Zhang, Z.; Fu, X. New understanding on water-oil displacement efficiency in a high water-cut stage. *Pet. Explor. Dev.* **2012**, *39*, 362–370. [CrossRef]
18. Yu, C. A New Prediction Method of Relative Permeability to Reflect Waterflooding Limitation. *Spec. Oil Gas Reserv.* **2014**, *21*, 4. [CrossRef]
19. Chen, D.; Li, J.; Zhu, W.; Xin, Z. Experimental Research on Reservor Parameters Variation After Water Flooding for Offshore Unconsolidated Sandstone Heavy Oil Reservoirs. *China Offshore Oil Gas* **2016**, *28*, 54–60. [CrossRef]
20. Wen, X.; Dai, Z.; Wang, H.; Zhang, X.; Li, H. Physical Properties of Marine Sandstone Reservoir after Long-Term Waterflooding. *Spetial Oil Gas Reserv.* **2017**, *24*, 157–161. [CrossRef]
21. Nie, F. Comparative study on seepage characteristics of core before and after high multiple water flooding in Different facies zones of Dongpu Oilfield. *Pet. Geol. Eng.* **2017**, *31*, 87–90.
22. Jiang, R.; Qiao, X.; Teng, W.; Xu, J.; Sun, Z.; Xie, L. Impact of physical properties time variation on waterflooding reservoir development. *Fault-Block Oil & Gas Field* **2016**, *23*, 768–771. [CrossRef]
23. Ma, K.; Cai, H.; Sun, Z. Nuclear Magnetic Resonance-based Experiment on the Effects of Displacement Velocity and Multiple on the Pore Throat Characteristics and Recovery Factor of Unconsolidated Sandstone Reservoirs. *China Offshore Oil Gas* **2019**, *31*, 86–91. [CrossRef]
24. Tao, S.; Wang, Y.; Tang, D.; Xu, H.; Lv, Y.; He, W.; Li, Y. Dynamic variation effects of coal permeability during the coalbed methane development process in the Qinshui Basin, China. *Int. J. Coal Geol.* **2012**, *93*, 16–22. [CrossRef]
25. Andersen, P.Ø.; Walrond, K.; Nainggolan, C.K.; Pulido, E.Y.; Askarinezhad, R. Simulation interpretation of capillary pressure and relative permeability from laboratory waterflooding experiments in preferentially oil-wet porous media. *SPE Reserv. Eval. Eng.* **2020**, *23*, 230–246. [CrossRef]
26. Dai, Z.; Jiang, J.; Li, H.; Cao, R.; Xin, J.; Luo, D. Wettability experiment of core and mico-mechanism during high-multiple water flooding in heavy oil reservoir. *Sci. Technol. Eng.* **2019**, *33*, 1671–1815.
27. Fang, Y.; Li, H.; Yu, C.; Fan, S.; Hou, X. Seepage characteristics of oil-water two-phase with high injection ratio. *Pet. Geol. Eng.* **2019**, *33*, 4.
28. Yu, H.; Wang, Y.; Nie, J.; Lv, C.; Cui, W.; Zhang, L. Sdudy and Application of a Correction Method for the Relative Permeability Curve of a High Water Injection Multiple. *Pet. Drill. Tech.* **1900**, *46*, 104–108. [CrossRef]
29. Yu, C.; Mi, L.; Wang, C.; Zhao, Y.; Jiang, H.; Tian, Y. Percolation characteristics investigation of microscopic remaining oil in water flooding reservoir with ultra-high water cut. *Fault-Block Oil Gas Field* **2016**, *23*, 592–594. [CrossRef]
30. Li, C.; Wang, S.; You, Q.; Yu, C. A New Measurement of Anisotropic Relative Permeability and Its Application in Numerical Simulation. *Energies* **2021**, *14*, 4731. [CrossRef]
31. Wang, S.; Yu, C.; Sang, G.; Zhao, Q. A new numerical simulator considering the effect of enhanced liquid on relative permeability. *J. Pet. Sci. Eng.* **2019**, *177*, 282–294. [CrossRef]
32. Gram, T.B.; Ditlevsen, F.P.; Mosegaard, K.; Fabricius, I.L. Water-flooding and consolidation of reservoir chalk–effect on porosity and Biot's coefficient. *Geophys. Prospect.* **2021**, *69*, 495–513. [CrossRef]

33. Cao, R.-Y.; Dai, Z.; Wang, Z.-K.; Wang, Y.-H.; Jiang, J.; Jia, Z.-H. Displacement behavior and mechanism of long-term water flooding in sandstone oil reservoirs. *J. Cent. South Univ.* **2021**, *28*, 834–847. [CrossRef]
34. Zhang, W.-Q.; Wang, Y.-N.; Liu, D.-W.; Deng, Y.; Xu, J.-C.; Gu, F.; Yang, Y.; Dong, R.-J. Water Flooding Characteristics of Carbonate Reservoirs with High Permeability Layer. In Proceedings of the International Field Exploration and Development Conference, Chengdu, China, 23–25 September 2020; Springer: Berlin/Heidelberg, Germany, 2020; pp. 3168–3175. [CrossRef]
35. Mohammad, R.S.; Tareen, M.Y.K.; Mengel, A.; Shah, S.A.R.; Iqbal, J. Simulation study of relative permeability and the dynamic capillarity of waterflooding in tight oil reservoirs. *J. Pet. Explor. Prod. Technol.* **2020**, *10*, 1891–1896. [CrossRef]
36. Mahmoudzadeh, A.; Fatemi, M.; Masihi, M. Microfluidics experimental investigation of the mechanisms of enhanced oil recovery by low salinity water flooding in fractured porous media. *Fuel* **2022**, *314*, 123067. [CrossRef]
37. Hu, J.; Li, A. Experimental Investigation of Factors Influencing Remaining Oil Distribution under Water Flooding in a 2-D Visualized Cross-Section Model. *ACS Omega* **2021**, *6*, 15572–15579. [CrossRef]
38. Sun, P.; Xu, H.; Zhu, H.; Jia, L.; Hu, X.; Fang, H.; Jiang, H.; Xu, Z.; Jiang, T.; Jiang, X. Investigation of pore-type heterogeneity and its control on microscopic remaining oil distribution in deeply buried marine clastic reservoirs. *Mar. Pet. Geol.* **2021**, *123*, 104750. [CrossRef]
39. Cheng, B.; Li, J.; Jiang, S.; Lu, C.; Su, H.; Yu, F.; Jiang, H. Pore-Scale Investigation of Microscopic Remaining Oil Variation Characteristic in Different Flow Rates Using Micro-CT. *Energies* **2021**, *14*, 3057. [CrossRef]
40. Ge, L.; Meng, Z.; Zhu, Z.; Zhu, X.; Wang, Y. Three-dimensional physical simulation experiment of reasonable initial oil recovery rate for the gas/edge water reservoirs. *China Offshore Oil Gas* **2019**, *31*, 99–105. [CrossRef]
41. Shi, D.; Yue, X.; Ling, S.; Dong, M.; Kong, B. Physical simulation on system of deep profile control and flooding on high water cut reservoir. *J. Shenzhen Univ. Sci. Eng.* **2018**, *35*, 179–186. [CrossRef]
42. Wu, Y. Experimental Research on Interlayer Interference of Thin Iterbed Reservoirs by 3-D Physical Simulation. *Res. Explor. Lab.* **2017**, *36*, 25–29.

energies

MDPI

Article

Heat Control Effect of Phase Change Microcapsules upon Cement Slurry Applied to Hydrate-Bearing Sediment

Guokun Yang [1], Tianle Liu [1,*], Hai Zhu [1], Zihan Zhang [1], Yingtao Feng [2], Ekaterina Leusheva [3] and Valentin Morenov [3]

[1] Faculty of Engineering, China University of Geosciences, Wuhan 430074, China; 1345781296@cug.edu.cn (G.Y.); zss9223@cug.edu.cn (H.Z.); cugzzh@cug.edu.cn (Z.Z.)
[2] Institute of Oilfield Chemistry, CNOOC Oil Service Co., Ltd., Sanhe 065201, China; fengyt@cosl.com.cn
[3] Department of Oil and Gas, Saint Petersburg Mining University, 199106 Saint Petersburg, Russia; leusheva_el@pers.spmi.ru (E.L.); morenov@spmi.ru (V.M.)
* Correspondence: yanggk@cug.edu.cn

Abstract: This study aims to develop a novel low-heat cement slurry using phase change microcapsule additives to reduce the decomposition of hydrate-bearing sediments during cementing. Microcapsules were prepared by coating mixed alkanes with polymethyl methacrylate, and lipophilic-modified graphite was incorporated to enhance the thermal conductivity of microcapsules. The effects of microcapsules upon the hydration heat, pore distribution, and compressive strength of the cement slurry/stone were studied through a variety of tests. The results showed that the phase-change temperature, thermal enthalpy, and encapsulation efficiency of the microcapsules were 8.99–16.74 °C, 153.58 Jg^{-1}, and 47.2%, respectively. The introduction of lipophilic-modified graphite reduced the initial phase-change temperature of microcapsules by 0.49 °C, indicating an improvement in their temperature sensitivity. The maximum hydration heat of cement slurry decreased by 41.3% with 7% dosage of microcapsules; the proposed microcapsules outperformed comparable low-heat additives. Moreover, the presence of microcapsules could reduce the number of large pores in (and thereby improve the compressive strength of) cement stone. The innovation of this study is that it comprehensively and intuitively confirms the feasibility of the application of low-heat cement slurry with MPCM as the key in hydrate sediments rather than just focusing on the reduction of hydration heat; furthermore, a self-made cementing device was developed to simulate the cementing process of hydrate deposition. The results show that the thermal regulation of microcapsules inhibited the temperature increase rate of the cement slurry, significantly reducing the damage caused to the hydrate. These findings should improve the safety and quality of cement in offshore oil and gas well applications.

Keywords: microencapsulated phase-change materials; heat control; temperature sensitivity; cementing; hydration heat; natural gas hydrate-bearing sediment

Citation: Yang, G.; Liu, T.; Zhu, H.; Zhang, Z.; Feng, Y.; Leusheva, E.; Morenov, V. Heat Control Effect of Phase Change Microcapsules upon Cement Slurry Applied to Hydrate-Bearing Sediment. *Energies* **2022**, *15*, 4197. https://doi.org/10.3390/en15124197

Academic Editor: Andrea Frazzica

Received: 16 May 2022
Accepted: 6 June 2022
Published: 7 June 2022

1. Introduction

In offshore oil and gas well cementing, natural gas hydrate-bearing sediment (GHBS) is often present at the well hole. The hydration heat released during the consolidation of cement slurry changes the stable phase equilibrium state of the hydrate (i.e., to a high pressure and low temperature) [1–3]. When this occurs, the gasified natural gas hydrate enters the cement slurry, which affects the sealing quality of the unhardened cement sheath [4]. In serious cases, the cementing operation may not succeed and a blowout accident could occur [5,6]. Therefore, to prevent accidents and ensure cementing quality in offshore cementing engineering, a cement slurry with a hydration-heat regulation function is urgently required [7,8]. One logical solution to this problem is to introduce phase-change materials (PCMs) as admixtures into the cement slurry, which would adjust its heat of hydration.

Owing to their ability to release or absorb heat during phase changes, and thereby maintain a stable system temperature, PCMs have been implemented in numerous other fields including thermal energy storage [9–11]. PCMs have been introduced as an additive to adjust the heat release rate of concrete with the aim of suppressing temperature cracks and achieved remarkable results [12,13]. However, the increase in fluidity of PCMs during phase change brings hidden danger to the safety of building gel materials [14,15]. Microencapsulated PCMs (MPCMs), which comprise a PCM (as the core material) encapsulated in a dense polymer or insoluble precipitation, are typically applied to address this problem, not only to retain a stable PCM shape during phase transitions but also to avoid incompatibility between the PCM and cement gel materials [16–18]. Usually, the phase change materials in MPCMs can be divided into two categories: inorganic hydrates and organic matter (such as alkanes, fatty alcohols, and fatty acids). Due to the characteristics of hydrated salt removing bound water step by step, when it is used as a PCM, it often has a wide temperature control range. For example, disodium hydrogen phosphate dodecahydrate can store heat in the temperature range of 25–75 °C [19]. In the process of offshore cementing, the cement slurry will experience the continuous changing temperature environment of 'seawater—GHBS—deep formation—GHBS', so the cement slurry additive is required to have excellent thermal stability. However, the well-known disadvantage of hydrate salt is that it is easy to precipitate and undercooled, which will damage its heat storage density in complex temperature environments. [20–22]. Compared with hydrated salts, organic PCMs exhibit better thermal stability [23–26]. Sari et al. [27] prepared MPCM by using heptadecane as a PCM, and tested its thermal stability. The results show that the latent heat of the prepared MPCM decreases by only 3.8% after 1000 phase transformations. Yu et al. [28] have also prepared PCMs with good thermal stability using octadecane as a PCM. However, their studies have pointed out that low thermal conductivity of organic compounds may reduce the sensitivity of MPCM to temperature. Furthermore, Liu et al. [29] microencapsulated paraffin with calcium carbonate as a shell to improve the temperature sensitivity of PCM by increasing the heating area of paraffin and the high thermal conductivity of calcium carbonate. Moreover, they tried to use MPCM to reduce the temperature rise rate of oil well cement slurry used in GHBS cementing. The results show that the hydration heat of cement slurry with 12 wt % MPCM decreases by 45.67%. To summarize, it seems feasible to introduce MPCMs into low-heat cement slurry as a more efficient heat-control additive, which might effectively avoid hydrate decomposition caused by high exothermic rates of cement slurry. In addition, to our knowledge, no clear research method has directly verified that the addition of MPCM will inhibit the damage of hydration heat of cement slurry on GHBS, which makes it obviously difficult to confirm the feasibility of MPCM application in this field.

In this study, a heat-control microcapsule (MPCM-1) with a mixed alkane core and polymethyl methacrylate (PMMA) shell was developed via in situ polymerization. The structures diagram of MPCM-1 is shown in Figure 1.

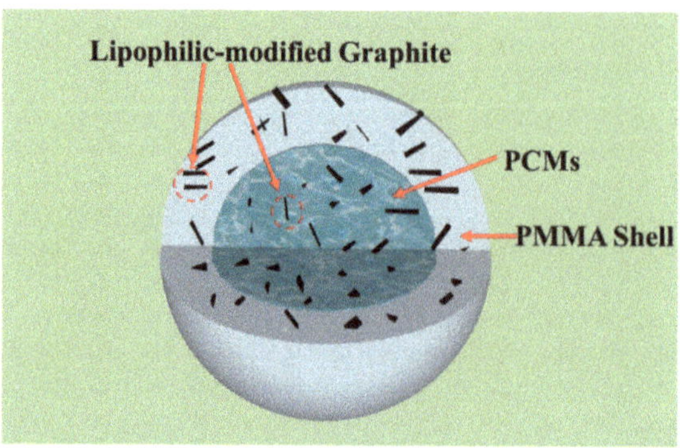

Figure 1. Structure diagram of MPCM-1.

Owing to the low thermal conductivity of alkanes and PMMA, lipophilic-modified graphite was introduced into MPCM-1 to improve its temperature sensitivity. Furthermore, the prepared MPCM-1 and other low-heat additive materials (e.g., fly ash, slag, and as-purchased MPCM) were added into the oil well cement slurry system to investigate their effects upon hydration temperature increase rates and maximum hydration heat. Based on the characteristic that gas channeling occurs easily during cementing in GHBS, the compressive strength and pore distribution of cement stone are studied, which are rarely focused on by other studies [29–31]. Finally, one of the innovations of this work is that a device to simulate the cementing process was designed, to intuitively evaluate the impact of low-heat cement slurry hydration upon the stability of hydrate-bearing sediments. This study verifies the feasibility of low-heat cement slurries containing phase-change microcapsules for safe cementing in hydrate-bearing sediments, which represents an important step forward in this area of research.

2. Experimental Section

2.1. Materials

N-tetradecane [analytical reagent grade (AR), 99%], n-hexadecane (AR, 98%), benzoyl peroxide (AR, 99%), polyvinyl alcohol (AR, 99%), and stearic acid (AR, 98%) were purchased from Aladdin Reagents (Shanghai, China) Co., Ltd. Aluminum chloride hexahydrate (AR, 97%) and absolute alcohol (AR, 99%) were purchased from Shanghai Meirel Chemical Technology Co., Ltd. (Shanghai, China). Microcrystalline graphite was purchased from Dingsheng Xin Chemical Co., Ltd. (Tianjin, China) Deionized water was self-made.

Class G oil-well cement and additives were provided by CNOOC (Sanhe, China) Co. Ltd. Fly ash and slag were purchased from Shanghai Weishen New Building Materials Co., Ltd. (Shanghai, China). The chemical compositions of Class G oil-well cement, fly ash, and slag are listed in Table 1.

Table 1. Chemical composition of cement, fly ash, and slag (%).

Sample	CaO	SiO_2	Al_2O_3	SO_3	K_2O	Na_2O	Fe_2O_3	TiO_2	CeO_2	LOI
Cement	64.10	20.10	3.67	3.54	0.62	0.29	4.89	0.31	-	3.10
Fly ash	6.69	52.45	29.35	0.74	1.07	0.89	5.93	1.24	0.20	4.41
Slag	34.66	30.04	13.54	0.83	0.31	0.33	0.31	0.57	0.14	0.39

In addition, several typical particle sizes are listed in Table 2.

Table 2. Particles size distribution of cement, fly ash, and slag (μm).

Sample	D_{10}	D_{50}	D_{90}
Cement	10.082	13.675	45.359
Fly ash	2.846	12.830	52.834
Slag	1.038	4.835	14.216

2.2. *Synthesis of MPCM-1*

2.2.1. Lipophilic Modification of Microcrystalline Graphite

First, 0.2 g stearic acid was dissolved in 20 mL absolute alcohol, and 0.22 g aluminum chloride hexahydrate was dissolved in 100 mL deionized water. The ethanol solution was then mixed with aqueous solution to form an emulsion, and 4 g microcrystalline graphite was added. The mixture was placed in a water bath with a temperature of 60 °C and stirred continuously at a rate of 100 r/min for 1.5 h. Finally, the modified microcrystalline graphite (with a rich lipophilic group on the surface) was obtained via suction filtration and drying. During this period, the filter material was not washed in order to prevent removal of the polymer material and suppression of its physical coating effect on the microcrystalline graphite during drying.

2.2.2. Synthesis of MPCM-1

First, 3 g polyvinyl alcohol was dissolved in 100 mL deionized water at 90 °C. Subsequently, 0.2 g benzoyl peroxide was dissolved in 10 mL methyl methacrylate (the inhibitor having been washed away by 1 wt% NaOH solution), and 0–0.8 g of lipophilic modified graphite (MG) was added and stirred evenly to produce a wall material solution (WM). Next, n-tetradecane and n-hexadecane were mixed and fused in a mass ratio of 1:4 to form 10 g of the phase change core material (i.e., the PCM). To prevent early polymerization of the methyl methacrylate (caused by high temperature or low temperature-induced solidification of PCM), the polyvinyl alcohol solution (100 mL), WM mixture (10 mL), and PCM solution (10 g) were kept at 50 °C for 10 min before being added together in a three-pot flask (500 mL). The water bath environment was then adjusted to 60 °C for 15 min (prepolymer process) and 80 °C for 1.5 h; the stirring speed was kept at 200 r/min. Finally, MPCM-1 was obtained by suction filtration, washing, and drying.

2.3. Characterization

The functional groups of the PCM, WM, and MPCM-1 were analyzed via FT-IR spectroscopy (Perkin Elmer Frontier, Perkin Elmer Co., Ltd., Waltham, USA) using the standard KBr disk method (scanning number: 32; optical range: 400–4000 cm^{-1}). The morphology of MPCM-1 without MG (MPCM-2) and MPCM-1 was observed using Zeiss stereoscopic optical microscopy (STEMI 508, Zeiss, Oberkochen, Germany). The NanoVoxel-3502E X-ray three-dimensional scanning system (micro-CT) (Sanying Precision Instrument Co., LT, Tianjin, China) was used to scan MPCM-1 (particle size > 200 mesh) placed in a cylindrical plastic mold. The core-wall structure of MPCM-1 (particle size > 200 mesh) was then assessed by analyzing the reconstructed MPCM-1 data using Voxel Studio Recon software (Sanying Precision Instrument Co., Ltd., Tianjin, China). A schematic diagram and digital photo of the micro-CT are shown in Figure 2.

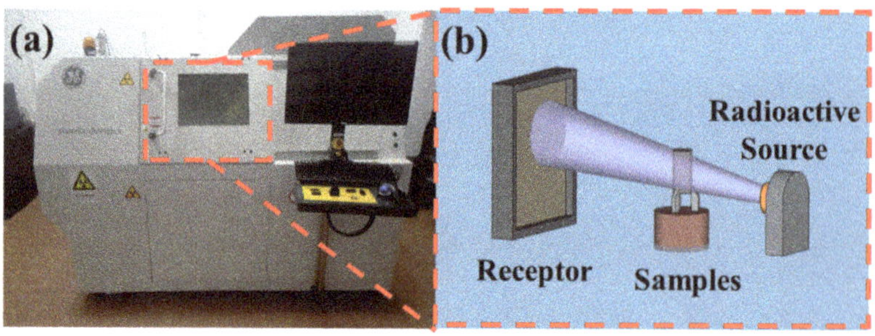

Figure 2. (**a**) Entity devices and (**b**) schematic diagram of micro-CT.

The microstructure of MPCM-1 (particle size < 200 mesh) and the element distribution upon its surface were observed using SEM (Carl Zeiss Sigma 300, Zeiss, Oberkochen, Germany) and energy-dispersive X-ray spectrometry (Smartedx, Zeiss, Oberkochen, Germany) after the MPCM-1 had been dispersed on a conductive adhesive and sprayed with gold powder. Using the transient plane heat source method, the thermal conductivity of WM (mass fractions: 0%, 2%, 4%, 6%, and 8% MG) fabricated into $5 \times 5 \times 0.05$ cm^3 films was analyzed by Hot Disk (TPS 2500S, Kegonas Instrument Trading Co., Ltd., Shanghai, China) at 10 °C. Here, three samples in each group were tested; the average of each group was then taken as the final test result. The phase change performances of MPCM-1, PCM, and MPCM-2 (i.e., the phase change temperature and latent heat) were analyzed using DSC (DSC3, ETTLER TOLEDO, Zurich, Switzerland) at a heating/cooling rate of 5 °C/min in the range 0–60 °C under a nitrogen environment. A thermal analyzer (TGA55, ETTLER TOLEDO, Zurich, Switzerland) was used to analyze the thermal stability of MPCM-1 and PCM at a heating rate of 10 °C/min in the range 20–800 °C under a nitrogen environment, after which the samples were maintained at 800 °C for 30 min. The MPCM-1 (5 g) and PCMs (5 g) were wrapped with dust-free blotting paper and placed in an electric blast drying oven. The samples were heated and cooled in 20 cycles over a temperature range of 0–60 °C. The samples were then weighed, and mass loss rate was calculated to estimate the thermal reliability of MPCM-1.

2.4. Preparation of Cement Slurry

First, the low-heat and low-density cement slurry applied to GHBS was prepared using the formula provided by the CNOOC Zhanjiang Branch (Sanhe, China); this was taken as the control group and denoted as "CS." Second, 10 wt% slag and 10 wt% fly ash were introduced into CS; the resulting samples were denoted as $CSS_{10\%}$ and $CSF_{10\%}$, respectively. Next, the 5 wt% phase change microcapsules (MPCM-3) purchased from Shanghai Feikang Products Factory were added into CS, and the result was denoted as $CSM_{5\%}$. Finally, 3.0 wt%, 5.0 wt%, and 7.0 wt% MPCM-1 were added into CS, and the resulting samples were denoted as $CSM1_{3\%}$, $CSM1_{5\%}$, and $CSM1_{7\%}$, respectively. The function of the additives and the design of the cement slurry system are shown in Tables 3 and 4.

Table 3. Functions of cement slurry additives.

Samples	X60L	F45L	G86L	BT5	P60
Function	Inhibit foam production	Improve rheological properties	Reduce water loss	Increase early strength	Reduce system density

Table 4. Mixture composition of cement slurry system (g).

Samples	Cement	Slag	Fly Ash	MPCM-1	MPCM	Water	X60L	F45L	G86L	BT5	P60
CS	600	-	-	-	-	1000	6	9	18	30	102
$CSS_{10\%}$	540	60	-	-	-	1000	6	9	18	30	102
$CSF_{10\%}$	540	-	60	-	-	1000	6	9	18	30	102
$CSM_{5\%}$	570	-	-	-	30	1000	6	9	18	30	102
$CSM1_{3\%}$	582	-	-	18	-	1000	6	9	18	30	102
$CSM1_{5\%}$	570	-	-	30	-	1000	6	9	18	30	102
$CSM1_{7\%}$	558	-	-	42	-	1000	6	9	18	30	102

The additives in all cement slurry samples were adjusted according to GB/T 19139-2012, and the density and rheological properties of the cement slurry are shown in Table 5.

Table 5. Physical properties of samples.

Samples	Density/g·cm^{-3}	Rheological Index	Consistency Coefficient (Pa·S^n)
CS	1.50	0.630	1.246
$CSS_{10\%}$	1.57	0.578	1.754
$CSF_{10\%}$	1.58	0.612	2.339
$CSM_{5\%}$	1.46	0.554	3.711
$CSM1_{3\%}$	1.48	0.598	1.874
$CSM1_{5\%}$	1.46	0.576	2.069
$CSM1_{7\%}$	1.45	0.561	2.154

2.5. Temperature and Hydration Heat of Cement Slurry

First, the cement slurry samples (shown in Table 4) were prepared and stirred at a high speed of 4000 r/min for 20 s and 12,000 r/min for 45 s. Before reaching the GHBS, the cement slurry passes through a low temperature (4 °C) region near the mud line. Hence, cement slurry samples were placed in a 4 °C curing tank for 15 min and transferred to a self-made semi-adiabatic calorimeter that met the GB/T12959-2008 standard and had a constant temperature of 8 °C. Fran bottles filled with the cement slurry were sealed with plasticine at the top, to reduce the effect of air temperature and to prevent water from entering the cement paste. Finally, the changes in cement slurry temperature and hydration heat release over 48 h were recorded.

2.6. Compressive Strength of Cement Stone

Following the GB/T 17671-2021 standard, the cement slurry samples were placed into the standard mold (40 × 40 × 40 mm^3) and cured at a constant temperature in a humidity curing box (humidity: 90%) and at temperatures of 8 °C, 10 °C, and 15 °C for 24 h. The average compressive strengths of the three samples in each group were then measured using a compression and flexural testing machine (ZCYA-W300C, Star Fire Testing Machine Co., Ltd., Jinan, China).

2.7. Spatial Distribution of Pores and MPCM-1 in Cement Stone

First, the cement slurry samples were placed into a standard mold (20 × 20 × 20 mm^3) and cured in a constant temperature and humidity curing box (humidity: 90%; temperature: 8 °C) for 24 h. Micro-CT was then used to scan the spatial distribution of pores and MPCM-1 in cement stone. The dates were recorded after reconstruction using VG Studio Max software (scanning accuracy: 15.63 μm; voltage: 120 kV; current: 110 μA).

2.8. Simulation of GHBS Damage

The self-made device used to simulate the cementing process in the GHBS is shown in Figure 3. First, the simulated GHBS was formed by injecting methane gas into a formation cavity preloaded with 30 g of tetrahydrofuran, 770 mL of deionized water, and 8 kg of sand particles under increasing pressure and decreasing temperature. The cement slurry samples (800 mL) were injected into the slurry cavity. The damage of the simulated GHBS during cement slurry hydration was estimated using temperature and pressure sensors.

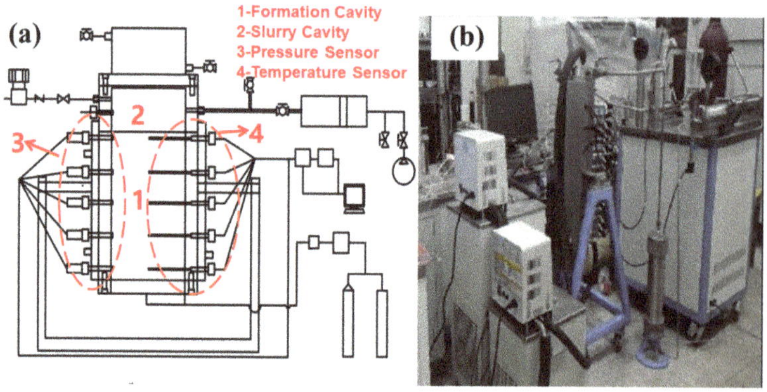

Figure 3. (**a**) Structure chart and (**b**) digital photo of device used to simulate cementing in GHBS.

3. Results and Discussion

3.1. Infrared Spectrum Analysis of MPCM-1

The infrared spectra of the PCM, WM, and MPCM-1 are shown in Figure 4.

For the FT-IR spectra of PCM, the absorption peak at 1464 cm^{-1} corresponds to the antisymmetric vibration of $-CH_3$, and the absorption peak at 720 cm^{-1} corresponds to the rocking vibration peak of $-(CH_2)_4$. In the infrared spectrum of WM, the strong absorption at 3429 cm^{-1} corresponds to the stretching vibration peak of −OH and −COOH dimer OH. The absorption peak at 1731 cm^{-1} corresponds to the umbrella bending vibration peak of $-CH_3$, and the absorption peak at 1270 cm^{-1} corresponds to the stretching vibration band of -O-C(O)-C [32]. Notably, the FT-IR spectrum of MPCM-1 contains all the characteristic absorption peaks of PCM and WM, and no new absorption peaks appear. Therefore, the preliminary results demonstrate that the energy storage density of the PCM has not been lost during microencapsulation, and MPCM-1 consists of PCM and WM.

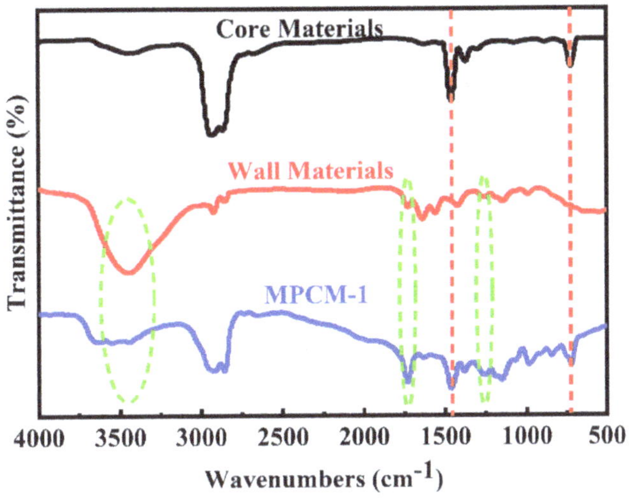

Figure 4. FTIR spectra of core materials (PCM), wall material (WM), and MPCM-1.

3.2. *Microstructural Analysis of MPCM-1*

Figure 5 shows the stereomicroscope views of MPCM-1.

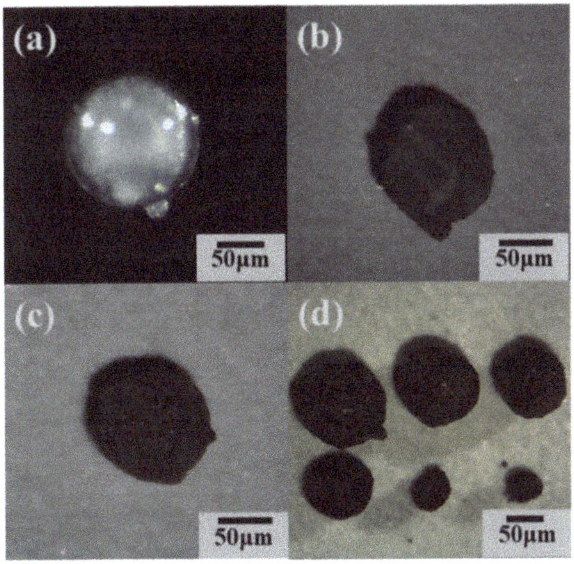

Figure 5. Stereomicroscope views of (**a**) MPCM-2, (**b**) MPCM-1 section, (**c**) MPCM-1, and (**d**) MPCM-1 of different sizes.

It can be observed from Figure 5 that the PCM was successfully encapsulated by the WM without the addition of MG. The 3D image model and slice images of MPCM-1 are presented in Figure 6. The images show that microcapsules combined with MG also exhibit an excellent core-wall structure; furthermore, the dense surface of WM helped to suppress PCM leakage during the phase transition.

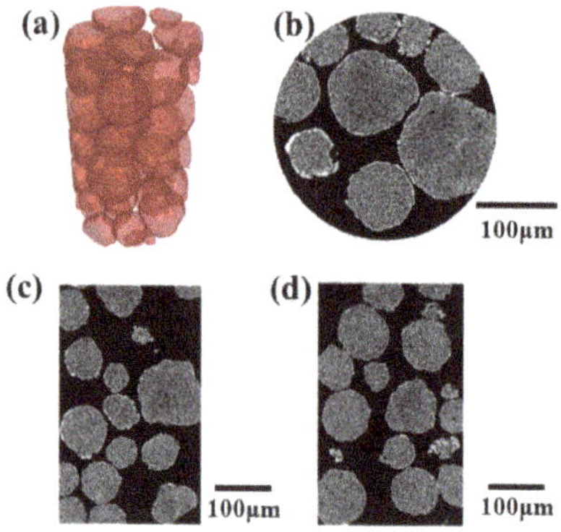

Figure 6. 3D image model and slice images of MPCM-1 (particle size > 200 mesh): (**a**) 3D image models, (**b**) XY plane slice, (**c**) YZ plane slice, and (**d**) XZ plane slice.

Besides this, the microcapsules prepared from the same batch with different particle sizes (Figure 5d) indicate that the mechanical strength of cement stone can be enhanced by varying the gradation relationship between the cement particles and MPCM-1, with suitable particle sizes isolated via screening [33–35]. The SEM micrographs of MPCM-1 (particle size < 200 mesh) are presented in Figure 7.

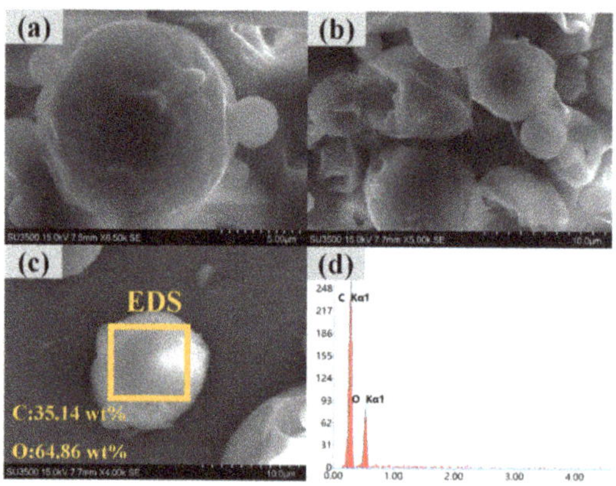

Figure 7. SEM micrographs of MPCM-1 (particle size < 200 mesh): (**a**) ×6500 and (**b**) ×5000. (**c**) EDS analysis. (**d**) Surface element content of MPCM-1.

The small-particle MPCM-1 appears easier to agglomerate but exhibits superior sphericity. The surface element content of MPCM-1 (shown in Figure 7c) demonstrates that the mass fractions of C and O are 35.14% and 64.86%, respectively; this conforms to the composition of C and O elements in the wall material (PMMA). Combined with the FT-IR analysis results, this provides strong evidence for the packaging effect of WM on PCMs, which allows them to satisfy design specifications.

3.3. Phase-Change Properties of MPCM-1

Thermal conductivity is an important factor in heat control; it affects temperature sensitivity and latent heat storage and release rates for MPCM-1 in the cement slurry. However, the thermal conductivity of organic PCMs is generally lower than that of cement slurry. In this study, the thermal conductivity of MPCM-1 was improved by adding microcrystalline graphite (which is inexpensive); furthermore, the lipophilic modification of graphite was performed to evenly disperse it in MPCM-1. Figure 8 shows the compatibility of MG with water before and after lipophilic modification.

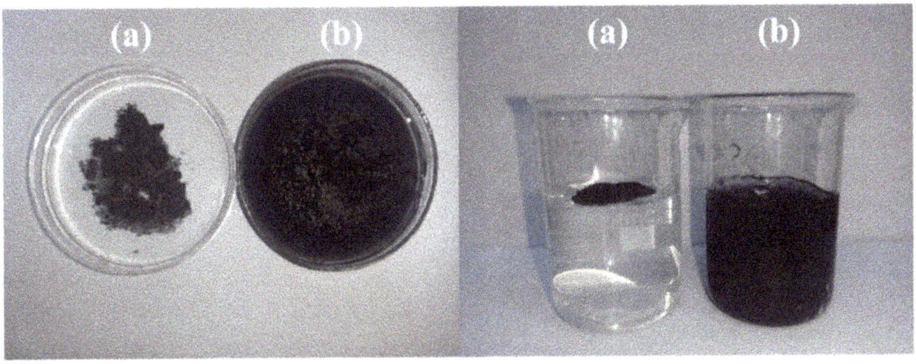

Figure 8. Comparison of hydrophobicity between (**a**) MG and (**b**) microcrystalline graphite.

The hydrophobicity of the MG increased markedly compared with that of the graphite, owing to the increase in the number of lipophilic groups on the surface. The improvement of MG lipophilicity was achieved via two stages: chemical modification and physical coating. The surface of the microcrystalline graphite contained hydroxyl groups [36,37] which could be esterified with stearic acid under aluminum chloride catalysis. By ester bonding, the carbon chain of stearic acid was partially distributed on the surface of the microcrystalline graphite, and the chemical modification was completed. During the drying process for microcrystalline graphite, the polymer (after esterification reaction) and stearic acid (without chemical reaction) were gradually precipitated out and coated on the surface of the microcrystalline graphite to produce a physical coating modification. Furthermore, the WM (with different mass fractions of MG) pressed into the films (as shown as Figure 9), and the hydrophobicity of the graphite after modification notably increased, allowing it to be uniformly dispersed in the WM.

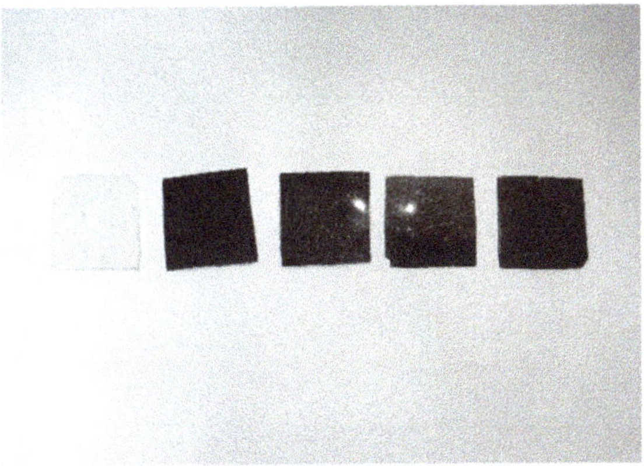

Figure 9. WM films at different MG dosages.

The thermal conductivity analysis of the WM is presented in Figure 10.

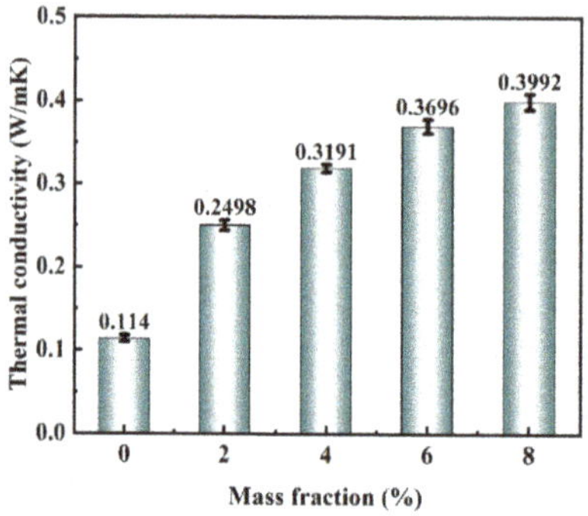

Figure 10. Thermal conductivity of WM films with modified MG.

Under the increase in MG mass fraction, the thermal conductivity of WM was considerably improved. However, MG entering the PCM reduced the PCM content per unit volume, lowering the thermal storage efficiency; therefore, a 4 wt% addition of MG was selected to optimally improve thermal conductivity.

The phase-change properties of PCM, MPCM-1, and MPCM-2 were tested, and the results are presented in Figure 11 and Table 6.

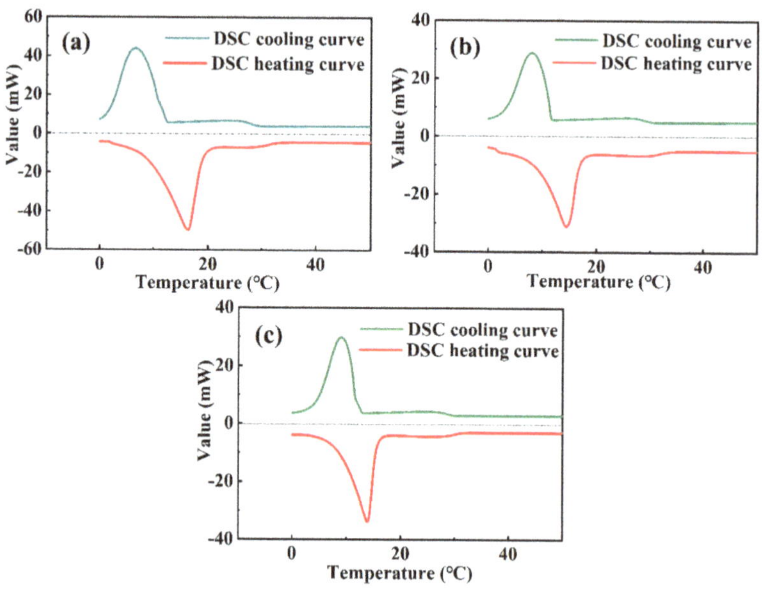

Figure 11. DSC curves of (**a**) PCM, (**b**) MPCM-1, and (**c**) MPCM-2.

Table 6. Phase-change behaviors.

Sample	Melting Process				Encapsulation Efficiency (%)
	T_o (°C)	T_p (°C)	T_e (°C)	ΔH_m (Jg^{-1})	
PCMs	9.01	15.32	18.82	325.34	
MPCM-1	8.99	13.77	16.74	153.58	47.2
MPCM-2	9.48	14.07	16.99	159.89	49.1

Both PCMs and MPCM-1 had high latent heat and a wide phase transition temperature range, owing to the mixed dissolution of n-tetradecane and n-hexadecane; this indicates that hydration heat was continuously absorbed by MPCM-1 over a large temperature range during the cement slurry hydration process. The phase-change temperature range, peak temperature, and latent heat of MPCM-1 were 8.99–16.74 °C, 13.77 °C, and 153.58 Jg^{-1}, respectively. Compared to MPCM-2, it is found that the initial phase-change temperature of MPCM-1 is 0.49 °C earlier, due to the addition of modified MG, and its temperature sensitivity was significantly enhanced. Moreover, only one phase transition peak appeared in each DSC curve, owing to the eutectic effect of the PCMs composed of n-tetradecane and n-hexadecane in this composition ratio during the phase transition process [20,38]. The formula for calculating the encapsulation efficiency E_e of MPCM-1 is [7,29]:

$$E_e = \frac{\Delta H_{m(MPCM-1)}}{\Delta H_{m(PCM)}} \times 100\% \tag{1}$$

where T_o, T_p, and T_e are the initial phase-change temperature, the peak phase-change temperature, and the end phase-change temperature in the melting process, respectively; ΔH_m is the latent heat during the melting process, for which $\Delta H_{m(MPCM-1)}$ and $\Delta H_{m(PCM)}$ are the latent heat of MPCM-1 and PCM during the melting process, respectively. Using this formula, the encapsulation efficiency of MPCM-1 is determined as 47.2% which may be lower than that of some other studies (the E_e is 67.8% and 66.5%, respectively) [7,29]; however, the lower encapsulation efficiency within a reasonable range ensures the ability of MPCM to resist external loads, which facilitates the mechanical strength of cement stone incorporated with *MPCM* [39].

3.4. Thermal Reliability of MPCM-1

Taking oil and gas resources in the South China Sea as an example, GHBS is typically present in the shallow surface of the mud line, with a water depth of 300–2000 m [40,41]. Cement slurry for deep-sea GHBS undergoes the following process: "seawater → GHBS → deeper formation → GHBS", during which the ambient temperature changes continually. In addition, MPCM-1 experiences frequent temperature variations during transport and storage on offshore platforms. The coating integrity of MPCM-1 under temperature change conditions is extremely important for ensuring its heat storage efficiency and the sealing performance of cement sheaths, which may be affected by PCM leakage. Hence, the thermal reliability of MPCM-1 was analyzed via TG and thermal cycles.

TGA and DTG curves of MPCM-1 and PCMs are shown in Figure 12.

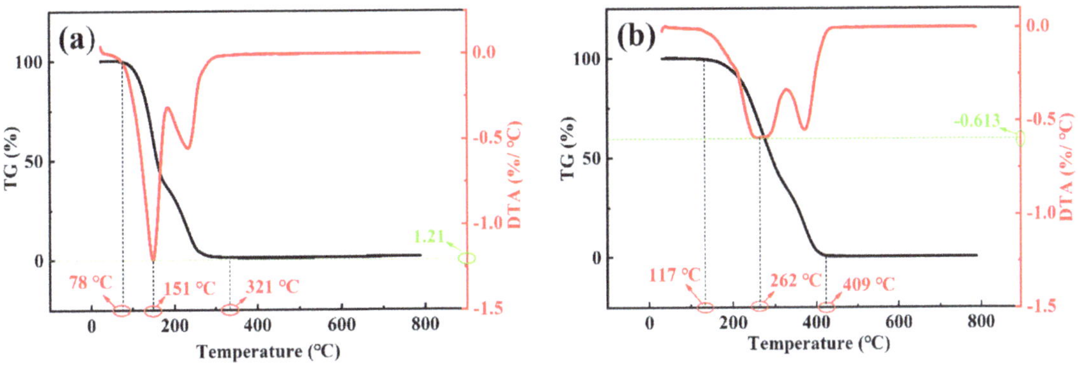

Figure 12. TGA curve and DTG curve of (**a**) PCM and (**b**) MPCM-1.

It can be observed that the initial and total mass loss temperatures of MPCM-1 were 117 °C and 409 °C, respectively; furthermore, when the temperature reached 262 °C, the mass loss rate was maximized at 0.613/°C. Compared with PCMs, the initial and maximum mass loss rates of MPCM-1 were higher and slower, respectively, because of the coating effect of the WM on the PCMs [42].

The samples of the thermal cycle experiment are shown in Figure 13, and the test results are shown in Table 7.

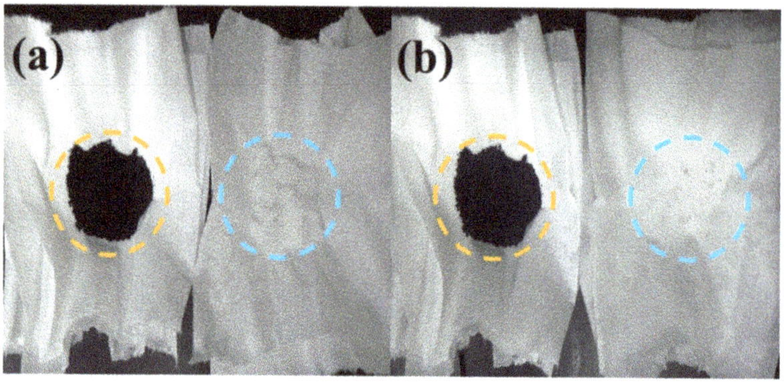

Figure 13. Digital photographs of (**a**) initial samples and (**b**) samples after thermal cycles.

Table 7. Mass loss of MPCM-1.

Sample	Initial Mass (g)	Mass after 5 Cycles (g); Mass Loss Rate (%)	Mass after 10 Cycles (g); Mass Loss Rate (%)	Mass after 15 Cycles (g), Mass Loss Rate (%)	Mass after 20 Cycles (g), Mass Loss Rate (%)
MPCM-1	5	4.5, 10	4.3, 14	4.2, 16	4.2, 16
PCM	5	0,100	0, 100	0, 100	0, 100

After five thermal cycling experiments, the mass loss rates of MPCM-1 and PCM were 10% and 100%, respectively. At this stage, the PCMs rapidly melted and leaked during phase change. Meanwhile, the quality of MPCM-1 also decreased significantly because several PCMs were not completely coated inside the microcapsule, and some were only embedded in the grooves on the WM surface. After twenty thermal cycles, the mass loss rate of MPCM-1 was 16%. The mass loss rate of MPCM-1 increased slightly because most of the remaining PCMs at this stage underwent a phase transformation in the WM, and the morphology of the PCMs remained essentially stable. These results further verify that WM can effectively prevent the leakage of PCMs and ensure their heat storage density.

3.5. Heat Control Effect of MPCM-1

Figures 14 and 15 show the temperature change and hydration heat release of each cement slurry sample, respectively, as obtained using the self-made semi-adiabatic calorimeter.

Table 8 shows the values corresponding to the curves.

Table 8. Temperature and heat of hydration data for cement slurry.

Sample	T_i (°C)	T_m (°C)	ΔT (°C)	t_m (h)	Q_{20h} (J)	Q_m (J)
CS	11.3	31.3	20	16.6	49,812	61,729
$CSS_{10\%}$	10.2	28.1	17.9	18.1	37,583	46,555
$CSF_{10\%}$	10.7	26.5	15.8	16.8	32,954	40,,384
$CSM_{5\%}$	10.2	24.2	14.0	16.6	32,608	40618
$CSM1_{3\%}$	10.2	24.6	14.4	18.9	30,535	40,463
$CSM1_{5\%}$	10.7	22.4	11.7	19.8	25,410	38,212
$CSM1_{7\%}$	10.2	21.6	11.4	22.5	19,905	36,220

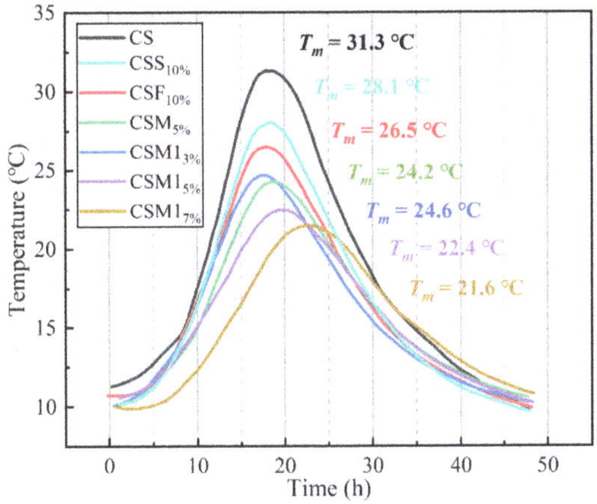

Figure 14. Hydration temperature of cement slurry samples.

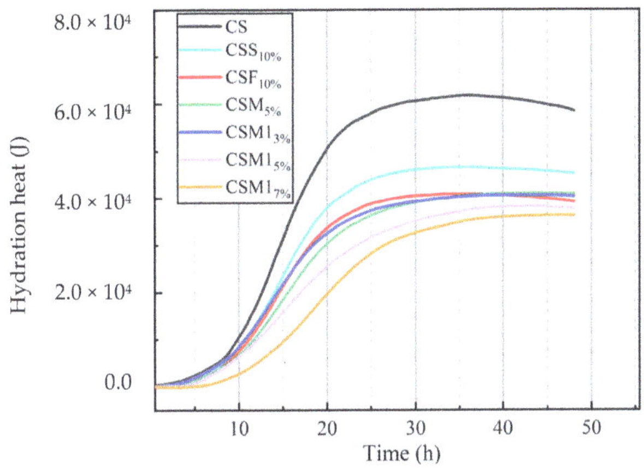

Figure 15. Hydration heat of cement slurry samples.

Compared to slag and fly ash, MPCM-1 has a more notable inhibitory effect on cement slurry temperature increase; this is reflected in the fact that the T_{max} of $CSM1_{7\%}$ was 6.5 °C and 4.9 °C lower than those of $CSS_{10\%}$ and $CSF_{10\%}$, respectively. Simultaneously, the T_{max} of $CSM1_{7\%}$ appeared later than those of $CSS_{10\%}$ and $CSF_{10\%}$. Similar results were also obtained in the research of Huo et al. [30]. They suggested that MPCM-1 not only replaced the active component of cement particles (similar to the function of slag and fly ash) but also absorbed heat released by the cement slurry during phase transformation. Compared to CSM, the maximum temperature and hydration heat of CSM1 were both lower, owing to the higher storage density of MPCM-1. In Figure 15 and Table 8, a significant positive correlation can be identified between the retarding effect of each low-heat material on the hydration heat release of the cement slurry; the effect of this on the temperature increase can also be seen. Moreover, the rate of increase in hydration heat for CSM1 was significantly faster than that for CS when the cement slurry temperature fell and approached the water bath temperature; this resulted from heat absorption during the liquid–solid phase transition of MPCM-1. Furthermore, compared with CS, the T_m and Q_m of $CSM1_{7\%}$ decreased by 9.7 °C and 25,509 J (41.3%), respectively, and T_m was extended by 5.9 h. The results show that MPCM-1 can effectively control the temperature increase and heat release rate of cement slurry, indicating that MPCM-1 has excellent prospects as a new additive for low-heat cement slurry. Huo et al. [43] conducted a similar study. The results showed that the peak temperature of cement slurry decreased by 6 °C with 7.5% MPCM (prepared from urea-formaldehyde resin coated paraffin) dosage. The reasons why the ability of MPCM in the study of Huo et al. to regulate the temperature of cement slurry was worse than that of this study, is that MPCM-1 can quickly sense the temperature change in cement due to the incorporation of high thermal conductivity materials, and that PCMs' phase transition occurs in time to complete the endothermic process [44]. Here, T_i, T_m, and ΔT represent the initial temperature, maximum temperature, and maximum temperature difference for the cement slurry, respectively; furthermore, t_m denotes the time taken for the cement slurry to reach the maximum temperature. Q_{20h} and Q_m denote the heat release in 20 h and the final heat release of the cement slurry.

3.6. Comprehensive Strength of Cement Stone

In this section, the effects of MPCM-1, low-heat inorganic materials, and as-purchased microcapsules upon the comprehensive properties of cement stone are studied. As GHBS is soft and the pressure window between fracture and formation-pore pressure gradients is narrow, a low-density cement slurry system is usually used to cement the well to prevent loss of circulation [45–47]. As shown in Table 5, the reduction effect of MPCM-1 upon the initial density of cement slurry was more notable than other low-heat materials. Thus, the MPCM-1 cementing system conforms to this design philosophy [48]. Compared with CS, the fluidity of CSM1 decreased slightly within the allowable range. Wang et al. [39] found that the reason for the decrease in fluidity of cement slurry is that the particle size of MPCM is generally larger than that of cement particles, which enhances the friction between particles in cement slurry. The compressive strength of cement increased under the increase in MPCM-1 dosages within a certain range (see Figure 16).

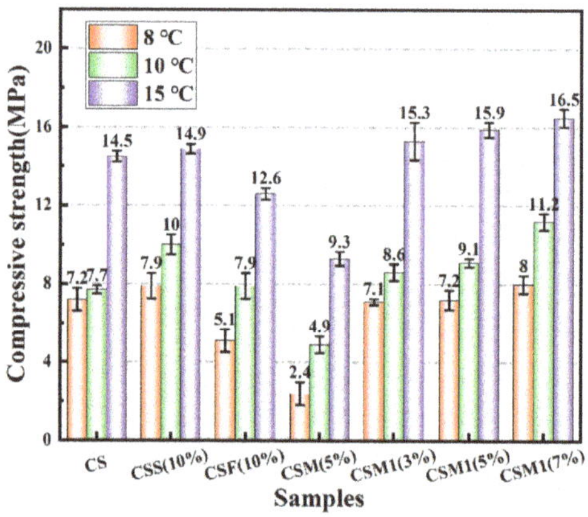

Figure 16. Compressive strength of cement stone after 24 h at 8 °C, 10 °C, and 15 °C.

Compared with CS, the compressive strength of $CSM1_{3\%}$, $CSM1_{5\%}$, and $CSM1_{7\%}$ cured at 15 °C for 24 h increased by 5.5%, 9.7%, and 13.8%, respectively. This occurs because PMMA (the shell of MPCM-1) is rigid; thus, it plays the role of an aggregate, which allows the cement stone of CSM1 to withstand higher external loads [49,50]. However, the results of Aguayo et al. [51] showed that a 5% dosage of microcapsules increased the number of pores in the cement stone, reducing its compressive strength by 12.5%. Ghods et al. [52,53] classified the pores inside the cement stone in terms of size and shape; they suggested that the macropores were the main factor affecting its compressive strength. Therefore, the reduction of large volume pores in cement stone may also represent an important factor when improving the compressive strength of CSM1. The spatial distribution of pores in the cement stone is studied in Section 3.7.

3.7. Distribution Analysis of MPCM-1 and Pores in Cement Stone

Figures 17 and 18 and Table 9 show the distribution of pores and MPCM-1 in cement stone, as obtained via micro-CT analysis.

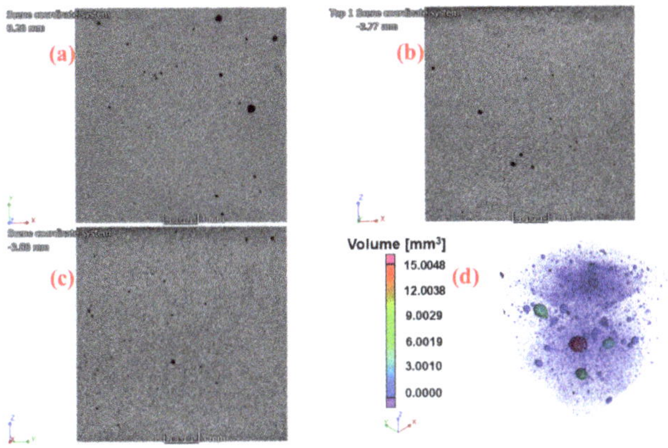

Figure 17. Slice images and 3D image model of the pores in CS.

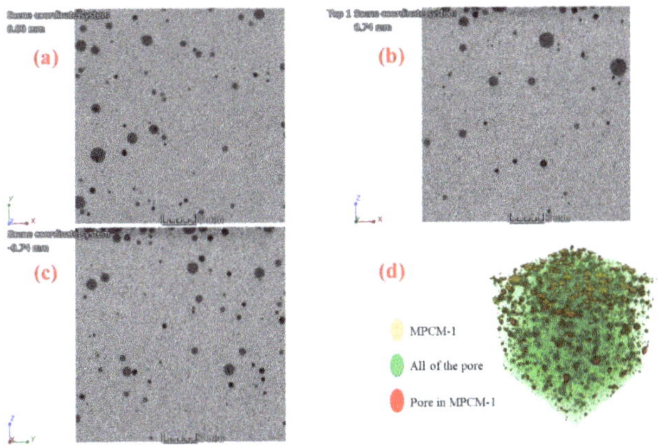

Figure 18. Slice images and 3D image model of the pores and MPCM-1 in $CSM1_{5\%}$.

Table 9. Total volume of MPCM-1, all pores, and the pores in MPCM-1.

Sample	CS	$CSM1_{5\%}$			
	Pore	MPCM-1	All Pores	Pores in MPCM-1	Pores in Cement
Total volume (mm^3)	62.4044	269.4997	98.7520	36.2646	62.4874

MPCM-1 is evenly distributed in the cement stone without agglomeration, which is amenable to the uniform heat control effect of MPCM-1 upon the cement slurry hydration process. Moreover, no clear arc-shaped pores (resulting from the incompatibility of MPCM-1 with cement slurry) were found around MPCM-1, and the porosity of cement in CSM1 was approximately equal to that of CS. Further, through reconstruction analyses, the volumes of MPCM-1 and its internal porosities were found to be 269.4997 mm^3 and 36.2646 mm^3, respectively. Calculated to be 13.4%, the porosity in MPCM-1 was already high, though it was lower than the true value, owing to accuracy limitations of the micro-CT scan.

Therefore, reducing the internal porosity of MPCM-1 should be considered in subsequent studies to improve its heat storage density and mechanical strength.

The voxel is the smallest volume unit that can be represented by a pixel after micro-CT reconstruction. The cumulative probability of voxel numbers in the pores inside the cement stone of CS and $CSM1_{5\%}$ is shown in Figure 19.

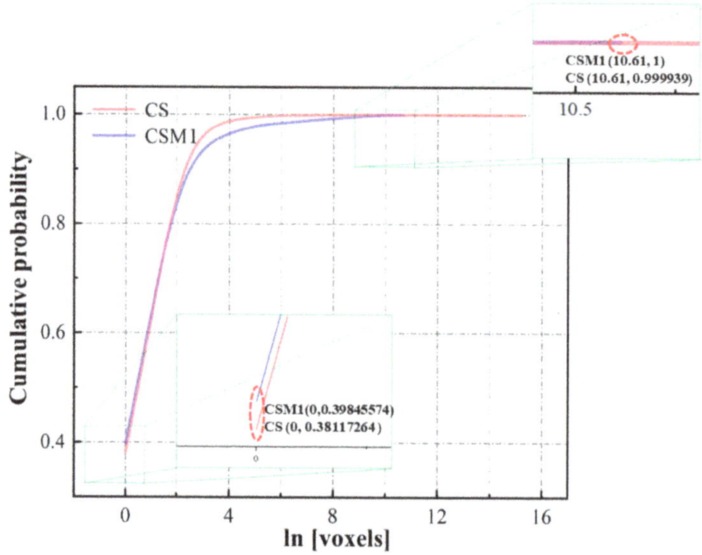

Figure 19. Cumulative probability of pore voxels in CS and $CSM1_{5\%}$ cement.

The voxel of the largest pore in $CSM1_{5\%}$ cement was 10.61 (i.e., 0.155 mm^3). Furthermore, it was calculated that 27 pores with a volume exceeding 0.155 mm^3 were present in CS, among which the maximum pore volume was 15.005 mm^3. Compared with CS, more small volume pores and fewer large volume pores were present in the cement stone of $CSM1_{5\%}$. Two factors may contribute to the reduction in the number of large volume pores: (1) the MPCM-1 increases the number of nucleation sites for the precipitation hydration products [39], and (2) the MPCM-1 (with a small particle size) plays a role in filling the pores. Therefore, MPCM-1 can (similar to defoamers) reduce the occurrence of large volume pores in the cement stone; this improves the mechanical strength of the cement stone [54,55]. Here, the side length of the unit voxel is 15.63 μm. Furthermore, micro-CT analysis at this precision shows that 447,743 and 501,657 pores are present in the CS and CSM1 cement stones, respectively.

3.8. Simulation of GHBS Damage

To further evaluate the protective function of a low-heat cement slurry system with MPCM-1 (to function as the key temperature control for GHBS stability), a device simulating the cementing operation of GHBS was designed and prepared (as shown in Figure 3). As the change in pressure during hydrate decomposition was considerably more drastic than the temperature change, the large pressure increase was selected as the standard by which to quantify the damage to GHBS stability.

The influence of the CS hydration heat on GHBS was observed, as shown in Figure 20.

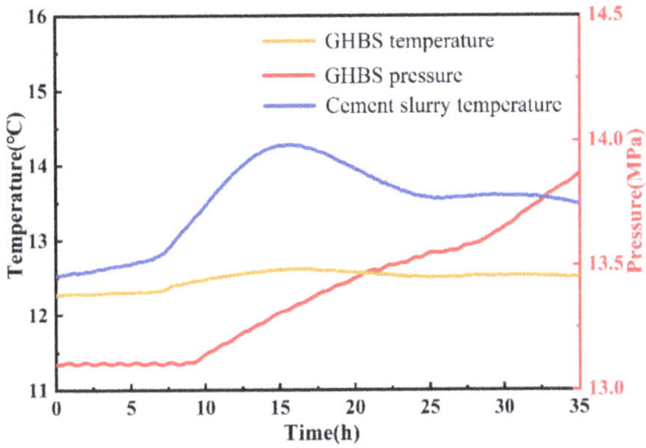

Figure 20. Damage of GHBS caused by hydration heat of CS.

In the first 7.5 h, owing to the slow increase of CS temperature in the initial hydration reaction [56], GHBS did not decompose under the thermal interaction between CS and GHBS. During this time, the temperature and pressure of GHBS remained essentially unchanged. Subsequently, the hydration process of CS entered an accelerated period [56], and the temperature increase in CS became more rapid. Owing to the enhanced heat exchange between CS and GHBS, the phase equilibrium condition for GHBS was destroyed, and the hydrate began to decompose. GHBS pressure increased significantly at 9.5 h, which continued until the end of the experiment, indicating that the stability of GHBS was completely damaged by the hydration heat release of CS.

As shown in Figure 21, under the hydration exothermic process of $CSM1_{5\%}$, the GHBS temperature and pressure began to increase significantly after 7 h via heat exchange between the CS and GHBS.

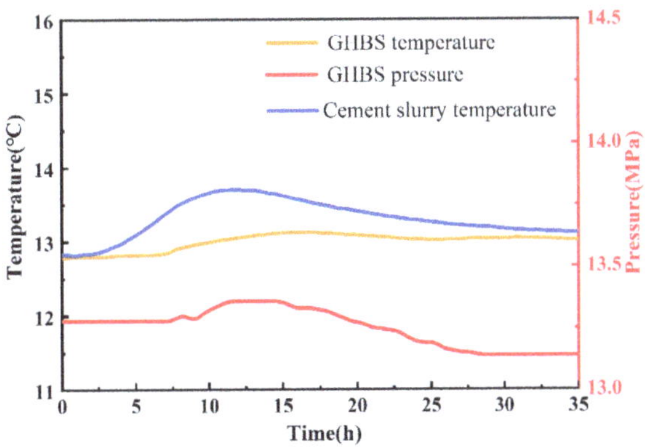

Figure 21. Damage of GHBS caused by hydration heat of $CSM1_{5\%}$.

When GHBS pressure reached the peak at 11.5 h, the ambient temperature and pressure were restored to the phase equilibrium via the pressure increase and temperature decrease caused by hydrate dissociation, after which the hydrate stopped decomposing. Meanwhile, the hydration exothermic rate of $CSM1_{5\%}$ gradually began to fall below the cooling rate, and the temperature of the cement slurry dropped. From 14.5 h, GHBS pressure began to decline until the GHBS reached a stable state at 28 h. At this stage, the GHBS pressure drop was attributable to re-formation of the lattice, in which the cavities that were trapped previously decomposed gas molecules and formed hydrate crystals. Compared to CS, the overall temperature increase rate of $CSM1_{5\%}$ was slower, and the temperature increased less, which was conducive to the phase equilibrium recovery within the GHBS [2]. These results indicate that the addition of MPCM-1 can successfully reduce the damage to GHBS during heat release in cement slurry.

4. Conclusions

In this work, a new low-heat cement slurry system incorporating prepared phase change microcapsules (MPCM-1) was developed for cementing in natural gas hydrate-bearing sediment. The following conclusions can be drawn:

(1) The addition of lipophilic-modified graphite reduced the initial phase transition temperature of MPCM-1 by 0.49 °C, indicating that the sensitivity of MPCM-1 to temperature change was improved.
(2) MPCM-1 was superior to fly ash and slag for temperature regulation in the cement slurry hydration process.
(3) The addition of MPCM-1 reduced the number of large volume pores in the cement stone, which helped to improve the compressive strength of the cement stone.
(4) The self-made cementing simulation device test results show that the addition of MPCM-1 inhibits the temperature increase rate of cement slurry, significantly reducing the damage to natural gas hydrate-bearing sediment during cementing. This also intuitively confirms the feasibility of low-heat cement slurry systems with MPCM-1 (a key objective in the field).

This work not only provides a comprehensive and intuitive proof of the feasibility of the application of MPCM-doped low-temperature cement slurry in hydrate sediment cementing, but also provides a new idea for inhibiting the formation of temperature cracks in mass concrete in civil engineering.

Author Contributions: Conceptualization, Writing—Original draft preparation, Methodology: G.Y. Project administration: T.L. Project administration: H.Z. Validation: Z.Z. Visualization: Y.F. Conceptualization, Data Curation: E.L. Writing—Review and Editing: V.M. All authors have read and agreed to the published version of the manuscript.

Funding: This work was financed by the National Key Research and Development Program of China [Grant No. 2016YFE0204300] and the National Natural Science Foundation of China [Grant No. 42072343].

Informed Consent Statement: Not applicable.

Data Availability Statement: The study did not report any data.

Conflicts of Interest: The authors declare that they have no known competing financial interests or personal relationships that could have appeared to influence the work reported in this paper.

Abbreviations

GHBS	Natural gas hydrate-bearing sediment
MPCM	Microencapsulated phase-change materials
MG	Lipophilic-modified graphite
WM	Wall materials
C-S-H	Calcium silicate hydrate
CS	Control group cement slurry
$CSF_{10\%}$	CS incorporated with 10 wt% fly ash
$CSM1_{3\%}$	CS incorporated with 3 wt% MPCM-1
$CSM1_{7\%}$	CS incorporated with 7 wt% MPCM-1
T_o	The initial phase-change temperature, °C
T_e	The end phase-change temperature, °C
$\Delta H_{m(PCM)}$	The latent heat of PCM, J/g
T_i	The initial temperature of cement slurry, °C

ΔT	The maximum temperature difference of cement slurry, °C
Q_m	The final heat release of the cement slurry, J
PCMs	Phase-change materials
MPCM-1	Prepared heat-control microcapsule
MPCM-2	Prepared MG-free MPCM-1
MPCM-3	Phase-change microcapsule purchased from Shanghai Feikang Products Factory
PMMA	Polymethyl methacrylate
$CSS_{10\%}$	CS incorporated with 10 wt% slag
$CSM_{5\%}$	CS incorporated with 5 wt% MPCM-3
$CSM1_{5\%}$	CS incorporated with 5 wt% MPCM-1
D_n	Particle size when cumulative particle size distribution percentage reaches n%, mm
T_p	The peak phase-change temperature, °C
$\Delta H_{m(MPCM\text{-}1)}$	The latent heat of MPCM-1, J/g
E_e	The encapsulation efficiency of MPCM-1,%
T_m	The maximum temperature of cement slurry, °C
Q_{20h}	The heat release in 20 h of the cement slurry, J
t_m	Time when cement slurry reaches maximum temperature, h

References

1. Wang, X.; Sun, B.; Wang, Z.; Gao, Y.; Li, H. Coupled heat and mass transfer model of gas migration during well cementing through a hydrate layer in deep-water regions. *Appl. Therm. Eng.* **2019**, *163*, 114383. [CrossRef]
2. Liang, H.; Duan, Y.; Pei, J.; Wei, N. Study on Hydrate Phase Equilibrium Diagram of Methane Containing System Based on Thermodynamic Model. *Front. Energy Res.* **2021**, *9*, 743296. [CrossRef]
3. Zheng, M.; Liu, T.; Gao, Z.; Wu, J.; Jiang, G.; Li, Q.; Li, Z.; Xie, L. Simulation of natural gas hydrate formation skeleton with the mathematical model for the calculation of macro-micro parameters. *J. Pet. Sci. Eng.* **2019**, *178*, 429–438. [CrossRef]
4. Wang, X.; Sun, B.; Gao, Y.; Wang, Z.; Li, H.; Chen, Y. Numerical simulation of the stability of hydrate layer during well cementing in deep-water region. *J. Pet. Sci. Eng.* **2019**, *176*, 893–905. [CrossRef]
5. Sun, Y.; Wang, Y.; Guo, W.; Jia, R.; Chen, G.; Zhang, P. Hole-bottom freezing technique based on phase change heat transfer for gas-hydrates sampling: Efficiency optimization of refrigeration change of phase. *Appl. Therm. Eng.* **2018**, *130*, 722–734. [CrossRef]
6. Cohen, E.; Klar, A.; Yamamoto, K. Micromechanical Investigation of Stress Relaxation in Gas Hydrate-Bearing Sediments Due to Sand Production. *Energies* **2019**, *12*, 2131. [CrossRef]
7. Huo, J.-H.; Peng, Z.-G.; Feng, Q.; Zheng, Y.; Liu, X. Controlling the heat evaluation of cement slurry system used in natural gas hydrate layer by micro-encapsulated phase change materials. *Sol. Energy* **2018**, *169*, 84–93. [CrossRef]
8. Huo, J.-H.; Peng, Z.-G.; Ye, Z.-B.; Zuo, G.-L. Preparation, characterization and investigation of low hydration heat cement slurry system used in natural gas hydrate formation. *J. Pet. Sci. Eng.* **2018**, *170*, 81–88. [CrossRef]
9. Yang, G.; Liu, T.; Aleksandravih, B.P.; Wang, Y.; Feng, Y.; Wen, D.; Fang, C. Temperature regulation effect of low melting point phase change microcapsules for cement slurry in nature gas hydrate-bearing sediments. *Energy* **2022**, *253*, 124115. [CrossRef]
10. Yang, Y.; Wu, W.; Fu, S.; Zhang, H. Study of a novel ceramsite-based shape-stabilized composite phase change material (PCM) for energy conservation in buildings. *Constr. Build. Mater.* **2020**, *246*, 118479. [CrossRef]
11. Amirahmad, A.; Maglad, A.M.; Mustafa, J.; Cheraghian, G. Loading PCM Into Buildings Envelope to Decrease Heat Gain-Performing Transient Thermal Analysis on Nanofluid Filled Solar System. *Front. Energy Res.* **2021**, *9*, 727011. [CrossRef]
12. Costa, J.A.C.; Martinelli, A.E.; Nascimento, R.M.D.; Mendes, A. Microstructural design and thermal characterization of composite diatomite-vermiculite paraffin-based form-stable PCM for cementitious mortars. *Constr. Build. Mater.* **2020**, *232*, 117167. [CrossRef]
13. Mahdaoui, M.; Hamdaoui, S.; Msaad, A.A.; Kousksou, T.; El Rhafiki, T.; Jamil, A.; Ahachad, M. Building bricks with phase change material (PCM): Thermal performances. *Constr. Build. Mater.* **2020**, *269*, 121315. [CrossRef]
14. Coppola, L.; Coffetti, D.; Lorenzi, S. Cement-Based Renders Manufactured with Phase-Change Materials: Applications and Feasibility. *Adv. Mater. Sci. Eng.* **2016**, *2016*, 7254823. [CrossRef]
15. Memon, S.; Lo, T.Y.; Shi, X.; Barbhuiya, S.; Cui, H. Preparation, characterization and thermal properties of Lauryl alcohol/Kaolin as novel form-stable composite phase change material for thermal energy storage in buildings. *Appl. Therm. Eng.* **2013**, *59*, 336–347. [CrossRef]
16. Wang, R.; Kang, Y.; Lei, T.; Li, S.; Zhou, Z.; Xiao, Y. Microcapsules composed of stearic acid core and polyethylene glycol-based shell as a microcapsule phase change material. *Int. J. Energy Res.* **2021**, *45*, 9677–9684. [CrossRef]
17. Erdem, S.; Gürbüz, E. Influence of microencapsulated phase change materials on the flexural behavior and micromechanical impact damage of hybrid fibre reinforced engineered cementitious composites. *Compos. Part B Eng.* **2019**, *166*, 633–644. [CrossRef]

18. Hu, Y.; Guo, R.; Heiselberg, P.K.; Johra, H. Modeling PCM Phase Change Temperature and Hysteresis in Ventilation Cooling and Heating Applications. *Energies* **2020**, *13*, 6455. [CrossRef]
19. Huang, K.; Li, J.; Luan, X.; Liu, L.; Yang, Z.; Wang, C. Effect of Graphene Oxide on Phase Change Materials Based on Disodium Hydrogen Phosphate Dodecahydrate for Thermal Storage. *ACS Omega* **2020**, *5*, 15210–15217. [CrossRef]
20. Xie, N.; Niu, J.; Zhong, Y.; Gao, X.; Zhang, Z.; Fang, Y. Development of polyurethane acrylate coated salt hydrate/diatomite form-stable phase change material with enhanced thermal stability for building energy storage. *Constr. Build. Mater.* **2020**, *259*, 119714. [CrossRef]
21. Yan, T.; Li, T.; Xu, J.; Chao, J. Understanding the transition process of phase change and dehydration reaction of salt hydrate for thermal energy storage. *Appl. Therm. Eng.* **2020**, *166*, 114655. [CrossRef]
22. Li, Z.; Yuan, J. Phase change microcapsules with high encapsulation efficiency using Janus silica particles as stabilizers and their application in cement. *Constr. Build. Mater.* **2021**, *307*, 124971. [CrossRef]
23. Wu, R.; Gao, W.; Zhou, Y.; Wang, Z.; Lin, Q. A novel three-dimensional network-based stearic acid/graphitized carbon foam composite as high-performance shape-stabilized phase change material for thermal energy storage. *Compos. Part B Eng.* **2021**, *225*, 109318. [CrossRef]
24. Wang, X.; Chen, Z.; Xu, W.; Wang, X. Capric acid phase change microcapsules modified with graphene oxide for energy storage. *J. Mater. Sci.* **2019**, *54*, 14834–14844. [CrossRef]
25. Chen, C.; Chen, J.; Jia, Y.; Topham, P.D.; Wang, L. Binary shape-stabilized phase change materials based on poly(ethylene glycol)/polyurethane composite with dual-phase transition. *J. Mater. Sci.* **2018**, *53*, 16539–16556. [CrossRef]
26. Khetib, Y.; Alahmadi, A.A.; Alzaed, A.; Sajadi, S.M.; Cheraghian, G.; Sharifpur, M. Numerical study of the effect of graphene nanoparticles in calcium chloride hexahydrate -based phase change material on melting and freezing time in a circular cavity with a triangular obstacle. *J. Energy Storage* **2021**, *43*, 103243. [CrossRef]
27. Sarı, A.; Saleh, T.A.; Hekimoğlu, G.; Tyagi, V.; Sharma, R. Microencapsulated heptadecane with calcium carbonate as thermal conductivity-enhanced phase change material for thermal energy storage. *J. Mol. Liq.* **2021**, *328*, 115508. [CrossRef]
28. Yu, S.; Wang, X.; Wu, D. Microencapsulation of n-octadecane phase change material with calcium carbonate shell for enhancement of thermal conductivity and serving durability: Synthesis, microstructure, and performance evaluation. *Appl. Energy* **2014**, *114*, 632–643. [CrossRef]
29. Liu, X.; Feng, Q.; Peng, Z.; Zheng, Y.; Liu, H. Preparation and evaluation of micro-encapsulated thermal control materials for oil well cement slurry. *Energy* **2020**, *208*, 118175. [CrossRef]
30. Huo, J.-H.; Peng, Z.-G.; Xu, K.; Feng, Q.; Xu, D.-Y. Novel micro-encapsulated phase change materials with low melting point slurry: Characterization and cementing application. *Energy* **2019**, *186*, 115920. [CrossRef]
31. Aguayo, M.; Das, S.; Maroli, A.; Kabay, N.; Mertens, J.C.; Rajan, S.D.; Sant, G.; Chawla, N.; Neithalath, N. The influence of microencapsulated phase change material (PCM) characteristics on the microstructure and strength of cementitious composites: Experiments and finite element simulations. *Cem. Concr. Compos.* **2016**, *73*, 29–41. [CrossRef]
32. Xie, W.; Guo, S.; Liu, Y.; Chen, R.; Wang, Q. Organic-inorganic hybrid strategy based on ternary copolymerization to prepare flame retardant poly(methyl methacrylate) with high performance. *Compos. Part B Eng.* **2020**, *203*, 108437. [CrossRef]
33. Zhu, B.; Huang, X.; Guo, Y. Influence of Cement Particle Size Distribution on Strength of Cement Paste. In *Advanced Engineering Materials*; Zeng, J.M., Li, T.S., Ma, S.J., Jiang, Z.Y., Yang, D.G., Eds.; Pts 1-32011; Trans Tech Publications: Zurich, Switzerland, 2011; pp. 1007–1011.
34. Qiao, D.; Cheng, W.; Xie, J.; Wang, J.; Huang, F.; Mo, Y.; Peng, J. Analysis of the influence of gradation on the strength of a cemented filling body and the cementation strength model. *Integr. Ferroelectr.* **2019**, *199*, 12–21. [CrossRef]
35. Althoey, F.; Zaid, O.; De-Prado-Gil, J.; Palencia, C.; Ali, E.; Hakeem, I.; Martínez-García, R. Impact of sulfate activation of rice husk ash on the performance of high strength steel fiber reinforced recycled aggregate concrete. *J. Build. Eng.* **2022**, *54*, 104610. [CrossRef]
36. Zhang, R.; Zhang, W.; Shi, M.; Li, H.; Ma, L.; Niu, H. Morphology controllable synthesis of heteroatoms-doped carbon materials for high-performance flexible supercapacitor. *Dyes Pigment.* **2022**, *199*, 109968. [CrossRef]
37. Tondro, H.; Zilouei, H.; Zargoosh, K.; Bazarganipour, M. Nettle leaves-based sulfonated graphene oxide for efficient hydrolysis of microcrystalline cellulose. *Fuel* **2020**, *284*, 118975. [CrossRef]
38. Zhou, D.; Zhou, Y.; Liu, Y.; Luo, X.; Yuan, J. Preparation and Performance of Capric-Myristic Acid Binary Eutectic Mixtures for Latent Heat Thermal Energy Storages. *J. Nanomater.* **2019**, *2019*, 2094767. [CrossRef]
39. Wang, Y.; Li, J.; Miao, W.; Su, Y.; He, X.; Strnadel, B. Preparation and characterizations of hydroxyapatite microcapsule phase change materials for potential building materials. *Constr. Build. Mater.* **2021**, *297*, 123576. [CrossRef]
40. Sha, Z.; Liang, J.; Zhang, G.; Yang, S.; Lu, J.; Zhang, Z.; McConnell, D.R.; Humphrey, G. A seepage gas hydrate system in northern South China Sea: Seismic and well log interpretations. *Mar. Geol.* **2015**, *366*, 69–78. [CrossRef]
41. Wu, X.; Liang, Q.; Ma, Y.; Shi, Y.; Xia, Z.; Liu, L.; Haeckel, M. Submarine Landslides and their Distribution in the Gas Hydrate Area on the North Slope of the South China Sea. *Energies* **2018**, *11*, 3481. [CrossRef]
42. Zhou, D.; Yuan, J.; Zhou, Y.; Liu, Y. Preparation and Properties of Capric–Myristic Acid/Expanded Graphite Composite Phase Change Materials for Latent Heat Thermal Energy Storage. *Energies* **2020**, *13*, 2462. [CrossRef]
43. Huo, J.-H.; Peng, Z.-G.; Feng, Q.; Zheng, Y.; Zhang, J. Controlling the heat evolution of cement slurry system using microencapsulated phase change materials. *Int. J. Energy Res.* **2018**, *42*, 4206–4220. [CrossRef]

44. Zhao, C.; Opolot, M.; Liu, M.; Bruno, F.; Mancin, S.; Hooman, K. Phase change behaviour study of PCM tanks partially filled with graphite foam. *Appl. Therm. Eng.* **2021**, *196*, 117313. [CrossRef]
45. Ravi, K.; Iverson, B.; Moore, S. Cement-Slurry Design to Prevent Destabilization of Hydrates in Deepwater Environment. *SPE Drill. Complet.* **2009**, *24*, 373–377. [CrossRef]
46. Yue, J.; Liu, Z.; Wang, J.; Sun, T.; Wu, Z.; Geng, Y. A Low-Cost and Low-Density Cement Slurry System Suitable for a Shallow Unconsolidated Stratum. *Adv. Mater. Sci. Eng.* **2020**, *2020*, 1628281. [CrossRef]
47. Zheng, M.; Liu, T.; Jiang, G.; Wei, M.; Huo, Y.; Liu, L. Large-scale and high-similarity experimental study of the effect of drilling fluid penetration on physical properties of gas hydrate-bearing sediments in the Gulf of Mexico. *J. Pet. Sci. Eng.* **2019**, *187*, 106832. [CrossRef]
48. Nikolaev, N.; Saint Petersburg Mining University; Leusheva, E. Low-Density Cement Compositions for Well Cementing Under Abnormally Low Reservoir Pressure Conditions. *J. Min. Inst.* **2019**, *236*, 194–200. [CrossRef]
49. Berodier, E.; Scrivener, K. Understanding the Filler Effect on the Nucleation and Growth of C-S-H. *J. Am. Ceram. Soc.* **2014**, *97*, 3764–3773. [CrossRef]
50. Nikolaev, N.I.; Haoya, L. Results of "Cement-to-Rock" Contact Study. *J. Min. Inst.* **2017**, *226*, 428–434.
51. Fernandes, F.; Manari, S.; Aguayo, M.; Santos, K.; Oey, T.; Wei, Z.; Falzone, G.; Neithalath, N.; Sant, G. On the feasibility of using phase change materials (PCMs) to mitigate thermal cracking in cementitious materials. *Cem. Concr. Compos.* **2014**, *51*, 14–26. [CrossRef]
52. Ghods, P.; Isgor, O.; Carpenter, G.; Li, J.; McRae, G.; Gu, G. Nano-scale study of passive films and chloride-induced depassivation of carbon steel rebar in simulated concrete pore solutions using FIB/TEM. *Cem. Concr. Res.* **2013**, *47*, 55–68. [CrossRef]
53. Li, L.; Liu, T.; Jiang, G.; Zheng, S.; Fang, C.; Sun, J.; Qu, B.; Zhu, Y. Interactive mechanism of manufacturing factors on the properties of microbial cementing slurry. *Constr. Build. Mater.* **2021**, *311*, 125308. [CrossRef]
54. Bu, J.; Tian, Z.; Zheng, S.; Tang, Z. Effect of sand content on strength and pore structure of cement mortar. *J. Wuhan Univ. Technol. Sci. Ed.* **2017**, *32*, 382–390. [CrossRef]
55. Li, L.; Liu, T.; Jiang, G.; Fang, C.; Sun, J.; Zheng, S.; Liu, H.; Leusheva, E.; Morenov, V.; Nikolaev, N. Field Application of Microbial Self-Healing Cement Slurry in Chunguang 17-14 Well. *Energies* **2021**, *14*, 1544. [CrossRef]
56. Bullard, J.W.; Jennings, H.M.; Livingston, R.; Nonat, A.; Scherer, G.; Schweitzer, J.S.; Scrivener, K.; Thomas, J. Mechanisms of cement hydration. *Cem. Concr. Res.* **2011**, *41*, 1208–1223. [CrossRef]

energies

Article

A Study on the Adaptability of Nonhydrocarbon Gas-Assisted Steam Flooding to the Development of Heavy Oil Reservoirs

Yong Huang [1], Wulin Xiao [2], Sen Chen [1], Boliang Li [3], Liping Du [3] and Binfei Li [3,*]

1 Oil Production Technology Research Institute, Petro China Xinjiang Oilfield Company, Karamay 834000, China; xjhy@petrochina.com.cn (Y.H.); chensen@petrochina.com.cn (S.C.)
2 Heavy Oil Development Company, PetroChina Xinjiang Oilfield Company, Karamay 834000, China; fcxwl@petrochina.com.cn
3 School of Petroleum Engineering, China University of Petroleum (East China), Qingdao 266580, China; s20020069@s.upc.edu.cn (B.L.); s20020130@s.upc.edu.cn (L.D.)
* Correspondence: libinfei@upc.edu.cn

Abstract: In view of the serious heat loss in the process of steam injection for heavy oil recovery, nonhydrocarbon gas combined with steam has attracted much attention in recent years to realize the efficient development of heavy oil. Due to the wide variety of nonhydrocarbon gases, their performance in pressurization, dissolution, viscosity reduction, and heat loss decrease is changeable. In this paper, four groups of one-dimensional physical simulation experiments on different nonhydrocarbon gas-assisted steam flooding methods were carried out, and the effect on oil displacement characteristics under high temperature and pressure conditions was studied. Moreover, the differences in N_2, CO_2, and flue gas in energy supplementation, heat transfer, and oil recovery efficiency were also analyzed. The results showed that the three nonhydrocarbon gas-assisted steam flooding methods could significantly improve the oil displacement efficiency, which was specifically embodied as a faster oil production rate and longer production period. Compared with pure steam flooding, the recovery was increased by 12.13%, 16.71% and 13.01%, respectively. The effects of N_2 in energy supplementation and heat transfer reinforcement were the greatest among the three nonhydrocarbon gases, followed by those of flue gas, and the CO_2 effects were the worst. The temperature at the end of the sandpack model increased by 14.3 °C, 8.8 °C and 13.1 °C, respectively. In addition, CO_2-assisted steam flooding had a prominent oil recovery effect, and the oil content of the sands in the front and middle of the model was significantly lower than that of other displacement methods. Most importantly, combined with the analysis of the remaining oil in the oil sands after displacement, we explained the contrasting contradictions of the three non-hydrocarbon gases in terms of recovery and energy supply/heat transfer, and further confirmed the gas properties and reservoir adaptability of the three non-hydrocarbon gases. The results may provide a theoretical basis for the selection of nonhydrocarbon gases for heavy oil reservoirs with different production requirements.

Keywords: heavy oil; nonhydrocarbon gas; steam flooding; reservoir adaptability

Citation: Huang, Y.; Xiao, W.; Chen, S.; Li, B.; Du, L.; Li, B. A Study on the Adaptability of Nonhydrocarbon Gas-Assisted Steam Flooding to the Development of Heavy Oil Reservoirs. *Energies* **2022**, *15*, 4805. https://doi.org/10.3390/en15134805

Academic Editors: Shu Tao, Dengfeng Zhang, Huazhou Huang, Shuoliang Wang and Yanjun Meng

Received: 30 May 2022
Accepted: 24 June 2022
Published: 30 June 2022

1. Introduction

With the development of conventional oil and gas resources entering the middle and late stages, many tricky problems have arisen at the sites, such as the high moisture content of the export liquid and depletion of natural energy in the formation, inducing higher development difficulty and reducing economic benefits at the same time. However, the demand for energy is increasing daily, and conventional oil resources obviously cannot meet the requirements of the current fast-developing industry. To ensure national energy security and the sustainable development of society, it is urgent to intensify the exploration of unconventional oil and gas resources and new energy. As one of the representatives, heavy oil is widely distributed in the world and has abundant reserves, accounting for more than 70% of the world's total oil reserves, and approximately 17% is recoverable, showing promising value and prospects for development [1,2]. Compared with conventional oil, the most notable features of heavy oil are high viscosity, high specific gravity, and the existence of many heavy components, such as sulfur, heavy metals, and asphaltenes, giving rise to its large flowing resistance in the formation, so the recoverable reserves through water flooding are extremely low, which is the fundamental reason for its limited production. Heavy oil has evident thermal expansion and temperature sensitivity; that is, when the temperature increases, the viscosity decreases, and the fluidity increases [3,4]. Therefore, thermal technologies have become an effective means of heavy oil development. Multiple conventional thermal methods of enhancing oil recovery, such as in situ combustion, hot water flooding and steam injection, which include steam huff and puff [5–7], steam flooding [8,9] and chemical assisted steam flooding [10], are being used by the oil and gas industry [11]. In addition, new measures, such as electromagnetic heating [12–14] and catalytic modification [15], also have significant potential exploration value.

Nevertheless, almost all thermal measures exhibit certain faults in technological aspects. In the process of steam flooding, the injected steam is prone to gas channeling along the high-permeability layer formed by the heterogeneity of the formation. The water breakthrough time of the oil well is greatly shortened, and the heat cannot comprehensively spread to the vast formation, so the ultimate recovery is located at a relatively low degree. On the other hand, when continuously injecting steam into the rock-oil-water system, a phase change (from liquid to steam phase) usually occurs, ultimately inducing unfavorable wettability and directly affecting the extent of heavy oil recovery [16–18]. Therefore, a series of measures to improve heat utilization, such as suppressing steam channeling, expanding the scope of heat spread, and reducing heat loss during the process, have been key to solving this problem. Flue gas, as a kind of non-condensable gas, is mainly formed in the process of producing steam. Because of its low thermal conductivity, it is widely used in thermal recovery processes to achieve the efficient development of crude oil and simultaneously reduce carbon emissions. Zhoujie Wang et al. [19] conducted steam condensation heat transfer experiments with the addition of flue gas and steam flooding experiments in a one-dimensional sandpack model. They believed that the flue gas can inhibit the condensation and heat release of steam in the front and middle parts of the reservoir and hinder the formation of condensing droplets, thereby promoting the expansion of the steam chamber into the deep reservoir. The adsorption and retention of nanoparticles in the reservoir can result in a significant plugging effect, forcing the direction of liquid flow to change and increasing the sweep coefficient. Osamah A. and Abdullah F. [20] synergized the recovery mechanisms of both EOR agents by injecting a hot hydrophilic nanofluid (HNF) slug, followed by superheated steam (SHS) in a second slug. The thermophysical properties of hydrophilic nanoparticles improve the thermal performance of SHS injection and increase oil mobility, which can substantially reduce steam consumption by up to 50% and reduce the costs of producing steam while also improving oil recovery through the utilization of nanotechnology. When passing through a narrow rock pore throat, the foam expands and deforms, generating additional flow resistance. With the continuous accumulation of foam, the resistance effect becomes increasingly obvious. Moreover, foam has the characteristics of blocking water but not oil and is often called an intelligent fluid. Yongqing Bai et al. [21] foamed a physically crosslinked clayey hydrogel Bent/PAM with low thermal conductance, high thermal stability and good mobility, which was synthesized by a one-pot process, enabling remarkable blockage of steam channeling. Zhanxi Pang et al. [22] selected a kind of foaming agent for thermal foam flooding and carried out many displacements in a sandpack, and the results showed that foam can effectively increase the displacement efficiency of steam flooding from 43.30% to 81.24% and that thermal foams can effectively improve the injection profile to restrain steam injection from gravity override and steam channeling in reservoirs. Changfeng Xi et al. [23] conducted 3D physical modeling experiments of steam flooding, CO_2-foam-assisted steam flooding, and CO_2-assisted steam flooding under different perforation conditions. The experimental results show that after the adjustment of perforation holes in the later stage of CO_2-assisted steam flooding, the steam chamber in the middle and lower part of the water injection well expand laterally, and the production and development mode of gravity drainage is formed in the top chamber of the production well.

In this paper, from the perspective of increasing the heat transfer range of steam and promoting the expansion of the steam chamber to the deep reservoir, the effects of different types of gases on the oil production rate, temperature field change, displacement pressure, etc., were researched based on a one-dimensional steam flooding experiment assisted by nonhydrocarbon gas. The oil displacement characteristics and the distribution of the remaining oil after displacement were also analyzed, yielding a certain guiding significance for the development of enhanced heavy oil recovery by injecting gas/steam.

2. Experimental Section

2.1. Materials

Removal of brine and gases from crude oil samples. The crude oil used in the experiment came from China's Shengli Oilfield. The viscosity of the crude oil was tested by a rheometer (Model MCR 302, Anton Paar, Austria). The relationship between viscosity and temperature is shown in Figure 1. The flue gas used in the experiment was a 1:4 mixture of CO_2 and N_2, which was similar to actual flue gas produced by steam generators in oil fields. The purity of CO_2 and N_2 were both 99.9 mol%, provided by China Tianyuan Company. Two types of silica sands were used to fill the sandbags: 80 mesh and 120 mesh, respectively. The water used in the laboratory to generate steam was distilled water. The parameters of the sand bag used in the oil displacement experiment are shown in Table 1. The physical properties of the sand bag were the same, which was consistent with actual field conditions.

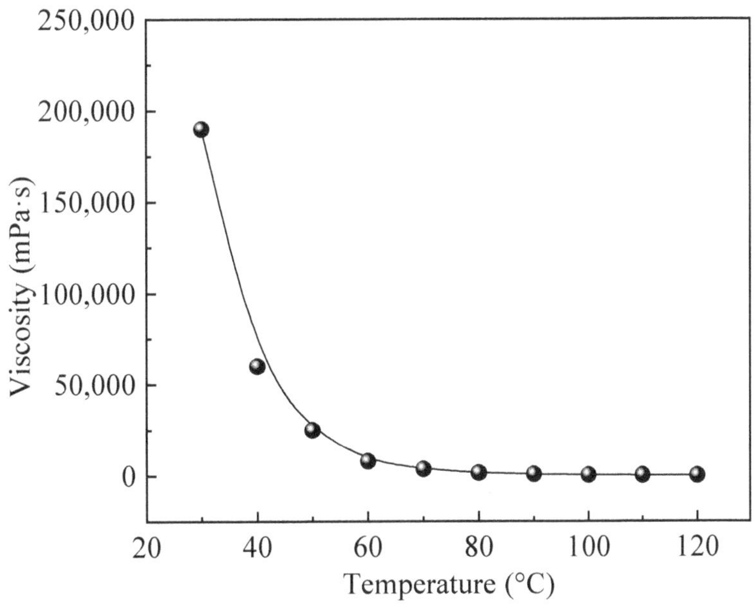

Figure 1. The relationship between the viscosity and temperature of the heavy oil used in the experiment.

Table 1. Parameters of the sandpack used in the flooding experiment.

Displacement	Length/cm	Diameter/cm	Porosity/%	Permeability/mD	Back Pressure/MPa
Steam flooding	60	2.54	33.1	1184	1
N_2-assisted steam flooding	60	2.54	34.17	1230	1
CO_2-assisted steam flooding	60	2.54	33.18	1152	1
Flue gas-assisted steam flooding	60	2.54	33.84	1203	1

2.2. Apparatus

The flow chart of the nonhydrocarbon gas-assisted steam flooding experiment is shown in Figure 2 and the involved experimental devices are listed in Table 2. A specific flow of flue gas was injected into the experimental model through a gas flow controller. Distilled water pressurized by a high-precision syringe pump was heated into steam by a steam generator. Water or heavy oil was injected from the intermediate vessel into the sand bag with three temperature test points. The temperature was measured by thermocouples. The schematic diagram and real picture of the sandpack model are shown in Figure 3.

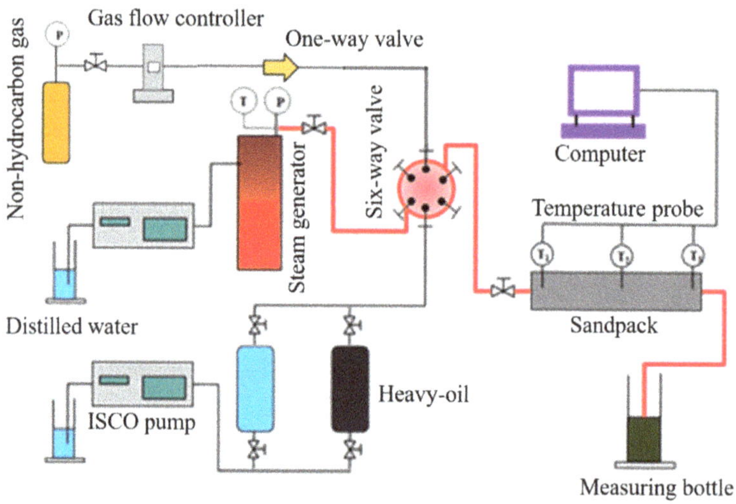

Figure 2. Schematic diagram of the experimental setup in the laboratory.

Table 2. Relevant parameters of experimental equipment.

Device	Type
Gas flow controller	Model Sla58550, Brooks, United States, flow rate range of 0–30 mL/min under standard conditions
High-precision syringe pump	Model 100DX, Teledyne ISCO Company, Teledyne Co., Ltd., USA, flow accuracy of ± 0.25 μL/min and pressure accuracy of ± 0.5%
Steam generator	Model GL-1, Haian Petroleum Equipment Company, temperature range of 100–350 °C and pressure range of 0.1–25 MPa
Thermocouples	Model K, Haian Petroleum Equipment Company, temperature accuracy of ± 0.1 °C

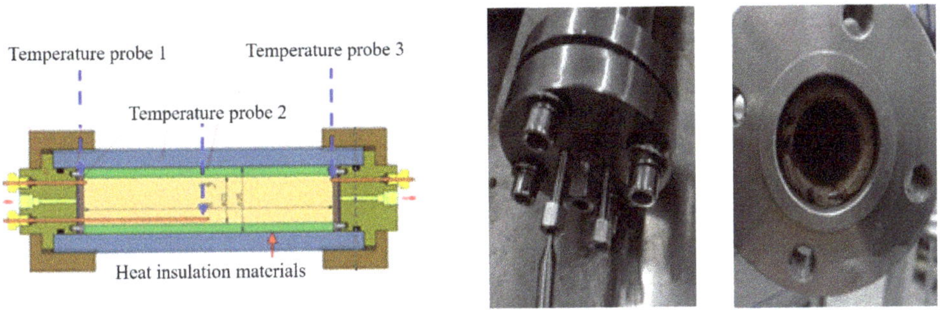

(a) (b)

Figure 3. The sandpack with three temperature probes: (**a**) is a schematic diagram and (**b**) is a real picture.

2.3. Experimental Procedures

(1) Fill the sandpack with sand of different diameters, and then test the air tightness of the sandpack.

(2) Measure the dry weight of the sandpack. Then, test the wet weight of the sandpack saturated with salt water. The porosity is calculated from the weight difference, and the permeability is calculated from the Darcy equation.

(3) Place the sandpack in a 100 °C oven for 2 h until the temperature of the sandpack stabilizes. Then, crude oil is injected into the sandpack at a flow rate of 0.5 mL/min to achieve oil saturation.

(4) The steam generator is pre-heated at a temperature of 250 °C. When the temperature of the sandpack and the steam generator stabilize, steam or steam with flue gas is injected into the sandpack according to the designed experimental parameters. The steam injection flow rate is set to the equivalent of condensate water.

(5) During the experiment, the temperature of each temperature measurement point in the sandpack is monitored and recorded in real time, and the water production and crude oil production are recorded.

(6) When the temperature of the sandpack is stable and the proportion of water in the produced fluid reaches 98%, the oil displacement experiment is stopped.

(7) The oil sand samples located in different positions of the model after the end of the displacement are selected to analyze the distribution of the remaining oil and evaluate the oil displacement effect of different displacement methods. The injection parameters of various displacement experiments are shown in Table 3.

Table 3. Injection parameters of various displacement experiments.

Steam Flow Rate/(mL·min^{-1})	Steam Dryness	Gas Flow Rate/(mL·min^{-1})	Back Pressure/MPa
1	0.7	1	1

3. Results and Discussion

3.1. Variation in the Oil Displacement Parameters

In the experiments, temperature changes at three locations within the sandpack were used to determine the heat transfer performance of steam flow in porous media. The temperature changes of the three thermocouples in the sandpack under different injection parameters were obtained. After the temperature of the sandpack was stable in the experiment, the data of each temperature measurement point were recorded. Taking the one-dimensional pure steam flooding as the standard control group, a total of 4 experiments were carried out with the addition of N_2, CO_2 and flue gas-assisted steam flooding. The oil characteristics of gas-assisted steam flooding and their mechanisms for enhancing heavy oil production were explored through the curves of apparent oil displacement parameters changing with time.

The liquid production characteristic curve included three parameters: recovery, water cut and oil production rate. For sandstone, the surface of the formation rock was mostly hydrophilic, and the seepage resistance of steam in the formation mainly came from the capillary force generated by the two-phase flow. In Figure 4, when the steam injection volume was approximately 0.3 PV, water started to appear at the outlet, and at this moment, the oil production rate reached a peak value of 0.81 mL/min. After that, due to the gradual formation of the dominant flow channel, the steam began to channel, and the water cut curve rose rapidly, corresponding to the rapid decrease in the oil production rate. During the subsequent displacement process, the flow of steam in the sandpack reached a steady state, the oil production rate fluctuated around the equilibrium value of 0.1 mL/min, and the recovery and water cut increased steadily as the remaining oil in the reservoir decreased. The ultimate recovery was near 48%, which was relatively low in terms of the thermal recovery of heavy oil.

From Figure 4b–d, it could be found that after adding nonhydrocarbon gas, the variation law of oil displacement parameters was similar to that of pure steam flooding; that is, there were two stages: a high-producing period and a low-producing period. The high-producing period appeared in the early stage of displacement, and the oil production rate increased sharply at first and then decreased rapidly. The low-producing period appeared in the middle and late displacement stages, and the production rate decreased slowly and gradually stabilized. The maximum oil production rates of N_2-assisted steam flooding, flue gas-assisted steam flooding, and CO_2-assisted steam flooding were 1.2 mL/min, 1.17 mL/min (since the main component of flue gas was N_2 and only a small amount of CO_2 existed, the maximum oil production rate of flue gas-assisted steam flooding was very close to the former), and 1.06 mL/min, respectively, which were higher than that of pure steam flooding. In Figure 5, the steam mixed with the nonhydrocarbon gas flowed faster than pure steam in the formation, and the peak value was higher due to the lower seepage resistance of the gas, thereby improving the oil displacement efficiency.

Figure 5 shows that after adding nonhydrocarbon gas, the ultimate recovery was improved to different degrees. It is not difficult to understand that since the main composition of flue gas was also N_2, the recovery of N_2-assisted steam flooding was very close to that of flue gas-assisted steam flooding. Although the maximum oil production rate of CO_2-assisted steam flooding was relatively lower, the high production period lasted for a longer time, so the final recovery degree was the highest.

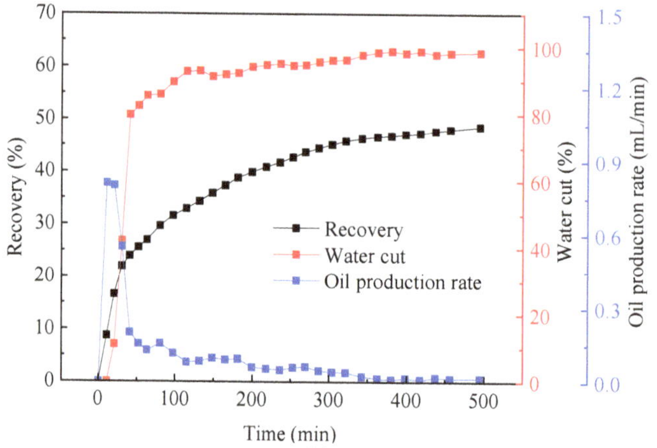

Figure 4. *Cont.*

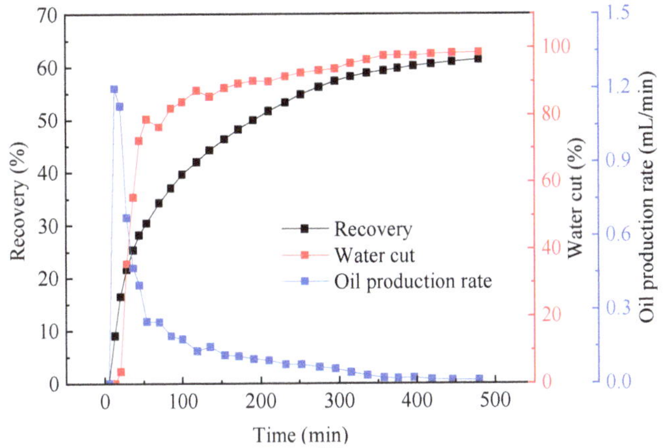

(b)

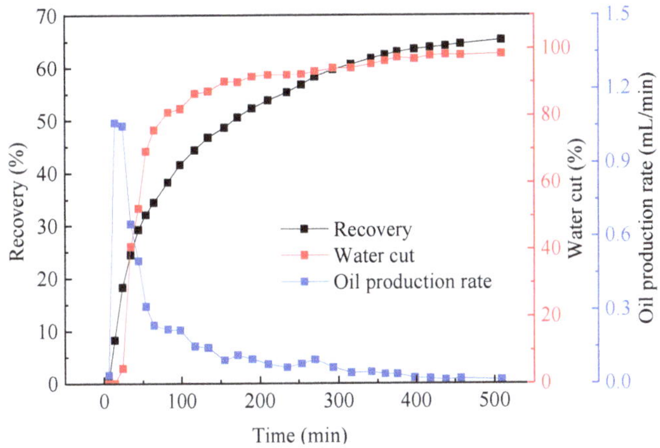

(c)

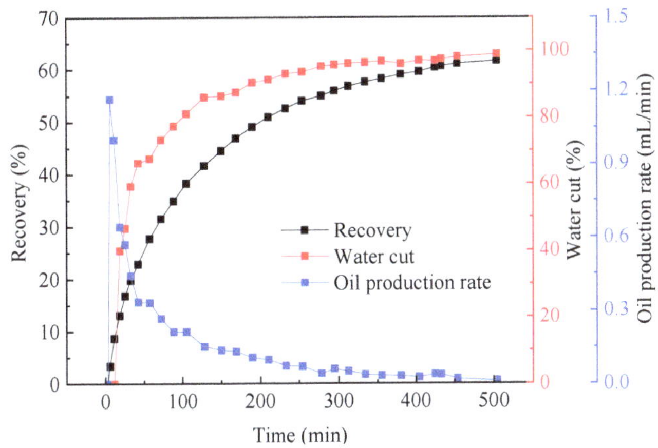

(d)

Figure 4. The liquid production characteristic curves under different flooding methods: (**a**) pure steam flooding, (**b**) N_2 + steam flooding, (**c**) CO_2 + steam flooding, and (**d**) flue gas + steam flooding.

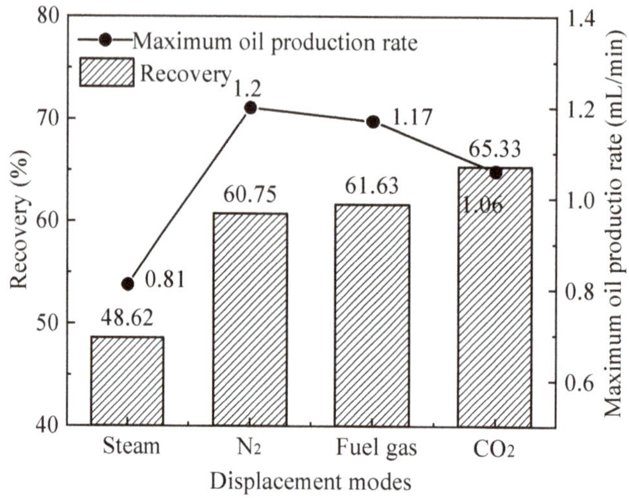

Figure 5. Maximum oil production rate and recovery under different displacement methods.

3.2. Variation in the Pressure Difference during Displacement

In Figure 6, the variation in the pressure difference of pure steam flooding and non-hydrocarbon gas-assisted steam flooding was similar, and the curve trends were basically the same. For a period of time when the steam was injected, the fluid flowed among the sand grains to open up the dominant channel for seepage. At this stage, no water was seen at the outlet, and the injection pressure increased rapidly, providing momentum for the flow of heavy oil. As the heavy oil along the dominant steam channel was continuously removed, high-permeability channels formed. After that, the flow resistance of steam/nonhydrocarbon gas dropped greatly, and a large amount of injected gas was channeled. Therefore, the displacement pressure suddenly dropped and finally remained stable with a small fluctuation. The heavy oil lying in the corners gradually expanded under the action of heat, occupying the large pores of fluid seepage, or the unstably filled sand particles formed displacement under the scouring of the steam/nonhydrocarbon gas, which slightly plugged the high-permeability channels. The pressure difference curves fluctuated in waves under this dynamic balance mechanism.

Further interpretation of the curves showed that the maximum displacement pressure difference of steam flooding was 2.28 MPa, and the maximum displacement pressure differences of N_2-assisted steam flooding, flue gas-assisted steam flooding and CO_2-assisted steam flooding were 3.16 MPa, 3.11 MPa and 2.86 MPa, respectively (Figure 7). After adding gas, the reasons why the maximum displacement pressure difference of each displacement mode was more than that of pure steam were as follows: First, in the high temperature and pressure conditions, the thermal motion of gas molecules was intensified, and the mixing degree of the two was higher. The dryness of the steam decreased, and the flow resistance of the mixed steam increased. Second, the non-condensate gas could form an insulating gas film on the surface of the rock particles, which prevented the direct contact between the steam and rock for heat exchange, and the viscosity of the heavy oil in the formation decreased slightly, resulting in a reduction in the flow capacity. In the later stage of displacement, due to the high gas mobility, the flow resistance of the mixed steam was reduced, so the stable pressure difference was lower than that of pure steam flooding. Compared with CO_2, N_2 was not easy to compress and had poor solubility in crude oil. Therefore, the maximum displacement pressure difference of N_2-assisted steam flooding was significantly greater than that of the other two gases, which also showed that N_2 possessed more ascendancies in supplementing the formation energy.

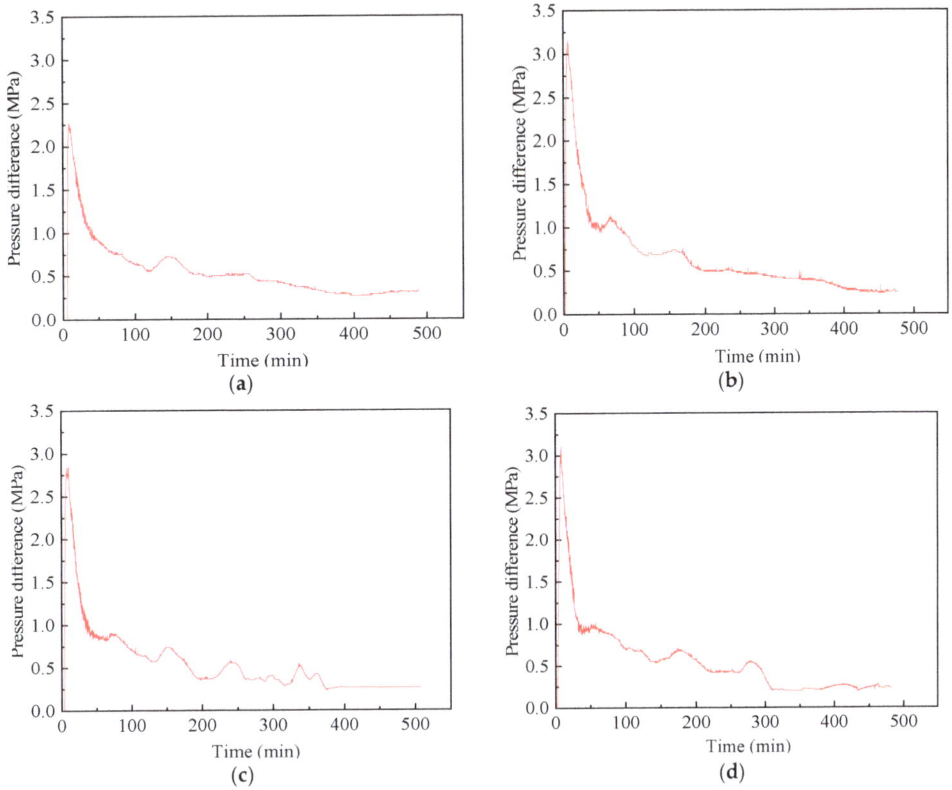

Figure 6. Variation in the displacement pressure difference under different displacement modes: (**a**) pure steam flooding, (**b**) N_2 + steam flooding, (**c**) CO_2 + steam flooding, and (**d**) flue gas + steam flooding.

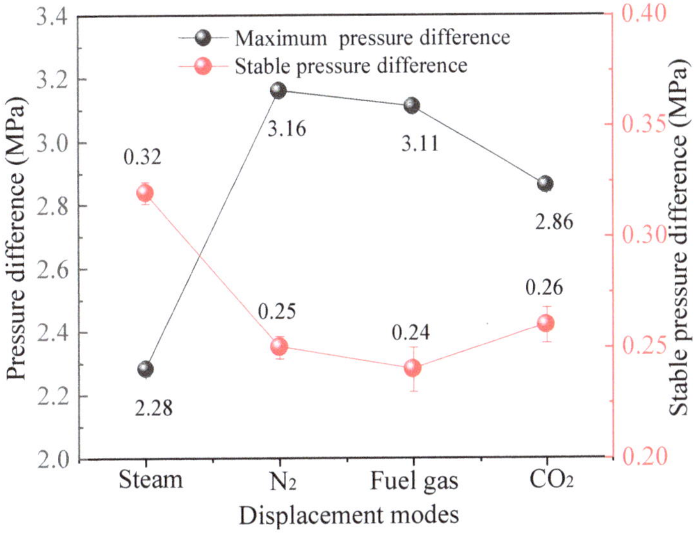

Figure 7. Maximum pressure difference and stable pressure difference under different displacement methods.

3.3. Variation in the Temperature Field

Figure 8 shows the temperature change process of the sandpack model under different displacement methods. The commonality was that as the steam front continued to advance into the deep reservoir, each temperature measurement point started to heat in sequence. The inlet first sensed the temperature change, and the curve increased nearly vertically and quickly reached a stable temperature. The temperature of the measurement points closer to the outlet end had a slower temperature rise, a longer heating time, and a lower stable temperature. It was determined that without changing the environmental conditions and injection parameters, the steam continuously exchanged heat with the formation rocks and fluids during the flow process, and with liquefaction and heat dissipation, its own temperature was lowered, inducing the heat it carried to be gradually reduced. Under the same displacement mode, the difference between the stable temperatures of adjacent temperature measurement points increased, indicating that the heat loss of the steam was severe along the way.

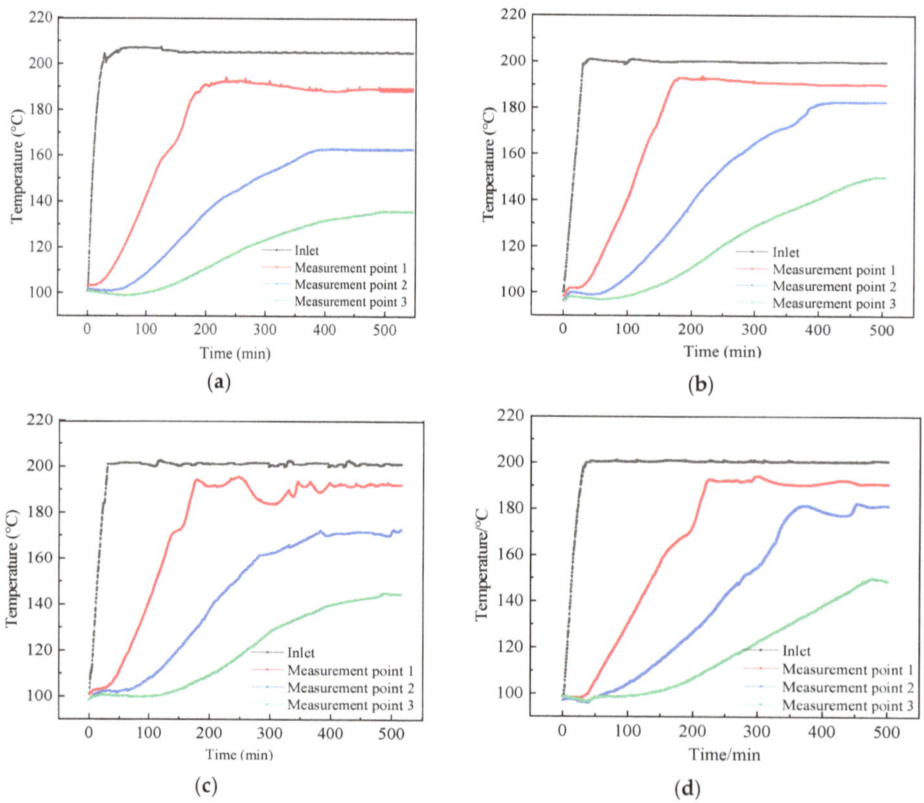

Figure 8. Temperature variation at temperature measurement points under different flooding modes: (**a**) pure steam flooding, (**b**) N_2 + steam flooding, (**c**) CO_2 + steam flooding, and (**d**) flue gas + steam flooding.

The stable temperature of each temperature measurement point under different displacement modes is shown in Figure 9 and Table 4. In pure steam flooding, the end temperature of the model was 135.9 °C, while it increased by 14.3 °C, 8.8 °C, and 13.1 °C in the steam flooding model assisted by N_2, CO_2, and flue gas, respectively. On the one hand, the addition of nonhydrocarbon gases improved the flow rate of the mixed thermal fluid. On the other hand, nonhydrocarbon gas would be enriched in the front edge of the flow, prompting heat exchange resistance between steam and rocks. Under the combined action of the two, the nonhydrocarbon gas spurred the steam to flow rapidly and reduced the heat loss in the low oil saturation region, thereby bringing more heat into the deep reservoir. The best effect of preventing the heat transfer mode was obtained with N_2-assisted steam flooding, followed by flue gas, and CO_2 was the worst because N_2 was not easily compressed with a low thermal conductivity, and the energy-providing effect was outstanding, while CO_2 had better dissolution and compressibility.

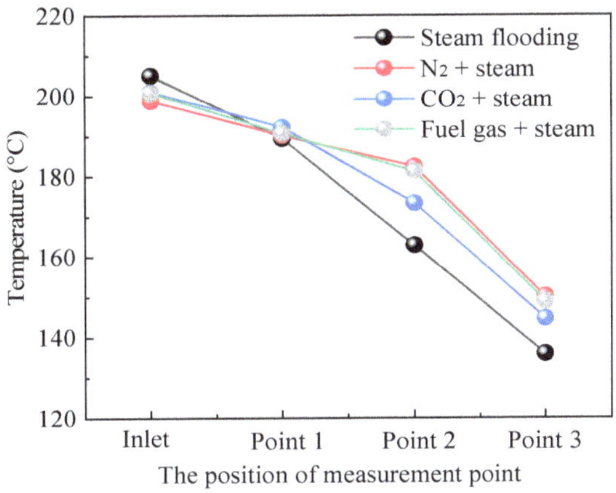

Figure 9. The comparative curves of the stable temperature of the measurement points under different displacement modes.

Table 4. The stable temperature of the measurement points under different displacement modes.

Displacement Mode	Inlet/°C	Measurement Point 1/°C	Measurement Point 2/°C	Measurement Point 3/°C
Steam	205.1	189.4	162.9	135.9
N_2 + steam	198.8	190.2	182.5	150.2
CO_2 + steam	201	192.3	173.4	144.7
Flue gas + steam	200.8	190.8	181.3	149

3.4. Distribution of the Remaining Oil

After flooding, samples were taken at different positions of the sandpack model to observe the distribution of the remaining oil, as shown in Figure 10. From the inlet to the outlet, the color of the oil sand became darker, and the oil content gradually increased. Due to the limited heat of the injected steam, the heavy oil near the inlet had the most obvious effect of heating up and reducing viscosity and had the highest recovery degree with an oil content of 5.15% (Table 5). Then, as the steam gradually liquefied and released heat, the heat brought to the rear of the reservoir dropped, and the oil content at the outlet was 12.88%. At the inlet and the middle position, the oil content of CO_2-assisted steam flooding was lower than that of N_2-assisted steam flooding, indicating that although CO_2-assisted steam flooding was not as good as N_2-assisted steam flooding in promoting the development of steam chambers, the oil recovery efficiency in the affected range was higher than that of N_2-assisted steam flooding (Figure 11). Compared with steam flooding assisted by nonhydrocarbon gas, it was found that the color of the oil sand was obviously lighter, and the remaining oil content was reduced, indicating that the addition of nonhydrocarbon gas significantly improved the overall oil displacement efficiency. In the area affected by steam heat, the result of multiple mechanisms, such as temperature rise and viscosity reduction, thermal expansion, and CO_2 extraction to light components, was that the oil recovery efficiency was greatly enhanced.

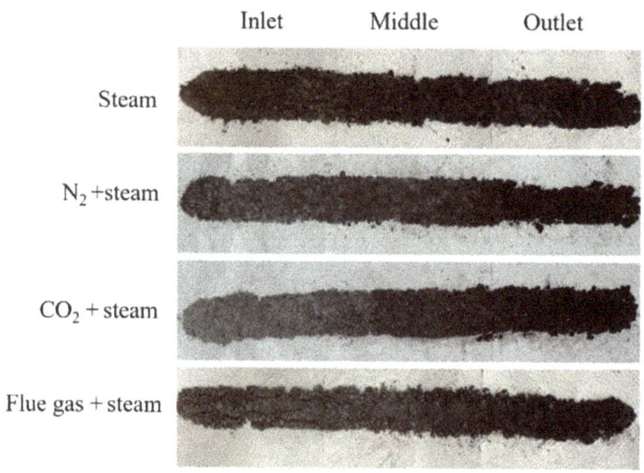

Figure 10. The distribution of the remaining oil under different displacement methods.

Table 5. The oil content of oil sands at different positions after displacement.

Oil Content/%. Displacement Modes	Inlet	Position Middle	Outlet
Initial oil sands	25.77	25.77	25.77
Steam	5.15	9.86	12.88
N_2 + steam	4.37	6.15	11.05
CO_2 + steam	3.76	5.72	11.95
Flue gas + steam	4.05	6.1	11.69

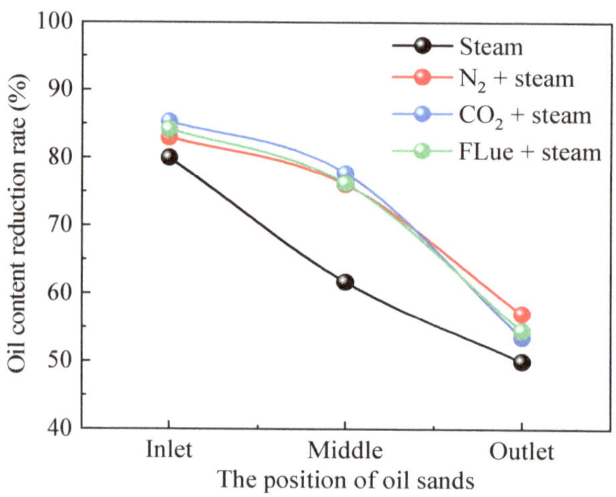

Figure 11. Comparison of oil content reduction rate of oil sands after displacement.

3.5. Analysis of the Reasons for the Effects Induced by Different Gases

From the above experimental results, some rules were determined. The mixing of the non-condensate gas with the steam increased the mobility of the displacement medium, thus prolonging the high production period and increasing the maximum oil recovery rate, because the gas was unable to liquefy into water when contacting cold objects, emitting much heat during the flow. The gas had a large molecular distance, and the viscous force between the molecules was mild, and the force generated by mutual shearing was significantly smaller than that of the liquid, so that it could quickly channel flow in the tiny pores, opening up the main path for the liquid flow and increasing the range of thermal spreading.

In terms of suppressing the heat transfer effect of condensation, the advantages of N_2 were more obvious than those of CO_2. After investigation, N_2 had low thermal conductivity and low compressibility. Because of lower density and better mobility, the injected N_2 could form a thermal isolation layer at the top of the reservoir. It not only inhibited steam override, but also prevented heat loss from the overlying rock, thus increasing thermal efficiency and sweep efficiency [24]. This was determined by the structure and physical properties of the gas molecules.

In terms of supplementing formation energy, compared with CO_2, N_2 was less compressible and insoluble in crude oil. Under the condition of the same injection amount, it could occupy a larger space, and better supplement formation energy, which provided higher displacement power and accelerated the advancement of thermal fluid to the deep formation, then expanding the swept range. This feature also positively affected the expansion of the heat spread.

In the aspect of enhancing oil recovery, CO_2 could dissolve in crude oil under certain conditions and reduce its viscosity, which not only supplements crude oil energy, but also increases fluid pressure in porous media. At the same time, the two further formed a miscible phase, which greatly reduced the interfacial tension, thereby decreasing the seepage resistance. As a kind of common acid gas, CO_2 also possessed a certain acid etching effect on rock particles that dissolved the cement between particles from the micro perspective, expanding the fluid flow space, and improving the flow capacity of the heavy oil originally located in the corner.

The research content of this paper was based only on the one-dimensional sandpack model of steam flooding. There are certain limitations in the testing and evaluation of oil production parameters, injection pressure and other indicators that affect the reservoir development effect, and the dynamic changes in the actual development process cannot be accurately predicted, but can provide a certain theoretical basis.

4. Conclusions

(1) Steam flooding assisted by three nonhydrocarbon gases, N_2, CO_2 and flue gas, could significantly improve the oil displacement efficiency, which was manifested as accelerated oil recovery and a prolonged high production period. Compared with that of pure steam flooding, the recovery factor was increased by 12.13%, 16.71% and 13.01%, respectively.

(2) The effect of enhancing the heat transfer of N_2-assisted steam flooding was the best among the three nonhydrocarbon gases, and the CO_2 effect was the worst. Compared with steam flooding, N_2-, flue gas- and CO_2-assisted steam flooding increased the temperature at the end of the sandpack by 14.3 °C, 13.1 °C and 8.8 °C, respectively. That is, N_2 could bring more heat into the deep formation under the same conditions.

(3) N_2 showed more evident assets in supplementing the formation energy. The maximum displacement pressure differences during steam flooding assisted by N_2, CO_2 and flue gas were 3.16 MPa, 2.86 MPa and 3.11 MPa, respectively.

(4) The oil content of the sands after undergoing CO_2-assisted steam flooding was lower than that of N_2-assisted steam flooding for the inlet and middle points of the sandpack, not the outlet, which indicated that CO_2 had a more remarkable effect on oil displacement.

Author Contributions: Y.H. and W.X. mainly participated in the investigation of the research background and the integration of experimental data. S.C. and B.L. (Boliang Li) participated in the entire experimental process, including designing the experimental plan, recording the experimental detection data, etc. L.D. and B.L. (Binfei Li) mainly participated in the writing of the article. All authors have read and agreed to the published version of the manuscript.

Funding: The project was financially supported by National Natural Science Foundation of China (No. ZX20210203) and project: Research on the Mechanism and Parameter Optimization Design of Preliminary Non-hydrocarbon Gas Injection in Heavy Oil Old Areas.

Institutional Review Board Statement: Not applicable.

Informed Consent Statement: Not applicable.

Data Availability Statement: Not applicable.

Acknowledgments: We sincerely appreciate the researchers from the Shandong Engineering Research Center for Foam Application in Oil and Gas Field Development and Xinjiang Oilfield Engineering Technology Research Institute, Xinjiang Oilfield Heavy Oil Development Company for their assistance in this study. The valuable comments made by the anonymous reviewers are also sincerely appreciated. The authors would like to thank the reviewers for their helpful suggestions.

Conflicts of Interest: The authors declare that they have no known competing financial interests or personal relationships that could have appeared to influence the work reported in this paper.

References

1. Pratama, R.A.; Babadagli, T. A review of the mechanics of heavy-oil recovery by steam injection with chemical additives. *J. Pet. Sci. Eng.* **2022**, *208*, 109717. [CrossRef]
2. Dong, X.; Liu, H.; Chen, Z.; Wu, K.; Lu, N.; Zhang, Q. Enhanced oil recovery techniques for heavy oil and oilsands reservoirs after steam injection. *Appl. Energy* **2019**, *239*, 1190–1211. [CrossRef]
3. Taylor, S.E. Interfacial chemistry in steam-based thermal recovery of oil sands bitumen with emphasis on steam-assisted gravity drainage and the role of chemical additives. *Colloids Interfaces* **2018**, *2*, 16. [CrossRef]
4. Xu, X.; He, Y. Blockchain application in modern logistics information sharing: A review and case study analysis. *Prod. Plan. Control* **2022**, 1–15. [CrossRef]
5. Cokar, M.; Kallos, M.S.; Gates, I.D. Reservoir simulation of steam fracturing in early-cycle cyclic steam stimulation. *SPE Reserv. Eval. Eng.* **2012**, *15*, 676–687. [CrossRef]
6. Butler, R.; Yee, C. Progress in the in situ recovery of heavy oils and bitumen. *J. Can. Pet. Technol.* **2002**, *41*, 31–40. [CrossRef]
7. Xu, X.; Wang, C.; Zhou, P. GVRP considered oil-gas recovery in refined oil distribution: From an environmental perspective. *Int. J. Prod. Econ.* **2021**, *235*, 108078. [CrossRef]
8. Feng, G.; Li, Y.; Yang, Z. Performance evaluation of nitrogen-assisted steam flooding process in heavy oil reservoir via numerical simulation. *J. Pet. Sci. Eng.* **2020**, *189*, 106954. [CrossRef]
9. Pang, Z.; Wang, L.; Yin, F.; Lyu, X. Steam chamber expanding processes and bottom water invading characteristics during steam flooding in heavy oil reservoirs. *Energy* **2021**, *234*, 121214. [CrossRef]
10. Irani, M.; Gates, I. Understanding the Convection Heat-Transfer Mechanism in the Steam-Assisted-Gravity-Drainage Process. *SPE J.* **2013**, *18*, 1202–1216. [CrossRef]
11. Sivakumar, P.; Krishna, S.; Hari, S.; Vij, R.K. Electromagnetic heating, an eco-friendly method to enhance heavy oil production: A review of recent advancements. *Environ. Technol. Innov.* **2020**, *20*, 101100. [CrossRef]
12. Hu, L.; Andy Li, H.; Babadagli, T.; Ahmadloo, M. A semianalytical model for simulating combined electromagnetic heating and solvent-assisted gravity drainage. *SPE J.* **2018**, *23*, 1248–1270. [CrossRef]
13. Hu, L.; Li, H.A.; Babadagli, T.; Ahmadloo, M. Experimental investigation of combined electromagnetic heating and solvent-assisted gravity drainage for heavy oil recovery. *J. Pet. Sci. Eng.* **2017**, *154*, 589–601. [CrossRef]
14. Paz, P.Z.; Hollmann, T.H.; Kermen, E.; Chapiro, G.; Slob, E.; Zitha, P.L. EM Heating-Stimulated Water Flooding for Medium–Heavy Oil Recovery. *Transp. Porous Media* **2017**, *119*, 57–75. [CrossRef]
15. Wang, X.; Li, X.; Liu, S.; Zhou, H.; Li, Q.; Yang, J. Cis-9-Octadecenylamine modified ferric oxide and ferric hydroxide for catalytic viscosity reduction of heavy crude oil. *Fuel* **2022**, *322*, 124159. [CrossRef]
16. Zhao, Y. Laboratory experiment and field application of high pressure and high quality steam flooding. *J. Pet. Sci. Eng.* **2020**, *189*, 107016. [CrossRef]
17. Pratama, R.A.; Babadagli, T. Reconsideration of Steam Additives to Improve Heavy-Oil Recovery Efficiency: Can New Generation Chemicals Be a Solution for Steam-Induced Unfavorable Wettability Alteration? *Energy Fuels* **2020**, *34*, 8283–8300. [CrossRef]
18. Pratama, R.A.; Babadagli, T. Wettability state and phase distributions during steam injection with and without chemical additives: An experimental analysis using visual micromodels. *SPE Reserv. Eval. Eng.* **2020**, *23*, 1133–1149. [CrossRef]
19. Wang, Z.; Li, S.; Li, Z. A novel strategy to reduce carbon emissions of heavy oil thermal recovery: Condensation heat transfer performance of flue gas-assisted steam flooding. *Appl. Therm. Eng.* **2022**, *205*, 118076. [CrossRef]
20. Alomair, O.A.; Alajmi, A.F. A novel experimental nanofluid-assisted steam flooding (NASF) approach for enhanced heavy oil recovery. *Fuel* **2022**, *313*, 122691. [CrossRef]
21. Bai, Y.; Lian, Y.; Zhao, J.; Cao, Z.; Sun, J.; Zhang, H. Thermal-insulation and temperature-resistant foamed gel for thermal management of heavy oil steam flooding. *J. Mol. Liq.* **2022**, *359*, 119304. [CrossRef]

22. Pang, Z.; Liu, H.; Zhu, L. A laboratory study of enhancing heavy oil recovery with steam flooding by adding nitrogen foams. *J. Pet. Sci. Eng.* **2015**, *128*, 184–193. [CrossRef]
23. Xi, C.; Qi, C.; Zhang, Y.; Liu, T.; Shen, D.; Mu, H.; Dong, H.; Li, X.; Jiang, Y.; Wang, H. CO_2 assisted steam flooding in late steam flooding in heavy oil reservoirs. *Pet. Explor. Dev.* **2019**, *46*, 1242–1250. [CrossRef]
24. Wu, Z.; Wang, L.; Xie, C.; Yang, W. Experimental investigation on improved heavy oil recovery by air assisted steam injection with 2D visualized models. *Fuel* **2019**, *252*, 109–115. [CrossRef]

energies

Article

Multiple-Level Tectonic Control of Coalbed Methane Occurrence in the Huaibei Coalfield of Anhui Province, China

Zhigen Zhao [1,2] and Sheng Xue [2,3,*]

1 School of Earth and Environment, Anhui University of Science and Technology, Huainan 232001, China; zhgzhao@aust.edu.cn
2 Institute of Energy, Hefei Comprehensive National Science Center, Hefei 230031, China
3 School of Safety Science and Engineering, Anhui University of Science and Technology, Huainan 232001, China
* Correspondence: sheng.xue@aust.edu.cn

Abstract: The Huaibei coalfield is an important coal base and one of the hot spots of coalbed methane development in China. Therefore, a detailed understanding of gas occurrence in the Huaibei coalfield is of great significance. This paper analyzes the gas occurrence from the perspective of multiple-level tectonic control, i.e., the regional tectonic level, the coalfield tectonic level, the mining area tectonic level, and the coal mine tectonic level. This study deduces that gas occurrence in the Huaibei coalfield is characterized by multiple-level tectonic control. At the regional level, the Huaibei coalfield is located in the southeast margin of the North China plate, affected by the tectonic evolution of the North China plate and by the evolution of the Dabie–Tanlu–Sulu orogenic belt. Therefore, the regional geological tectonic is complex, leading to the high gas content and serious gas hazard. At the coalfield level, gas occurrence in the Huaibei coalfield is controlled by east–west faults, NNE faults, and the Xuzhou–Suzhou arc nappe tectonic, which results in the highest gas occurrence in the Suxian mining area, followed by the Linhuan mining area and the Suixiao mining area, while the lowest amount of gas occurs in the Guoyang mining area. At the mining area level, considering the Suxian mining area as an example, the gas occurrence is controlled by the distance from the Tancheng–Lujiang fault zone and the intensity of tectonic compression, i.e., coal mine gas in the east is the highest, followed by coal mines in the south, while coal mine gas in the west is the lowest. At the coal mine level, gas occurrence is controlled by the buried depth of the coal seam, the tensional normal fault, magmatic activity, and uplift and erosion of strata. Finally, the findings of this study may help in the prevention of gas hazard and the exploration and development of coalbed methane in the Huaibei coalfield and other coalfields of similar geological characteristics.

Keywords: Huaibei coalfield; tectonic evolution; gas occurrence; multiple-level tectonic control; coalbed methane

Citation: Zhao, Z.; Xue, S. Multiple-Level Tectonic Control of Coalbed Methane Occurrence in the Huaibei Coalfield of Anhui Province, China. *Energies* **2022**, *15*, 4977. https://doi.org/10.3390/en15144977

Academic Editor: Reza Rezaee

Received: 9 May 2022
Accepted: 4 July 2022
Published: 7 July 2022

1. Introduction

Gas explosions and coal and gas outbursts are potential hazards in high-gas coal mines [1,2]. Methane is the main component of coal mine gas, which is a clean and efficient energy source [3,4]. China is seriously affected by coal mine gas, and it is also actively engaged in the development and utilization of coalbed methane [5–7]. It is, therefore, important to further investigate the occurrence characteristics and gas distribution patterns of coal mines.

In the long-term study of gas geology, it is well recognized that the generation, migration, and occurrence of a certain area's coal mine gas are controlled by the geotectonic evolution of this area [8–10]. The multiple-level tectonic control of gas occurrence is proposed [11]. A macroscopic study of the multiple-level tectonic control of gas occurrence in China reveals that there are 29 areas of coal mine gas occurrence, including 16 high- and outburst-gas areas and 13 low-gas areas [12]. There are also some studies based on the multiple-level tectonic control of gas occurrence from the perspective of a certain coalfield, mining area, or coal mine. It is important to mention that in existing studies, multiple-level tectonic control may be expressed as "tectonic-level control" [12], "tectonic level-by-level control" [13], "stepwise tectonic control" [14,15], "tectonic gradual control" [16], and "step by step control" [17].

The Huaibei coalfield is rich in coal and coalbed methane resources. It features thick coal seams with a moderate depth and a relatively high gas content [18]. Therefore, the Huaibei coalfield is an important coal base in China and faces serious safety challenges in mitigating gas risk during coal mining operation [19,20]. As this coalfield is rich in coalbed methane resources [21,22], it is a hot spot for the development of coalbed methane resources. The coalfield has been explored and developed for its coal resources since 1955, and a number of surface test boreholes have been created in several pilot studies of coalbed methane development. A wealth of data has been obtained on the coalfield geology, gas geology and coalbed methane resources of the Huaibei coalfield. These data have substantially improved the understanding of the occurrence and distribution of coalbed methane. In this paper, the Huaibei coalfield is considered a target area in order to study the gas occurrence from the perspective of multiple-level tectonic control.

2. Geological Background

Anhui Province is located in eastern China. It is part of the Yangtze River Delta region, with a well-developed economy and a strong energy demand. Its west, north, and east are bound by Henan, Shandong, and Jiangsu Provinces [23]. The Huaibei coalfield is located in the northern part of Anhui Province and spreads across Huaibei City, Suzhou City, and Bozhou City of Anhui Province, with a coal-bearing area of approximately 4100 km^2 [24]. In the Huaibei coalfield, two large coal-producing enterprises mainly exist: the Huaibei Mining (Group) Co., Ltd., headquartered in Huaibei City of Anhui Province, China, and Wanbei Coal-electricity Group Co., Ltd., headquartered in Suzhou City of Anhui Province, China. In addition, at least 23 underground coal mines have been developed [18].

The Huaibei coalfield is in the southeastern margin of the North China plate. It belongs to Xusu depression in the south–central part of the Luxi-Xuhuai uplift area [25,26]. Regionally, the Huaibei coalfield is located between the nearly east–west Feng–Pei uplift and the Bengbu uplift and lies adjacent to the He-Huai subsidence area on the west and the Tancheng–Lujiang fault zone on the east [26,27]. The Huaibei coalfield is in north–south orientation. Its fault structure has a grid structure pattern. From north to south, the main east–west orientation faults are the Feng–Pei fault, the Subei fault, and the Banqiao fault. From east to west, the main north–south orientation faults are the Guzhen–Changfeng fault, the Fengxian–Kouziji fault, and the Xiayi–Gushi fault, as shown in Figure 1. According to the characteristics of tectonic development, the Huaibei coalfield can be subdivided into the Suixiao, Linhuan, Suxian, and Guoyang mining areas [23,24,27]. The Suixiao mining area is located north of the Subei fault, while the Suxian, Linhuan, and Guoyang mining areas are located between the Subei fault and the Banqiao fault. The Nanping fault is the boundary between the Suxian and Linhuan mining areas, while the Fengxian–Kouziji fault is the boundary between the Linhuan and Guoyang mining areas.

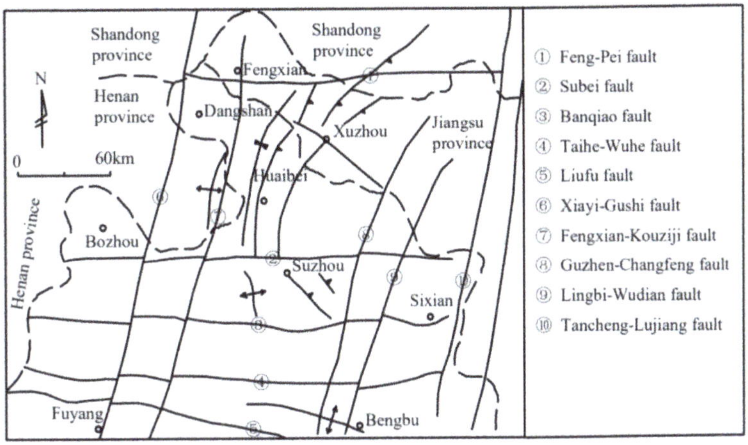

Figure 1. Tectonic outline map of the Huaibei coalfield. Modified from [28].

The strata in this study area in ascending order consists of the Archean, Qingbaikou system and Sinian, Cambrian and Ordovician, Carboniferous and Permian, Triassic, Jurassic and Cretaceous, Paleogene, Neogene, and Quaternary formations [23,26,27]. The Carboniferous and Permian are the coal-bearing formations, including in ascending order, the Benxi, Taiyuan, Shanxi, Lower Shihezi, Upper Shihezi, and Sunjiagou formations. The coal-bearing strata has a thickness of nearly 920 m. It is formed by shallow marine deposition, shore coastal deposition, fluvial delta deposition, and alluvial plain deposition [29]. The Benxi and Taiyuan formations contain several thin coal seams that are not minable, the Shanxi, Lower Shihezi, and Upper Shihezi formations are major coal-bearing strata, while the Sunjiagou Formation contains no coal seams. In total, 25 coal seams exist in the Huaibei coalfield. Among them, 12 coal seams are minable, with an average total thickness of almost 14.25 m, and the main minable coal seams are nos. 3_2, 7_2, 8_1, 8_2, and 10_2 [18,23,30].

3. Multiple-Level Tectonic Control of Gas Occurrence

The multiple-level tectonic control of gas occurrence in the Huaibei coalfield is analyzed at four levels: the regional tectonic level, the coalfield tectonic level, the mining area tectonic level, and the coal mine tectonic level.

3.1. Regional Tectonic Level

The basement of the coal-bearing strata in the Huaibei coalfield is complete and continuous, with a flat terrain and no large tectonic differentiation in this area. The whole crust subsides in Late Paleozoic. In addition, deposition occurs on basin basement, forming continuous and stable coal accumulation in Carboniferous Permian [31,32]. After the formation of coal-bearing strata, the evolution of coal seams' buried depth and temperature in the Huaibei coalfield can be subdivided into three major stages: Triassic to Middle Jurassic, Late Jurassic to Cretaceous, and Paleogene onward. From Triassic to Middle Jurassic, the buried depth of the Permian strata rapidly increased and reached almost 3000 m, with a maximum temperature range of 140–180 °C. From Late Jurassic to Cretaceous, the Permian strata uplifted, the overlying strata suffered from erosion, and the maximum erosion thickness reached 1800–2650 m. From Paleogene onward, the deposition in the area continued and the temperature of the coal-bearing formation was kept between 27 and 50 °C for a period of time [33,34]. The coal seam temperature was mainly controlled by its buried depth of the coal seam. In this case, the deep burial metamorphism of coal is important to the coalification process. In addition, the Yanshanian tectonic-thermal event superimposed magmatic thermal metamorphism on the coal seams in the Huaibei coalfield, which led to an increase in coal rank to coking coal, and even natural coke in some areas [35–37].

From the perspective of plate tectonics, since the formation of the Carboniferous Permian coal-bearing strata, the North China plate where the Huaibei coalfield is located, was mainly pushed by the Siberian plate from north to south and by the Yangtze block from south to north during the Indosinian period, forming wide and flat folds and faults in nearly east–west orientation. The North China plate was then mainly pushed by the subduction and collision of the Pacific Kula plate during the Yanshanian period, which formed a series of large-scale uplift and depression in NNE and NE directions, accompanied by intense magmatism. On the west side of the Tancheng–Lujiang fault zone, the Xuzhou–Suzhou arc nappe tectonic was formed by the subduction and collision of the Yangtze plate to the North China plate [11]. In summary, the tectonic evolution of the Huaibei coalfield has gone through three stages: the east–west trending tectonic development stage in the Indosinian period (257–205 Ma), the NEE trending tectonic development stage in the Yanshanian period (205–65 Ma), and the extensional tectonic development stage in the Himalayan period (65 Ma–) [38].

The Huaibei coalfield is located in the southeast margin of the North China plate. It lies adjacent to the Tancheng–Lujiang fault zone on the east, and therefore it is affected by the tectonic evolution of the North China plate and the evolution of the Dabie–Tanlu–Sulu orogenic belt. Consequently, the geological tectonic of the Huaibei coalfield is complex, the magmatic activity is frequent, and the faults and folds are well developed [26,39]. The tectonic in the Huaibei coalfield is controlled by east–west faults, NNE faults, and the Xuzhou–Suzhou arc nappe tectonic. The fault structure of the whole coalfield presents a grid structure pattern, and the NNE trending faults formed in the Yanshanian period transformed the east–west trending faults in the Indosinian period. Under the influence of the Dabie orogenic belt and the Tancheng–Lujiang fault zone, the Huaibei coalfield generally shows the tectonic pattern of pushing in the east part and sinking in the west part. More precisely, the east part is the Xuzhou–Suzhou arc nappe tectonic, while the west part is the Guoyang–Linhuan fault depression belt [26,40], as shown in Figure 2.

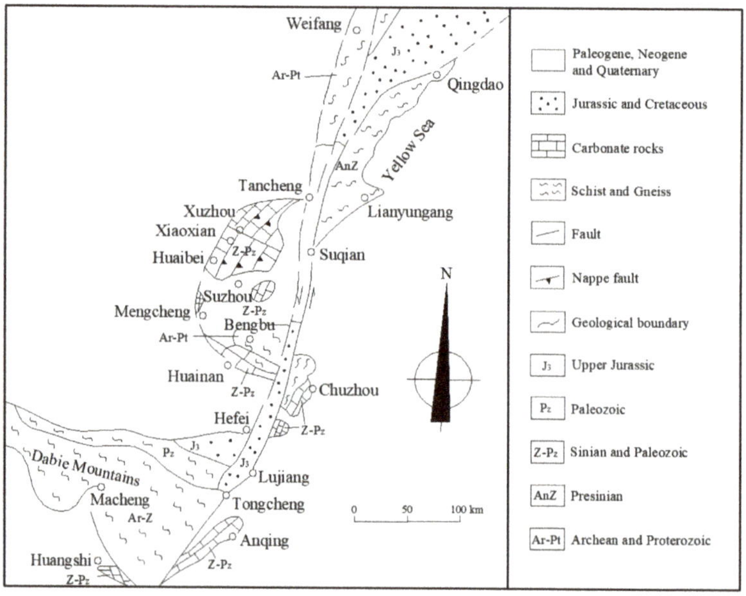

Figure 2. Geological background of the Xuzhou–Suzhou arc nappe tectonic. Modified from [41,42].

The regional tectonic evolution history controls the gas occurrence in the Huaibei coalfield. The complex regional geological tectonic leads to the high gas content and serious gas hazard in the Huaibei coalfield [12,43,44]. Li and Wang (2009) mention that all the 14 coal mines in Huaibei Mining (Group) Co., Ltd. are of high content. In addition, 7 of them, namely Luling, Haizi, Qinan, Zhuxianzhuang, Shitai, Taoyuan, and Tongting mines, are prone to coal and gas outbursts [43]. Since the development of coal resources in the Huaibei coalfield, 33 outburst incidences have occurred, with a total coal ejection of 11,339 t and gas emission of 1,404,000 m^3 [43]. Zhang and Wu (2013) mention that the Huaibei coalfield has 13 coal- and gas-outburst-prone coal mines and 9 high-gas coal mines. Moreover, more than 140 incidences of coal and gas outbursts have occurred. The largest outburst in the Huaibei coalfield occurred at the Luling coal mine, with a coal ejection of 10,500 t and gas emission of 1,230,000 m^3 [12].

3.2. Coalfield Tectonic Level

As previously mentioned, the Huaibei coalfield is controlled by the Xuzhou–Suzhou arc nappe tectonic (cf Figure 3). Therefore, the stress of the arc structure mainly controls the distribution characteristics of gas enrichment in the coalfield. The stress of the arc structure has two states, divided by the neutral plane of the arc structure. The stress state is compressive in the inward arc direction and tensile in the outward arc direction. In general, the curvature of the inner arc zone is greater than that of the outer arc. Therefore, the stress of the inner arc zone is concentrated and the compression is serious, which results in the development of more reverse faults. Thus, the gas content in the middle part of the arc is higher than that in the two wings of the arc, and the gas content in the inner arc zone is greater than that in the outer zone [45]. The Suxian mining area, which is located on the inner side of the arc structure and close to the Tancheng–Lujiang fault zone, is of high stress concentration due to compression. It has many reverse faults, and it is conducive to the accumulation and preservation of coal seam gas. Therefore, the Suxian mining area has a high gas content and faces serious gas hazard in coalmining operations [38]. The Linhuan mining area, which is located in the middle of the arc structure and slightly far away from the Tancheng–Lujiang fault zone, is in the area of transformation from compressive stress to tensile stress and has many tensile normal faults. This area is relatively conducive to the migration and release of coal mine gas. Therefore, although the Linhuan mining area has a high gas content, the gas hazard risk in the area is lower than that in the Suxian mining area. Although the Suixiao mining area is also located in the middle of the arc structure, it is on the wing of the arc structure rather than at the center of the arc structure. Therefore, it is in the area of tensile stress, and many tensile normal faults are developed. This area is relatively unfavorable to the accumulation and preservation of coal seam gas. Therefore, the Suixiao mining area has generally a low gas content. The Guoyang mining area is located on the outside of the arc structure and farthest from the Tancheng–Lujiang fault zone. Therefore, the influence of the Xuzhou–Suzhou arc nappe tectonic has largely disappeared. Thus, the Guoyang mining area has the lowest gas content in the Huaibei coalfield.

Besides the control of the Xuzhou–Suzhou arc nappe tectonic, the uneven subsidence of the Huaibei coalfield also affects the distribution characteristics of coal seam gas. The uneven subsidence of the Huaibei coalfield that occurred in the Himalayan period causes subsidence and sedimentation in Linhuan and Suxian mining areas located in the south of the Subei fault and the east of the Fengxian–Kouziji fault, which provides favorable conditions for gas preservation. The uneven subsidence has also caused uplift and erosion in the Suixiao mining area in the north of the Subei fault and the Guoyang mining area in the west of the Fengxian–Kouziji fault, which provides favorable conditions for gas migration and release and therefore results in low gas contents in the areas [25].

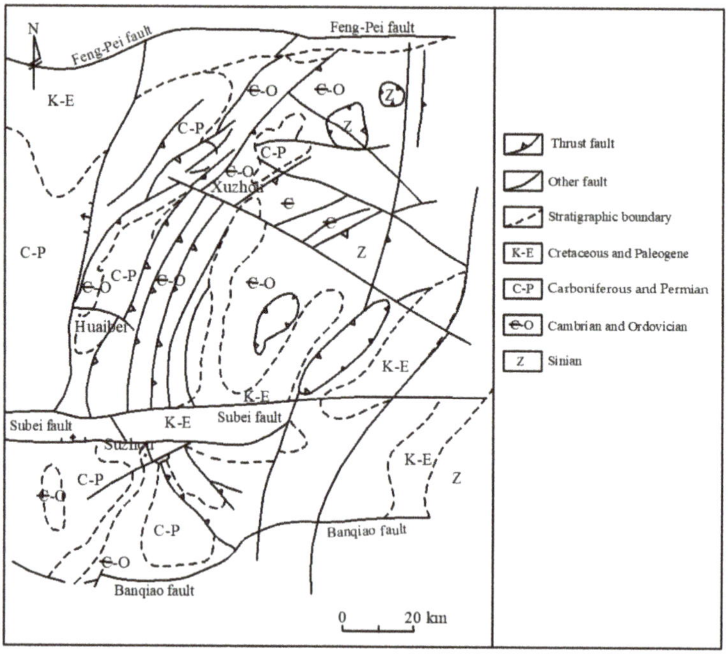

Figure 3. Geological map of the Xuzhou–Suzhou arc nappe tectonic. Modified from [46].

Thus, the gas distribution in the Huaibei coalfield presents a distribution pattern of "high in the south and low in the north," "high in the east and low in the west," and "the highest in the southeast" [25,40]. "High in the south and low in the north" denotes that the gas content and the gas hazard in the Suxian mining area and the Linhuan mining area in the south of the Subei fault are greater than those of the Suixiao mining area in the north of the Subei fault. "High in the east and low in the west" denotes that there are 3 mining areas in the south of the Subei fault: the Suxian, Linhuan, and Guoyang mining areas. The gas content and the gas hazard in the Suxian mining area in the east are the highest, followed by those in the Linhuan mining area in the middle, while the Guoyang mining area in the west has the lowest gas content and gas hazard risk. "The highest in the southeast" denotes that among the four mining areas in the study area, the gas content and the gas hazard in the Suxian mining area in the southeast of the Huaibei coalfield are the highest.

Previous studies have confirmed this understanding, as shown in Table 1. For instance, Zhou and Wang (2000) report that the gas content in the east of the Suxian mining area is the highest, averaging at more than 8 m^3/t. This is followed by the gas content in the south of the Suxian mining area, while the gas content in the Guoyang mining area is the lowest, averaging at less than 4 m^3/t. The average gas contents of no. 8 coal seam in the Suxiang, Linhuan, Suixiao, and Guoyang mining areas are 7.3, 6.1, 5.2, and 3.4 m^3/t, respectively [47]. Qu et al. (2008) report that the gas content of the Suixiao mining area north of the Subei fault is only 2–12 m^3/t, which is far less than that of mining areas in the south of the Subei fault. In addition, from the west to the east of the Huaibei coalfield, the gas content in the Guoyang mining area is 2–8 m^3/t, which is less than the 6–16 m^3/t in the Linhuan mining area and the 6–24 m^3/t in the Suxian mining area [25]. Wang (2015) reports that the gas contents of the Suxian mining area and the Linhuan mining area are, respectively, 12–15 m^3/t and 8–12 m^3/t, which is higher than the 8–10 m^3/t in the Suixiao mining area and less than the 8 m^3/t in the Guoyang mining area [45]. Hu et al. (2014) report that 75.8% outburst incidences occurred in the Suxian mining area, 21.2% in the Linhuan mining area, 3.0% in the Suixiao mining area, and none in the Guoyang mining area as no active coal production exists in this area at the time of reporting. Among the incidences, the Luling coal mine in the Suxian mining area is the most serious, and the number of outburst incidences accounts for 66.7% of the total in the Huaibei coalfield [40].

Table 1. The gas characteristics of the Huaibei coalfield.

Characteristic	Suxian Mining Area	Linhuan Mining Area	Suixiao Mining Area	Guoyang Mining Area	Source
Average gas content of no. 8 coal seam (m^3/t)	7.3	6.1	5.2	3.4	[47]
Gas content (m^3/t)	6–24	6–16	2–12	2–8	[25]
Gas content (m^3/t)	12–15	8–12	8–10	<8	[45]
Outburst incidence (%)	75.8	21.2	3.0	0.0	[40]

3.3. Mining Area Tectonic Level

This section analyzes the gas occurrence in these four mining areas from the perspective of the mining area tectonic level.

3.3.1. Suxian Mining Area

The main coal mines in the Suxian mining area include the Luling coal mine and the Zhuxianzhuang coal mine in the east, the Taoyuan coal mine, the Qinan coal mine and the Qidong coal mine in the south, and the Qianyingzi coal mine and the Zouzhuang coal mine in the west (cf Figure 4). The gas hazards these coal mines pose are serious. However, the characteristics of coal seam gas occurrence and the risk of coal and gas outburst are different for specific coal mines due to their different tectonic positions in the Suxian mining area.

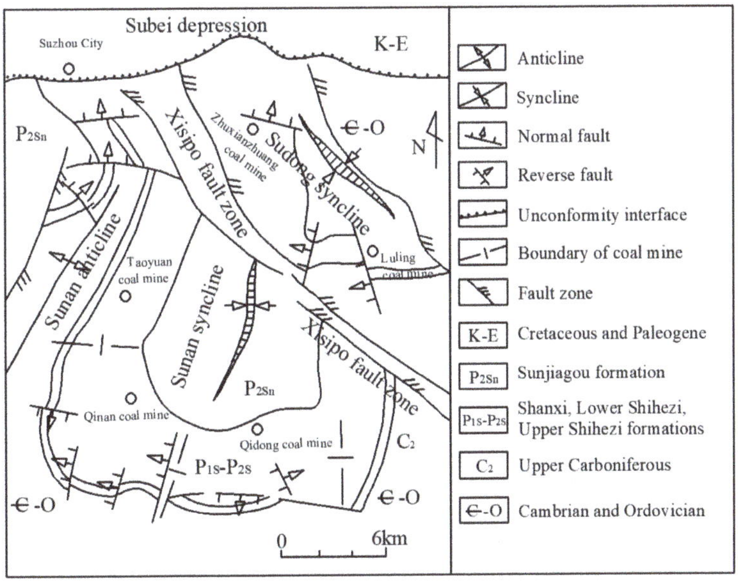

Figure 4. Tectonic outline map in the Suxian mining area. Modified from [38].

As for the Sudong syncline in the east of the mining area, it is in the hanging wall of the NW trending the Xisipo thrust fault (cf Figure 5). In addition, the tectonic compression and shear deformation are strong, and therefore the tectonic coals in coal seams are well developed, with large thickness and continuous distribution, which is the main reason why the Luling coal mine and the Zhuxianzhuang coal mine in the Sudong syncline are highly prone to coal and gas outburst [48,49].

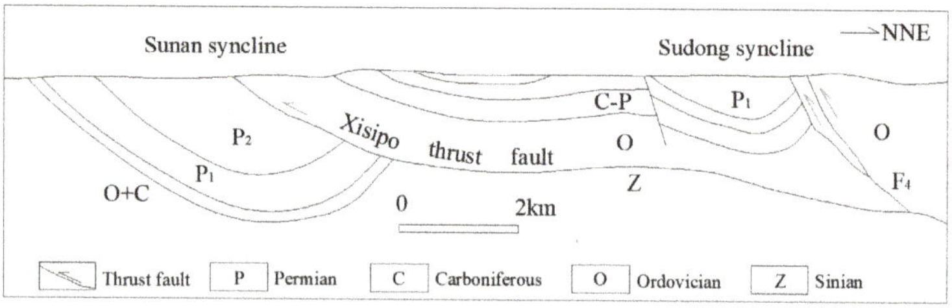

Figure 5. Geological section of the Xisipo thrust fault and its surrounding area. Modified from [50].

As for the Sunan syncline in the south of the mining area, it is in the footwall (i.e., a passive moving wall) of the NW trending Xisipo thrust fault (cf Figure 5). The tectonic compression and shear deformation are smaller than those of the Sudong syncline in the hanging wall. The south end of the Sunan syncline is close to the east–west direction's Bengbu uplift, and the east–west tectonics are relatively well developed, such as the Weimiao fault, which creates good conditions for gas release [3,51]. Therefore, the risk of coal and gas outburst at Taoyuan, Qinan, and Qidong coal mines in the Sunan syncline is smaller than that at the Luling and Zhuxianzhuang coal mines in the Sudong syncline.

As for the Sunan anticline in the west of the mining area, the Qianyingzi and Zouzhuang coal mines are on the west wing of the Sunan anticline. The tectonic compression and shear deformation of this location are further smaller than those of the Sudong syncline and the Sunan syncline. Furthermore, this location is divided by two high-angle normal faults (i.e., the Nanping normal fault and the Shuangduiji normal fault), which is conducive to gas release. In the long process of geological evolution, a large amount of coal seam gas has been released. Therefore, the risk of coal and gas outburst is much smaller than that of the Sudong syncline and the Sunan syncline [52,53].

3.3.2. Linhuan Mining Area

Affected by the superposition of multi-stage tectonic movements, many faults have been formed in the Linhuan mining area. The openness of these faults is fairly good, which is conducive to the release of coal seam gas, resulting in a significant reduction in gas content compared with the adjacent Suxian mining area. There are more than 10 coal mines in the Linhuan mining area, i.e., Haizi, Linhuan, Tongting, Zhangyoufang, Wugou, Jiegou, Xutuan, Renlou, Suntuan, and Yangliu coal mines, many of them distributed around the Tongting anticline. At the core of the Tongting anticline, due to the uplift of coal-bearing strata, the coal seams suffer from weathering and erosion, which forms gas release channels. The gas of coal seams the nearby weathering area can continuously release through the channels, thus reducing the gas pressure and gas content of these coal seams. Therefore, in general, the gas content of coal seams around the Tongting anticline is small [54].

However, the northwest area of the Tongting anticline where the Tongting coal mine, the Linhuan coal mine, and the Haizi coal mine are located is the superimposed intersection of the Tongting anticline formed in the early Yanshan period and the Yuandian syncline, the Qingtuan anticline, and the Haizi syncline formed in the Indosinian period. Consequently, the compressive stress is concentrated in this area. Moreover, in this area, the magmatic intrusion is serious. Due to the concentrated stress and serious magmatic intrusion, the coal seam structure is damaged and tectonic coal and even a soft layer in coal seams are developed. Thus, this is an area having a high gas pressure and gas content of coal seam, is a potential gas hazard, and could lead to serious coal and gas outburst in the Linhuan mining area [55]. For instance, the Haizi coal mine is a coal-and-gas-outburst coal mine, its maximum absolute gas emission rate is 55.80 m^3/min, and its maximum relative gas emission rate is 28.79 m^3/t. In addition, Xutuan and Renlou coal mines are located in the southeast edge of the Linhuan mining area. They are in the transition zone from the Suxian mining area, dominated by a compressive fault, to the Linhuan mining area, dominated by a tensile fault. It is also a stress concentration zone, which is conducive to gas accumulation and preservation. Thus, the gas pressure of the main coal seams at Xutuan and Renlou coal mines is greater than that in the middle and west of the Linhuan mining area. In addition, the coal seam here has a certain risk of coal and gas outburst. For instance, the Xutaun coal mine is a coal-and-gas-outburst coal mine, its maximum absolute gas emission rate is 58.34 m^3/min, and its maximum relative gas emission rate is 15.82 m^3/t.

3.3.3. Suixiao Mining Area

The Suixiao mining area is the earliest mining area in the history of coal resources development in the Huaibei coalfield. More than 10 coal mines are distributed around Zhahe syncline, i.e., the Yangzhuang, Zhuzhuang, Zhangzhuang, Daihe, Shitai, Shuoli, Yuanzhuang, Shenzhuang, Mengzhuang, and Maoyingzi coal mines [56]. The Suixiao mining area is located at the north of the Subei fault, which is relatively far from the Bengbu uplift. Therefore, it is less affected by the Bengbu uplift and is mainly controlled by the Xuzhou–Suzhou arc nappe tectonic. The Suixiao mining area is located in the wing of the arc structure rather than at the center of the arc structure. Therefore, it is relatively unfavorable to the accumulation and preservation of coal seam gas. In addition, the uneven subsidence of the Huaibei coalfield that occurred in the Himalayan period causes uplift and erosion of Triassic strata in the Suixiao mining area, which provides favorable conditions for gas migration and release. Therefore, the gas content in the Suixiao mining area is relatively low and several low gas coal mines exist [57]. These coal mines in the west of the Suixiao mining area, such as the Baishan coal mine, Liuqiao no. 1 coal mine, and the Hengyuan coal mine, are all low-gas coal mines due to the strong weathering and erosion of the coal seams.

When coal seams are located in the conditions of complex tectonic, strong tectonic compression, or deep burial depth, they may have high gas contents and even have the risks of coal and gas outburst. The magmatic activity may lead to secondary hydrocarbon generation, which results in uneven gas distribution in the area and sudden enrichment of gas in parts of the area. This location extruded by magmatic activity can be easily enriched by coal seam gas, and it is prone to coal and gas outbursts in mining activities. For instance, in general, the Maoyingzi coal mine has a low gas content and a low risk of coal and gas outburst. Its maximum absolute gas emission rate is only 4.9 m^3/min, and its maximum relative gas emission rate is only 6.4 m^3/t. However, there exists the risk of coal and gas outburst in the zones affected by magmatic intrusion and the zones with stress concentration, and an incidence of coal and gas outburst did occur in the mine [57].

3.3.4. Guoyang Mining Area

The Guoyang mining area is located on the west of the Linhuan mining area, bound by the Subei fault in the north, the Banqiao fault in the south, the Xiayi–Gushi fault in the west, and the Fengxian–Kouziji fault in the east. The Guoyang mining area is the most recently explored and developed among the four mining areas in the Huaibei coalfield. In the reported year [58], there were only two coal mines in the Guoyang mining area (Guobei and Liudian), while the others were at exploration stages. Compared with other mining areas in the Huaibei coalfield, the Guoyang mining area is the farthest away from the Tancheng–Lujiang fault zone and has been far away from the tectonic active zone on the plate edge. Therefore, most of the tectonic stress formed by the strong rock deformation at the plate edge has been diminished in the area [59]. In addition, the fault tectonic in the Guoyang mining area is relatively well developed and the main faults in the area are the east–west trending and north–south trending normal faults, which result in poor continuity of coal bearing strata and easy gas release. Therefore, the gas content and the gas hazard in this area are far less than those at the other three mining areas in the Huaibei coalfield [60].

However, it is important to mention that in the zones having a high gas content, the risk of coal and gas outburst may occur under the condition of an increased coal seam buried depth, complex tectonic, magmatic intrusion, and high compressive stress [61].

3.4. Coal Mine Tectonic Level

There are many coal mines in each of the mining areas. A total of four coal mines, selected respectively from each of the four mining areas, are used in this study.

3.4.1. Qidong Coal Mine in the Suxian Mining Area

There are two distinct characteristics of the tectonic control of the Qidong coal mine in terms of gas occurrence. The first is the differentiation between the north part and the south part of this coal mine. The Weimiao fault in the east–west direction is a major controlling fault of the Qidong coal mine, which divides it into a north part and a south part. The Mawan syncline and the Weidong anticline, with an axial nearly east–west direction, are developed around the Weimiao fault, as shown in Figure 6. Due to the fact that the Weimiao fault is a tensional normal fault system, it provides a good channel for gas release. In addition, a large part of the coal-bearing strata in the core of the Mawan syncline and the Weidong anticline has been eroded, making gas release favorable. Therefore, it is deduced that the gas content in the north part is 2–3 times higher than that in the south part at the same burial depth [62]. The second is the effect of the burial depth. For a given coal seam, the gas content and the risk of coal and gas outburst increase with the burial depth of the coal seam [62,63].

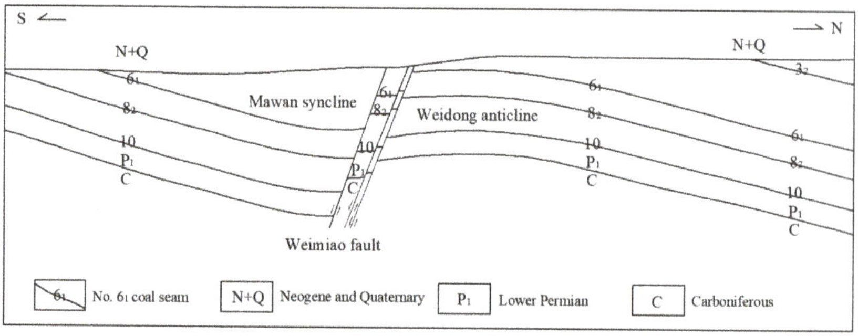

Figure 6. Geological section of no. 24 exploration line at the Qidong coal mine. Simplified from [64].

3.4.2. Tongting Coal Mine in the Linhuan Mining Area

The Zhaokou fault and the Mengji fault are two large normal faults at the boundary of the Tongting coal mine. The coal seams are cut and controlled by the Mengji fault in the shallow part and the Zhaokou fault in the deep part, as shown in Figure 7. Consequently, the occurrence characteristics of coal seam gas are largely controlled by these two faults. It is widely accepted in gas geology that the gas content and pressure in coal seams increase with the increase in the burial depth of coal seams. In addition, the gas emission and the risk of coal and gas outburst will be high when the seams are mined. However, an abnormal phenomenon occurs at the Tongting coal mine. That is, with the increase in the coal seam burial depth, the gas content, the gas pressure, and the gas emission first increase and then decrease [65,66], which is mainly due to the fact that the distribution of the coal seam gas is controlled by the Zhaokou fault in the deep part. The Zhaokou fault is a tensional normal fault in the deep part of this coal mine, which is characterized by a large fault drop, a wide fracture zone, and extended distance. During the long geological time, this normal fault has become a good channel for gas release, which results in a reduction in the gas content in the coal seams around the Zhaokou fault.

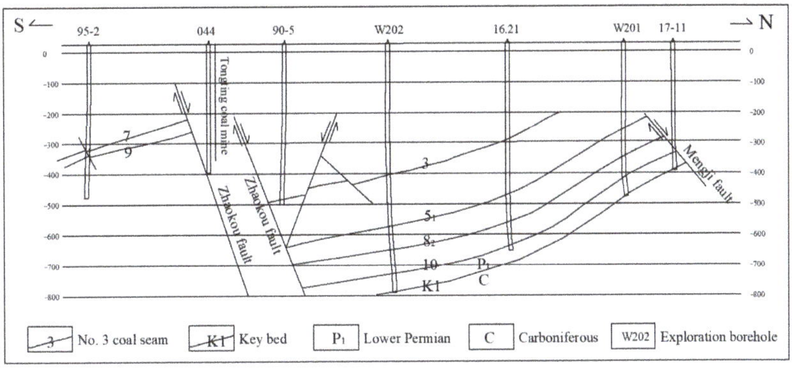

Figure 7. Geological section of no. w20 exploration line at the Tongting coal mine. Simplified from [67].

3.4.3. Qianling Coal Mine in the Suixiao Mining Area

Several coal and gas outbursts have occurred at the Qianling coal mine due to its complex geological tectonic [68]. The magmatic intrusion is serious in this coal mine. The magmatic rock has been found in 27 of a total 32 exploration boreholes. Based on the statistics of gas emissions of the working faces at the Qianling coal mine, it is deduced that the gas emission from the working faces affected by magmatic intrusion is greater than that from the unaffected working faces, and so is the risk of coal and gas outburst. This is due to the secondary hydrocarbon generation of the coal seam by magmatic intrusion at a high temperature and the overlying sill of the coal seam, which prevents gas from escaping from the coal seam [69].

3.4.4. Guobei Coal Mine in the Guoyang Mining Area

The overall tectonic of the Guobei coal mine is a monocline with nearly north–south strike and a near-west inclination. The south boundary is the F9 fault and the F9-1 fault, while the north boundary is the Liulou fault. Two intersecting faults (F22 and F26) divide this coal mine into four blocks, as shown in Figure 8. The tectonic characteristics in this coal mine are that the faults are mainly developed, while the folds are not well developed and the magmatic activities are not serious. The five normal faults are most important: F9, F9-1, Liulou, F22, and F26. In addition, 54 faults, i.e., 51 normal faults and 3 reverse faults, are found in this coal mine. These tensional normal faults provide favorable conditions for gas migration and gas release, which results in low gas pressure, low gas content, low gas emission, and low risk of coal and gas outburst [70,71]. In addition, the uplift and erosion of the Triassic strata that occurred in the Himalayan period also provide favorable conditions for gas migration and release. The data of geological exploration and coal mine construction confirm this understanding. The data of geological exploration reveal that the gas pressure (elevation from −442.10 to −806.40 m) of this coal mine is 0.20–0.60 MPa, the gas content (elevation from −439.62 to −878.71 m) is 0.01–6.96 m^3/t, the relative gas emission rate (elevation from −457.56 to −685.00 m) is 0.38–4.54 m^3/t, and the absolute gas emission rate (elevation from −457.56 to −685.00 m) is 0.37–5.72 m^3/min. The data in the coal mine construction reveal that the gas pressure in nos. 8_1 and 8_2 coal seams at the main shaft and the auxiliary shaft is 0.96 Mpa and 0.85 Mpa, respectively. In addition, the gas contents in nos. 8_1 and 8_2 coal seams at the main shaft are respectively 5.81 m^3/t and 5.23 m^3/t, and the gas content in no. 8_1 coal seam at the auxiliary shaft is 4.32 m^3/t [71]. The coal seams generally have a low gas content. However, the gas content of the seams may increase with the increase in the buried depth of the seam and tectonic complexity.

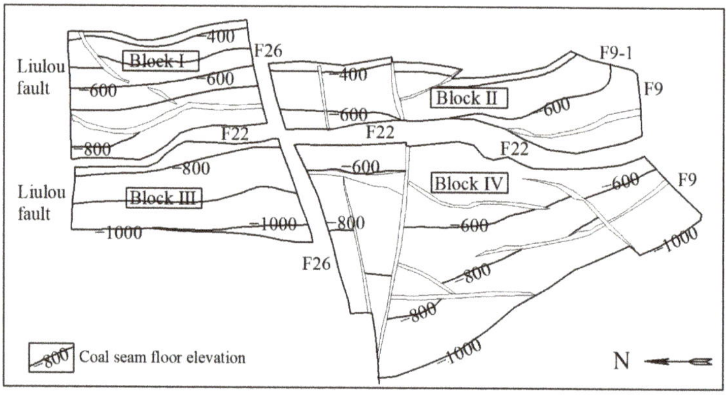

Figure 8. Tectonic map of no. 8 coal seam at the Guobei coal mine. Modified from [72].

Although these four coal mines considered in this study may not represent all the characteristics of all the coal mines in the Huaibei coalfield, the above analysis shows that in general, the gas occurrence characteristics are largely controlled by the buried depth of the coal seam, the tensional normal fault, magmatic activity, as well as the uplift and erosion of strata.

4. Discussion

Whether it is for the prevention and control of gas hazard or for the development of coalbed methane resources, the Huaibei coalfield is one of the areas for intensive study in China. However, these studies lack a systematic approach on gas occurrence in the Huaibei coalfield from the perspective of multiple-level tectonic control. Since the discovery of coal resources in the Huaibei coalfield in 1955, rich data have accumulated on regional geology, coalfield geology, and gas geology after a large period of coal resource exploration, coal mine construction and operation, and coalbed methane drainage [26,36,73]. This provides a solid foundation for the study of the multiple-level tectonic control of gas occurrence in the Huaibei coalfield. Based on the previous research results and the analysis of the rich data, this paper studied the multiple-level tectonic control of coal mine gas occurrence in the Huaibei coalfield from four levels: the regional level, the coalfield level, the mining area level, and the coal mine level.

Several geological factors affect the occurrence of coal seam gas. For instance, the coal maceral and coal rank may affect the generation of gas. The permeability of the coal seam and the hydrogeological condition of the coal mine may affect the migration of gas. The burial depth of the coal seam and the porosity of surrounding rock strata may affect the preservation of gas. The development characteristics and the evolution history of geological tectonic are important factors controlling the generation, migration, and preservation of gas [12,38,39]. The multiple-level tectonic control of gas occurrence in the Huaibei coalfield is helpful to understand the gas distribution characteristics at the regional level, the coalfield level, the mining area level, and the coal mine level. It is also useful in the prevention and control of gas hazard in coal mines and in the selection of favorable belts for coalbed methane development.

The Huaibei coalfield is one of the key areas of coal resource prospecting in Anhui Province [74,75]. In order to ensure sustainable and stable coal production, improve the guarantee capacity of coal resources, and promote the development of coalbed methane, it is required to continue the exploration and development of coal resources, coalbed methane resources, and related scientific studies. This study provides a reference for the future exploration and development in the Huaibei coalfield and for other coalfields or mining areas having a similar geological tectonic evolution history or gas geological occurrence. Finally, any new geological data related to the Huaibei coalfield or the research results of other coalfields or mining areas will also help to enrich and improve the understanding of the Huaibei coalfield.

5. Conclusions

Gas occurrence in the Huaibei coalfield is characterized by multiple-level tectonic control, as well as a tectonic level controls gas occurrence in a certain range. At the regional level, the Huaibei coalfield is located at the southeast margin of the North China plate. It is affected by the tectonic evolution of the North China plate and by the evolution of the Dabie–Tanlu–Sulu orogenic belt. Therefore, the regional geological tectonic is complex, which leads to a high gas content and serious gas hazard in the Huaibei coalfield. At the coalfield level, gas occurrence in the Huaibei coalfield is controlled by east–west faults, NNE faults, and the Xuzhou–Suzhou arc nappe tectonic, which results in the overall pattern of gas distribution. In other words, the Suxian mining area has the highest gas content, followed by the Linhuan mining area, the Suixiao mining area, and the Guoyang mining area. At the tectonic level of mining areas, considering the Suxian mining area as an example, the gas distribution of the coal mines is controlled by the distance from the Tancheng–Lujiang fault zone and the intensity of tectonic compression. That is, the highest gas content occurs in mines in the east, followed by mines in the south, and mines in the west. At the tectonic level of coal mines, the gas occurrence characteristics in each mining area are different, which shows the tectonic control characteristics as the buried depth of coal seams, the tensional normal fault, magmatic activity, and uplift and erosion of strata.

This study provides a reference for the subsequent exploration and development of coal and coalbed methane resources and the prevention and control of gas hazard in the Huaibei coalfield. It may also provide a reference for studies on other coalfields or mining areas having the same geological characteristics.

Author Contributions: Conceptualization, Z.Z.; methodology, Z.Z. and S.X.; software, Z.Z.; validation, S.X. and Z.Z.; formal analysis, Z.Z.; investigation, Z.Z.; resources, Z.Z. and S.X.; data curation, Z.Z.; writing—original draft preparation, Z.Z.; writing—review and editing, S.X.; visualization, Z.Z.; supervision, S.X.; project administration, S.X.; funding acquisition, S.X. All authors have read and agreed to the published version of the manuscript.

Funding: This work was supported by the Institute of Energy, Hefei Comprehensive National Science Center (No. 21KZS218).

Institutional Review Board Statement: Not applicable.

Informed Consent Statement: Not applicable.

Data Availability Statement: Not applicable.

Acknowledgments: The authors would like to thank the other members of the CO_2-ECBM team of the Institute of Energy, Hefei Comprehensive National Science Center, for their constructive advice. The authors also express their gratitude to Shuangying Tan and Xiang Fu from the Wanbei Coal-electricity Group Co., Ltd. for their kind help.

Conflicts of Interest: The authors declare no conflict of interest.

References

1. Xue, S.; Yuan, L. The use of coal cuttings from underground boreholes to determine gas content of coal with direct desorption method. *Int. J. Coal Geol.* **2017**, *174*, 1–7. [CrossRef]
2. Sachsenhofer, R.F.; Privalov, V.A.; Panova, E.A. Basin evolution and coal geology of the Donets Basin (Ukraine, Russia): An overview. *Int. J. Coal Geol.* **2012**, *89*, 26–40. [CrossRef]
3. Zhao, Z.G.; Sui, F.T.; Yan, J.P. Prediction of methane content of deep coal seams in the Sunan mining area in Anhui Province, China. *Int. J. Oil Gas Coal Technol.* **2020**, *23*, 351–364. [CrossRef]
4. Mendhe, V.A.; Bannerjee, M.; Varma, A.K.; Alka, D.K.; Subhashree, M.M.; Bhagwan, D.S. Fractal and pore dispositions of coal seams with significance to coalbed methane plays of East Bokaro, Jharkhand, India. *J. Nat. Gas Sci. Eng.* **2017**, *38*, 412–433. [CrossRef]

5. Qin, Y.; Moore, T.A.; Shen, J.; Yang, Z.B.; Shen, Y.L.; Wang, G. Resources and geology of coalbed methane in China: A review. *Int. Geol. Rev.* **2018**, *60*, 777–812. [CrossRef]
6. Hon, C.L.; Li, H.Y.; Huang, S. Challenges and opportunities of coalbed methane development in China. *Energy Fuels* **2017**, *31*, 4588–4602.
7. Zhou, F.B.; Xia, T.Q.; Wang, X.X.; Zhang, Y.F.; Sun, Y.N.; Liu, J.S. Recent developments in coal mine methane extraction and utilization in China: A review. *J. Nat. Gas Sci. Eng.* **2016**, *31*, 437–458. [CrossRef]
8. Zhao, Z.G.; Yan, J.P.; Liu, X. Research on gas occurrence regularity in Fengcheng mining area of Jiangxi province. *Energy Explor. Exploit.* **2014**, *32*, 591–599. [CrossRef]
9. Kędzior, S.; Maciej, J.; Pękała, Z. Geology, spatial distribution of methane content and origin of coalbed gases in Upper Carboniferous (Upper Mississippian and Pennsylvanian) strata in the south-eastern part of the Upper Silesian Coal Basin, Poland. *Int. J. Coal Geol.* **2013**, *105*, 24–35. [CrossRef]
10. Fisne, A.; Esen, O. Coal and gas outburst hazard in Zonguldak Coal Basin of Turkey, and association with geological parameters. *Nat. Hazards* **2014**, *74*, 1363–1390. [CrossRef]
11. Zhang, Z.M. *Gas Geology*; China University of Mining and Technology Press: Xuzhou, China, 2009; pp. 43–118.
12. Zhang, Z.M.; Wu, Y. Tectonic-level-control rule and area-dividing of coal mine gas occurrence in China. *Earth Sci. Front.* **2013**, *20*, 237–245.
13. Luo, Y.; Zhou, X.J.; Shi, J.B.; Feng, X.Z.; Liu, Z.J.; Zhang, Q. Discussion on tectonic level-by-level control and tectonic indicator prospecting in Panzhaozhuang copper nickel mining area. *Nonferrous Met. Eng.* **2013**, *3*, 48–50.
14. Yan, J.W.; Zhang, Y.Z.; Wang, W. Characteristics of gas occurrence under stepwise tectonic control in Pingdingshan mining area. *Coal Geol. Explor.* **2015**, *43*, 18–23.
15. Jia, T.R.; Zhang, Z.M.; Wei, G.Y.; Tang, C.A. Mechanism of stepwise tectonic control on gas occurrence: A study in North China. *Int. J. Min. Sci. Technol.* **2015**, *25*, 601–606. [CrossRef]
16. Huang, Z.X.; Gong, Y.W.; Ma, S. Tectonic gradual control on the gas occurrence of Shuicheng coalfield. *China Min. Mag.* **2018**, *27*, 307–310.
17. Zhao, T.Y.; Yang, S.Q.; Wang, X.N.; Yang, P. Step by step control of gas geology and main controlling factor analysis for Zhaoguan coal mine. *Saf. Coal Mines* **2020**, *51*, 174–177.
18. Jiang, B.; Qu, Z.H.; Wang, G.G.; Li, M. Effects of structural deformation on formation of coalbed methane reservoirs in Huaibei coalfield, China. *Int. J. Coal Geol.* **2010**, *82*, 175–183. [CrossRef]
19. Zhao, Z.G.; Wu, J.W.; Li, Q.G. Research on the prediction model of the amount of gas emission of No. 2 mining district and No. 4 mining district of Taoyuan coal mine. *Energy Technol. Manag.* **2007**, *4*, 43–44.
20. Zhang, K.Z.; Wang, L.; Cheng, Y.P.; Li, W.; Kan, J.; Tu, Q.Y.; Jiang, J.Y. Geological control of fold structure on gas occurrence and its implication for coalbed gas outburst: Case study in the Qinan coal mine, Huaibei coalfield, China. *Nat. Resour. Res.* **2019**, *29*, 1375–1395. [CrossRef]
21. Wang, C.; Tang, S.H.; Wang, Z.Y.; Sun, J.J.; Li, W.J. Coalbed methane resources exploration potential evaluation in Huaibei mining area. *Coal Geol. China* **2014**, *26*, 22–26.
22. Song, Y.; Jiang, B.; Liu, J.G. Nanopore structural characteristics and their impact on methane adsorption and diffusion in low to medium tectonically deformed coals: Case study in the Huaibei coalfield. *Energy Fuels* **2017**, *31*, 6711–6723.
23. Zhao, Z.G. Evaluation of coalbed methane resources in the Huainan-Huaibei coalfield of Anhui province, China, using an analytic hierarchy process and grey cluster method. *Arab. J. Geosci.* **2021**, *14*, 1718. [CrossRef]
24. Li, Q.G.; Ju, Y.W.; Bao, Y.; Yan, Z.F.; Li, X.S.; Sun, Y. Composition, origin, and distribution of coalbed methane in the Huaibei coalfield, China. *Energy Fuels* **2015**, *29*, 546–555. [CrossRef]
25. Qu, Z.H.; Jiang, B.; Wang, J.L.; Li, M. Characteristics of tectonic evolution and its controlling effects on coal and gas in Huaibei area. *Coal Geol. China* **2008**, *20*, 34–37.
26. Zhang, W.Y.; Zhu, W.W.; Dou, X.Z.; Zhao, Z.Y.; Hu, G.Q.; Ding, H.; Yi, X.H. Research progress on coal measure natural gas exploration and development in Huaibei and Huainan coalfield. *Coal Sci. Technol.* **2018**, *46*, 245–252.
27. Sui, F.T.; Dou, X.Z. Systematic research and its significance of unconventional natural gas in coal measures in Huainan and Huaibei coal fields. *Shanxi Coal* **2016**, *36*, 18–21.
28. Xu, D.J.; Hu, B.L.; Hu, W. Geological structure controlling magmatic intrusions into coalbeds in the Wolong coal mine of Huaibei coalfield. *Coal Geol. Explor.* **2011**, *39*, 1–5.
29. Zhao, Z.G.; Tang, X.Y.; Li, B.F. Geochemistry of rare earth elements of coal in Huaibei coalfield. *Geochimica* **2000**, *29*, 578–583.
30. Tong, L.; Luo, Y.; Zhou, F.; Wang, Y.F.; Li, S.Y.; Jia, L.X.; Yue, T. Genetic mechanisms of coalbed methane in typical districts from Huaibei coalfield, Eastern China. *Geodin. Acta* **2018**, *30*, 241–248. [CrossRef]
31. Shang, G.X. *Late Paleozoic Coal Geology of North China Platform*; Shanxi Science and Technology Press: Taiyuan, China, 1997; pp. 5–12.
32. Cheng, A.G.; Wei, Z.D. Discussion on the relationship between stratigraphic sequence and coal accumulation in Late Paleozoic coal accumulation basin in North China. *Coal Geol. China* **2001**, *13*, 7–9.
33. Wu, Y.D.; Ju, Y.W.; Hou, Q.L.; Hu, S.B.; Pan, J.N.; Fan, J.J. Comparison of coalbed gas generation between Huaibei-Huainan coalfields and Qinshui coal basin based on the tectono-thermal modeling. *Sci. China Earth Sci.* **2011**, *54*, 1069–1077. [CrossRef]
34. Liu, X.; Li, Y.Y. The factors influencing on 72 coal seam gas occurrence in Zouzhuang coal mine. *Coal* **2015**, *24*, 7–10.

35. Yang, Q.; Pan, Z.G.; Weng, C.M.; Su, Y.C.; Wang, Z.P. Regional magmatic thermal metamorphism and its effect on coal quality in China. *Geoscience* **1987**, *1*, 123–130.
36. Zhao, Z.G.; Tang, X.Y.; Huang, W.H.; Zhao, J.Y.; Tan, S.Y. Analysis of coal quality change at Qidong coal mine and its major controlling factors. *J. Anhui Univ. Sci. Technol. Nat. Sci.* **2009**, *29*, 1–4.
37. Xie, C.L.; Yan, J.P. Magmatic intrusion and its impact on coal seams in Huagouxi minefield, Guoyang mining area, Huaibei. *Coal Geol. China* **2015**, *27*, 18–20.
38. Li, Y.B.; Jiang, B. Structural characteristics and controlling on CBM hosting in Suxian mining area. *Coal Geol. China* **2014**, *26*, 26–30.
39. Zhai, M.G. Tectonic evolution of the North China Craton. *J. Geomech.* **2019**, *25*, 722–745.
40. Hu, S.R.; Zhang, Y.; Yue, T.; Peng, J.C.; Zhang, X.Q.; Zhang, Y.X.; Yu, M.D.; Li, X.F. Structural features and main mine gas controlling factors in northeastern part of North China Plate south margin. *Coal Geol. China* **2014**, *26*, 29–33.
41. Shu, L.S.; Wu, J.Q.; Liu, D.Z. Thrust tectonics of Xuzhou-Suzhou region, eastern China. *J. Nanjing Univ.* **1994**, *30*, 638–647.
42. Wu, J.L.; Shang, D.F. Framework of the Xu-Su arc structural belt and its influence on the coal (rock) bed in its front. *Geol. Anhui* **2019**, *29*, 190–195.
43. Li, L.; Wang, N. Analysis of geological law of coal and gas outburst in Huaibei mining area. *Coal Technol.* **2009**, *28*, 127–129.
44. Kuang, C.B. Study on prevention and control measures of coal and gas outburst in Taoyuan coal mine. *Inner Mongolia Coal Econ.* **2016**, *13*, 134–135.
45. Wang, Q.X. Characteristics of gas controlling in arc structure belt. *J. Henan Polytech. Univ. Nat. Sci.* **2015**, *34*, 451–454.
46. Wang, P.P.; Wang, L.C.; Wang, J.L.; Li, L.; Wang, J.; Du, J.P. The influence mechanism of Xusu arc structure on Yuanzhuang mine structure. *Coal Geol. Explor.* **2012**, *40*, 18–22.
47. Zhou, R.F.; Wang, X.Z. Analysis of gas characteristics of coal seams in Huaibei coalfield. *Jiangsu Coal* **2000**, *3*, 5–7.
48. Zhang, B. Analysis of geological factors affected gas in coal seam of Luling mine in Huaibei mining area. *J. North China Inst. Sci. Technol.* **2011**, *8*, 13–15.
49. Chen, W.S.; Liu, Z. Study of gas pressure distribution rule of Zhuxianzhuang mine of Suxian mining area in Huaibei. *Coal Technol.* **2013**, *32*, 131–133.
50. Yang, X. The Gas-Geology Law and Gas Forecasting in Qianyingzi Mine. Master Dissertation, Henan Polytechnic University, Jiaozuo, China, 2010.
51. Tong, L.H.; Ma, Z.X.; Zhao, Z.G. Study on distribution characteristics of gas content and deep gas prediction in Sunan mining area. *Coal Mine Blasting* **2019**, *37*, 19–23.
52. Wei, J.G.; Cui, H.Q.; Ma, D.X.; Li, X.X. Study on gas geological law and outburst prediction of no. 32 coal seam at Qianyingzi coal mine. *Coal Mine Mod.* **2011**, *1*, 39–41.
53. Song, Y.D.; Gao, J.N.; Zhang, Y. Study on gas occurrence characteristics of no. 82 coal seam at Zouzhuang coal mine based on the division of gas geological units. *Min. Technol.* **2019**, *19*, 168–171.
54. Kong, Y.F. The control of geological tectonic on gas occurrence and coal and gas outburst. *Energy Technol. Manag.* **2013**, *38*, 41–43.
55. Zhang, W.Y.; Wu, J.W.; Xu, S.P. Geological tectonic characteristics of Suxian and Linhuan mining areas and their influence on gas occurrence. *Coal Eng.* **2010**, *10*, 75–77.
56. Liu, J.; Li, W.; Yan, J.P. Comprehensive management and land efficient utilization of coal mining subsidence area in Huaibei Zhahe mining area. *Mine Constr. Technol.* **2018**, *39*, 4–9.
57. Shi, X.W.; Zhang, Y.G.; Zhang, Z.M. Structure control analysis on coal and gas outburst in Maoyingzi mine. *Coal Sci. Technol.* **2007**, *35*, 55–57.
58. Chen, Y.C.; Yan, J.P.; Li, L. Comparative analysis study on prospecting and mining of large and medium-sized faults in Guoyang mining area. *Coal Sci. Technol.* **2019**, *47*, 235–239.
59. Lv, F.J.; Xu, S.P. Tectonic characteristics and genetic analysis of Guoyang mining area in the west part of Huaibei coalfield. *J. Huainan Inst. Technol.* **2002**, *22*, 13–16.
60. Xu, M.Z. Geological structure control of gas enrichment process in Huaibei coalfield. *Saf. Coal Mines* **2008**, *39*, 73–75.
61. Cao, S.L. Safety technology of cutting through coal seam of shaft at Guobei coal mine. *Min. Saf. Environ. Prot.* **2007**, *34*, 53–55.
62. Li, H.L.; Qin, Y.; Zhang, Y.G.; Jian, K. Structure control on gas deposition in Qidong minefield of Huaibei mining area. *Coal Sci. Technol.* **2014**, *42*, 42–46.
63. Zhao, Z.G.; Tang, X.Y. Characteristics of borehole distribution and its effect to methane content in coalbed. *Sci. Technol. Manag. Land Resour.* **2003**, *20*, 62–63.
64. Qidong Coal Mine of Anhui Hengyuan Coal & Electric Co., Ltd. *Production Geological Report of Qidong Coal Mine*; Qidong Coal Mine of Anhui Hengyuan Coal & Electric Co., Ltd.: Suzhou, China, 2019.
65. Ying, C.J.; Wang, L. Influence of open fault on methane occurrence in coal seam. *Coal Min. Technol.* **2012**, *17*, 87–89.
66. Wang, L.G.; Cheng, Y.P.; Wang, L.; Li, W. Research on the control Zhaokou fault on the coal seam gas occurrence. *Saf. Coal Mines* **2013**, *44*, 6–9.
67. Tongting Coal Mine of Huaibei Mining (Group) Co., Ltd. *Mine Geological Report of Tongting Coal Mine*; Tongting Coal Mine of Huaibei Mining (Group) Co., Ltd.: Huaibei, China, 2002.
68. Si, C.F.; Chen, Y.C.; Qian, X.S. Risk prediction of coal and gas outburst at Qianling coal mine. *Coal Min. Technol.* **1998**, *4*, 39–41.
69. Cai, C.C.; Cheng, Y.P.; Wang, L.; Li, W.; Lu, S.Q. Study on gas occurrence law in magma erosion area of Qianling coal mine. *Saf. Coal Mines* **2012**, *43*, 15–19.

70. Cao, S.L. Discussion on gas comprehensive treatment technology in fully-mechanized top coal caving mining face. *Miner. Eng. Res.* **2009**, *31*, 86–88.
71. Qi, Z.K.; Zheng, X.J. Determination of basic gas parameters of shaft coal uncovering at Guobei coal mine. *Inner Mongolia Coal Econ.* **2018**, *22*, 152–154.
72. Ji, H.J.; Li, Z.H.; Liu, Z.; Yang, Y.L.; Liu, J. Influence of geological structure on gas emission law in Guobei coal mine. *China Coal* **2011**, *37*, 26–29.
73. Wei, Z.D. *Occurrence Law of Coal Resources and Coal Prospecting Prediction in Anhui Province*; Geological Publishing House: Beijing, China, 2012.
74. Jiang, D.; Wang, H.Z. Analysis of coal prospecting prospect in the east of Suixiao mining area of Huaibei coalfield. *Shandong Coal Sci. Technol.* **2020**, *12*, 159–161.
75. Sun, G.; Zhan, R.; Sui, F.T.; Wang, S. Analysis of the guarantee capacity and prospecting direction of coal resources in Anhui province. *Geol. Anhui* **2021**, *31*, 103–105.

energies

MDPI

Article

Experimental Investigation of Pore Characteristics and Permeability in Coal-Measure Sandstones in Jixi Basin, China

Huazhou Huang [1,2,3,*], Yuantao Sun [2,3], Xiantong Chang [2,3], Zhengqing Wu [2,3], Mi Li [2,3] and Shulei Qu [2,3]

1 Jiangsu Key Laboratory of Coal-Based Greenhouse Gas Control and Utilization, China University of Mining and Technology, Xuzhou 221008, China
2 Key Laboratory of Coalbed Methane Resources and Reservoir Formation Process, China University of Mining and Technology, Ministry of Education, Xuzhou 221008, China
3 School of Resources and Geosciences, China University of Mining and Technology, Xuzhou 221008, China
* Correspondence: huazhouh@163.com

Abstract: The research of pore and permeability characteristics of tight sandstone reservoirs in coal-measure is critical for coal-measure gas development. In this study, the pore systems of tight sandstones were studied based on low-field nuclear magnetic resonance (LF-NMR) data. The permeability of tight sandstones was obtained by the tester based on the pulse transient method. The permeability variation with the effective stress, grains, and pore characteristics was analyzed. The results show that the tight sandstone reservoirs in the coal-measure have low total porosity (2.80–4.14%), low effective porosity (0.51–1.56%), and low permeability (0.351×10^{-6}–13.910×10^{-6} um^2). LF-NMR T_2 spectra of the testing sandstones show that the micropores are the most developed, but most of the micropores are immovable pores. The pore characteristics are significantly affected by the grain size of sandstones. The pore connectivity ranks from good to poor with decreasing sandstone particle size. The total porosity and effective porosity increase with the grain size. There is a near-linear negative relationship between permeability and effective stress when the effective stress is between 405 psi and 808 psi. The greater the number of movable pores and the larger the effective porosity, the bigger the permeability of the sandstone. The effective porosity of sandstones is a sensitive indicator for evaluating the permeability of tight sandstone reservoirs. The stress sensitivity coefficient of permeability (S_s) increases with the increase of the effective stress. The sandstone with lower permeability, smaller effective porosity, and finer grains has a higher S_s. The particle size of sandstone from coal-measure has a great influence on both permeability and S_s. The findings will provide a better understanding of the characterization of pore structure and permeability in the process the coal-measure gas extraction, which is useful for the efficient development of coal-measure gas.

Keywords: sandstone; permeability; pore characteristics; permeability stress sensitivity; Jixi Basin

Citation: Huang, H.; Sun, Y.; Chang, X.; Wu, Z.; Li, M.; Qu, S. Experimental Investigation of Pore Characteristics and Permeability in Coal-Measure Sandstones in Jixi Basin, China. *Energies* **2022**, *15*, 5898. https://doi.org/10.3390/en15165898

Academic Editor: Rouhi Farajzadeh

Received: 14 July 2022
Accepted: 11 August 2022
Published: 14 August 2022

1. Introduction

The coal-measure formation is rich in unconventional natural gas resources, which is called coal-measure gas, including coalbed methane in coal reservoirs, tight sandstone gas in sandstone reservoirs, and shale gas in shale reservoirs [1,2]. Coal-measure gas development has attracted global attention because of its huge resource potential, especially in China [1,3,4]. Coal seams are generally mixed with sandstones, mudstones, and shales in coal-measure [5]. Coal seams and shale in the coal-measure could be the main source rocks for coal-measure gas because they are favorable for hydrocarbon generation and gas accumulation, while the conditions, such as burial depth, time, and temperature, are appropriate [1,6,7]. Sandstones in the coal-measure are one kind of reservoir that could store gas charged by the above source rocks [4,6,8,9]. Most of the sandstones in coal-measure are tight sandstones because their porosity is less than 10%, and their permeability is less than 0.1 mD [1,6,10]. The reservoir characteristics among the coal, shale, and sandstone in the coal-measure vary greatly [11]. The pore and permeability characteristics of the coal-measure gas reservoir could affect the gas accumulation and production [12,13]. Many studies have focused on the reservoir characteristics of coal [14,15], while the research on the pore and permeability characteristics of tight sandstone in coal-measure in the study area has been less focused upon. The tight sandstones in coal-measure are characterized by smaller pores and complex pore throat connections [16], which lead to lower permeability. The research on pore and permeability characteristics of tight sandstone reservoirs is critical for coal-measure gas development [15].

Low field nuclear magnetic resonance (LF-NMR) is a non-destructive method to characterize the pore structure parameters of complex porous media on a wide scale compared with N_2 adsorption and mercury porosimetry [1,4,17–19]. The sandstone sample should be saturated with water when the pore characteristics of the sandstone are obtained by LF-NMR. The water-saturated sample is then placed in a strong static magnetic field generated by an external magnet, and a small radio frequency magnetic field is superimposed on the static magnetic field to excite the hydrogen nuclei in the water, thereby inducing nuclear magnetic resonance [20]. When the radio frequency field is turned off, a signal varying with time can be received [21]. The amplitude of the signal change can be described by the longitudinal relaxation time (T_1) and the transverse relaxation time (T2). For LF-NMR, T_2 carries similar information to that of T_1 [20,22,23]. Therefore, T_2 is generally applied to the analysis of the pore structure parameters of sandstone samples. Pore sizes, porosity, pore connectivity, and pore throat distribution for tight sandstone reservoirs could be characterized by the relaxation time T_2 distribution [1,14,19,24]. However, it cannot determine the permeability of the sandstone directly [1,16]. Tight sandstone reservoirs have complex, highly heterogeneous pore structures, and poor pore connectivity [1,25]. The pore size and pore distribution vary greatly for the samples with different lithology. There is a close relationship between permeability and pore characteristics including porosity, pore-throat size, and pore volume [14,15,26,27]. The permeability of tight sandstones cannot be assessed only by total porosity because of the sandstone's complex pore-throat geometry [26,28]. However, the movable fluid porosity attained by LF-NMR might have a certain indication on permeability [1,14].

The previous reports have studied the relationship between permeability and scaling in the water injection, pore fluid pressure, confining pressure, and effective stress [29–32]. Generally, permeability decreases with the rise of the confining pressure and the effective stress [31], the permeability declines dramatically under 4–8 MPa confining stress, and permeability reduces slightly for 8–12 MPa conditions [29–31]. In a deep coal-measure reservoir, stress increases with the burial depth, which results in low permeability. The extraction of coal-measure gas would increase the effective stress on pores and fractures, which results in the compression of pores and the closure of fractures [29,31,33–37]. The permeability of the reservoir gradually decreases with the increase of the effective stress in the process of coal-measure gas extraction [38]. This kind of phenomenon is known as the stress sensitivity of the reservoir [39]. The permeability loss by the stress sensitivity will significantly affect the water and gas production rate because the original permeability of the coal-measure reservoir is generally low [15,29,38,40,41]. The stress sensitivity coefficient of reservoir permeability (S_s) for sandstones [37,42] is an index of reservoir permeability in response to effective stress variation [38], which could be affected by the plastic mineral content, pore throat size, porosity, permeability, etc. [37]. For instance, a higher the plastic mineral content could result in a bigger S_s [37].

The effect of sandstone grains on pore characteristics, permeability, and S_s is still unclear, and the relevant research about that is of great significance to coal-measure gas production [26,37,38,43,44]. The objective of this study was to analyze the characteristics of pore systems and permeability in tight sandstones with different grains based on Coal-measure Gas Well HJD1 in Jixi Basin in Northeast China. The pore systems were studied based on LF-NMR data. The permeability was obtained by the tester based on the pulse transient method. The effects of pore characteristics on permeability variation were discussed. Then, permeability variation with effective stress and grain size of sandstones were analyzed in detail. The findings will provide a better understanding of the characterization of pore structure and permeability in the process of coal-measure gas extraction, and will be useful for the efficient development of coal-measure gas.

2. Geological Setting

Jixi Basin is mainly composed of the northern depression and the southern depression, which are separated by the central Mashan fault (Figure 1a). Large-angle normal faults are well developed in the basin. The strikes of the faults are mainly east–west and northeast.

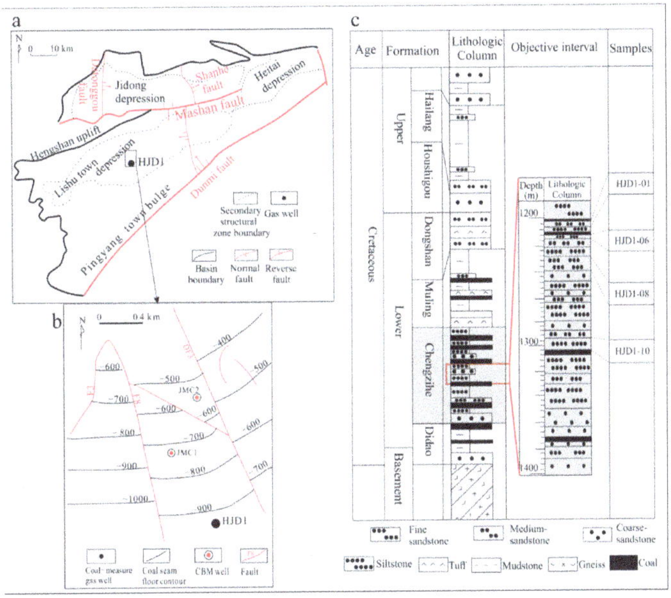

Figure 1. Geological background of the study area: (**a**) Location of Coal-measure Gas Well HJD1 in Jixi Basin; (**b**) geological structure of the area where Well HJD1 is located; (**c**) sampling stratigraphic column in Well HJD1.

The Quaternary in the Cenozoic and the Lower Cretaceous in the Mesozoic are the main strata in the research area where Well HJD1 is located (Figure 1). The Lower Cretaceous covers the Jixi Group and Huashan Group. The Jixi Group consists of Didao Formation, Chengzihe Formation, and Muling Formation from bottom to top (Figure 1c). The Muling Formation and Chengzihe Formation in the Lower Cretaceous are the main coal-bearing strata. Well HJD1 was constructed for coal-measure gas extraction from the Chengzihe Formation in the Jixi Basin, which is in eastern Heilongjiang province, in northeast China.

The Chengzihe Formation is the target strata in this research (Figure 1c). The buried depth of the Chengzihe Formation in Well HJD1 is 767.50–1739.05 m. The Chengzihe Formation is mainly composed of siltstone, fine sandstone, and coal seam. The sandstone samples used in this study are taken from the middle section of the Chengzihe Formation. The main coal seams involved in the sampling section are Numbers 22, 23, 25, and 28, respectively. In addition, most sandstone layers are between 2–8 m in thickness, as the sandstone layers are separated by thin mudstone layers and coal seams. The grains of sandstone also vary greatly.

3. Samples and Experimental Methods

3.1. Sample Collection and Processing

There are 4 sandstone samples collected from Well HJD1: medium sandstone (HJD1-01), fine sandstone (HJD1-06 and HJD1-08), and siltstone (HJD1-10). The 4 sandstone samples were lithologically classified according to the rock flake identification standard SY/T5368-2016 issued by the National Energy Administration in China. The grain size, and the percentage of clastic constituents and clay of the four samples can be seen in Table 1. The appearance of the rock samples was gray white sandstone without visible cracks.

Table 1. Rock thin section identification results of sandstone samples.

Sample No.	Clastic Constituents/%				Matrix/%	Grain Size/mm	Lithology
	Quartz	Feldspar	Detritus	Mica	Clay		
HJD1-01	56	12	13	3	-	0.30–0.50	Medium sandstone
HJD1-06	32	28	13	2	23	0.05–0.10	Fine sandstone
HJD1-08	55	18	11	1	10	0.06–0.12	Fine sandstone
HJD1-10	15	11	-	-	72	0.03–0.06	Siltstone

The collected sandstones were made into cylinders for permeability testing with a diameter of 50 mm and a height of 30 mm–100 mm. In addition, the flatness of the top and bottom surfaces of the sandstone samples was within 0.05 mm. Then, these samples were washed and dried before the test, in accordance with the requirements of Practices for Core Analysis (SY/T5336-2006).

HJD1-10 (silts sandstone), HJD1-06 (fine sandstone), and HJD1-01 (medium sandstone) are sandstones with different grain sizes. They were selected to undergo LF-NMR experiments. The LF-NMR experiments were conducted to characterize the pore system of the sampling sandstones and illustrate the effect of pore characteristics on the permeability of the sandstone.

3.2. Permeability Experiment

The permeability was tested by a Pulse Decay Permeameter-PDP-200 (American Core Lab Company, Tulsa, OK, USA, Figure 2), which was based on the pulse transient method. The permeability could be obtained based on the pressure decay curve. Practices for Core Analysis (SY/T5336-2006) was the reference method for this experiment, and the test gas was nitrogen. Permeability experiments were conducted by increasing the confining pressure and fixed pore pressure. This simulates the increasing effective stress during the coal-measure gas development.

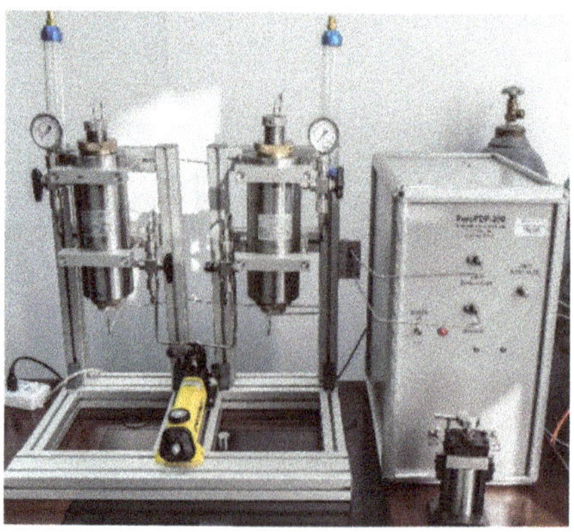

Figure 2. Pulse Decay Permeameter-PDP-200.

In this experiment, the test sample was put into a closed container, then the initial pressure was set, and time allowed for the initial pressure to reach the equilibrium state. Then, a pulse pressure (ΔP) was applied to the top of the sample. The gas percolated through the sample under the pressure ΔP, and this process resulted in the decrease of the pressure of the top container and the increase of the pressure of the bottom container. Finally, another equilibrium state (P_f) was reached between the top container and the bottom container; there was no pressure difference attenuation.

The intake nitrogen gas pressure, which is also known as pore pressure, was set to 700 psi, and the average pore pressure stabilized between 682 psi and 704 psi. The confining pressure for each sample was set to 1100 psi, 1200 psi, 1300 psi, 1400 psi, and 1500 psi to obtain the permeability. The temperature in the permeability tests was the indoor atmospheric temperature (about 25 °C).

3.3. LF-NMR Experiments

The LF-NMR experiments were performed by the MesoMR23-060H-I instrument with a main frequency of 23.400 MHz (Figure 3). The temperature of the laboratory where the NMR experiments were performed was 23.5 °C, the magnet temperature was constant at (35 ± 0.02) °C, the echo spacing was 0.1 ms, the numbers of scans were 32, and the echo numbers were 8000.

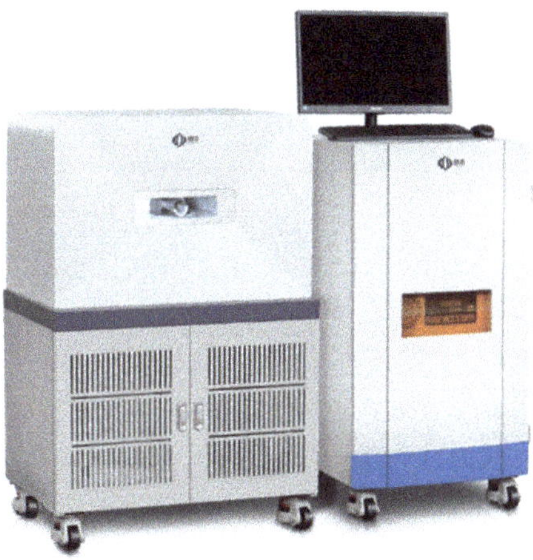

Figure 3. MesoMR23-060H-I instrument for LF-NMR experiments.

First, the columnar sandstone samples were dried. The sample was put into a 101 electric heating blast box for nearly 24 h until its weight remained constant. In addition, the samples were vacuumed for 2 h, and then they were saturated with water for nearly 24 h in the 2XZ-4B vacuum-saturation device. The quality difference of the sandstone sample before and after saturation was less than 0.05%. Moreover, the T_2 spectrum distribution of the saturated sandstone sample was obtained by the NMR test instrument. After that, the movable water within the saturated sandstone sample was removed by the TCL-21M desktop high-speed refrigerated centrifuge. Finally, the T_2 spectrum distribution of the centrifuged sandstone samples was tested by the MesoMR23-060H-I instrument.

In an LF-NMR test, the number of hydrogen atoms present within a fluid in a porous medium can be detected by the transverse relaxation time (T_2) [1,14], and the T_2 spectrum is generally used to characterize the physical parameters of the rock, such as pore size distribution and connectivity. It is supposed that longer T_2 corresponds to larger pores, while shorter T_2 corresponds to the smaller pores [4,28]. The amplitude of the T_2 distribution reflects the number of pores within a certain size. The higher the amplitude, the greater the number of pores [20,21].

The application of NMR in the study of pore types in sandstone is based on the fact that T_2 is positively correlated with pore size. This relationship can be expressed as [20,28,45]:

$$\frac{1}{T_2} = \rho_2\left(\frac{S}{V}\right) \tag{1}$$

where T_2 is the transverse relaxation time resulted from surface interactions, ms; and ρ_2 is a constant representing the transverse relaxation strength, μm/ms; S is the surface area of pores (cm^2); and V is the pore volume (cm^3).

4. Results

4.1. Permeability

The permeability of testing sandstones was between 0.351×10^{-3}–13.910×10^{-3} mD under the confining pressure of 1100–1500 psi. The permeability of the medium sandstone was the highest, followed by the fine sandstone, and the siltstone was the smallest under similar pore pressure and confining pressure.

For the medium sandstone (HJD1-01), its permeability was between 11.160×10^{-4}–13.910×10^{-3} mD under the confining pressure of 1100–1500 psi, with an average of 10.118×10^{-3} mD (Table 2).

Table 2. Permeability results.

Sample Number	Test Gas	Confining Pressure/psi	Average Pore Pressure/psi	Permeability/ $\times 10^{-3}$ mD	Lithology	Sampling Depth/m
HJD1-01	N_2	1100	695	13.910	Medium sandstone	1201.60–1201.88
	N_2	1200	700	13.410		
	N_2	1300	699	12.860		
	N_2	1400	704	12.110		
	N_2	1500	704	11.160		
HJD1-06	N_2	1100	687	1.188	Fine sandstone	1300.12–1300.54
	N_2	1200	695	1.081		
	N_2	1300	697	0.891		
	N_2	1400	697	0.738		
	N_2	1500	703	0.689		
HJD1-08	N_2	1100	691	1.060	Fine sandstone	1301.88–1302.5
	N_2	1200	700	0.932		
	N_2	1300	705	0.788		
	N_2	1400	705	0.656		
	N_2	1500	704	0.594		
HJD1-10	N_2	1100	693	0.643	Siltstone	1380.95–1381.82
	N_2	1200	707	0.571		
	N_2	1300	682	0.489		
	N_2	1400	698	0.410		
	N_2	1500	699	0.351		

For the fine sandstone (HJD1-06, HJD1-08), the permeability was between 0.594×10^{-3}–1.188×10^{-3} mD. The permeability of HJD1-06 was between 1.188×10^{-3}–0.689×10^{-3} mD and the average permeability was 0.917×10^{-3} mD under 1100–1500 psi confining pressure. The permeability of HJD1-08 between 1.060×10^{-3}–0.594×10^{-3} mD and the average value was 0.806×10^{-3} mD under the condition of 1100–1500 psi confining pressure (Table 2).

For the siltstone (HJD1-10), its permeability was between 0.351×10^{-3}–0.643×10^{-3} mD, with an average of 0.493×10^{-3} mD (Table 2).

The grain size of sandstone has an obvious effect on permeability. The permeability of the sampling sandstone increases with the increase of grain size. The permeability of the medium sandstone (HJD1-01) with an average of 10.118×10^{-3} mD was higher than that of the other samples. However, the permeability of the fine sandstone (HJD1-06 and HJD1-08) and the siltstone (HJD1-10) was relatively close. The permeability of the fine sandstone was bigger than that of the siltstone under similar confining pressure and pore pressure.

The permeability and confining pressure are negatively correlated. The permeability gradually decreases as the confining pressure increases for the medium sandstone, the fine sandstone, and the siltstone (Figure 4). The decrease in permeability for HJD1-01, HJD1-06, HJD1-08, and HJD1-10 while the confining pressure increased from 1100 to 1500 psi was 2.75×10^{-3} mD, 0.499×10^{-3} mD, 0.466×10^{-3} mD, and 0.292×10^{-3} mD, respectively.

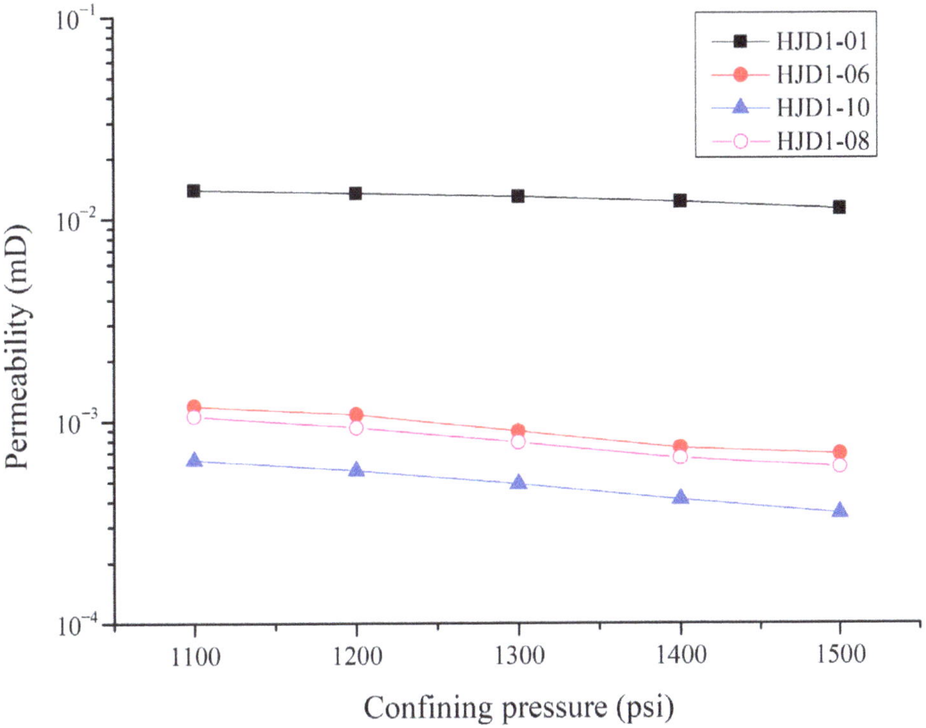

Figure 4. Relationship between confining pressure and permeability.

4.2. LF-NMR Pore Characteristics

The pore size distribution, total porosity, effective porosity, and the T_{2C} value could be obtained by the T_2 spectrum [1,14,46]. The T_2 spectra of the sampling sandstones under the water-saturated condition and the centrifugal condition is shown in Figure 5a–c. T_2 spectra for three sandstones range from 0.01 to 1889.65 ms. T_2 spectra of HJD1-01, HJD1-06, and HJD1-10 show a wide bimodal distribution with a main peak and a subpeak. There is a significant difference in T_2 distribution between the three sandstones (Figure 5d). The main peak of HJD10 is separate from the subpeak; however, the main peak of HJD1-01 and HJD1-06 is interconnected with the subpeak. The main peaks of HJD1-01, HJD1-06, and HJD1-10 are within 0.13–11.09 ms, 0.13–9.66 ms, and 0.01–1.38 ms. The subpeaks of HJD1-01, HJD1-06, and HJD1-010 are within 11.09–1889.65 ms, 9.66–666.99 ms, and 12.75–622.25 ms.

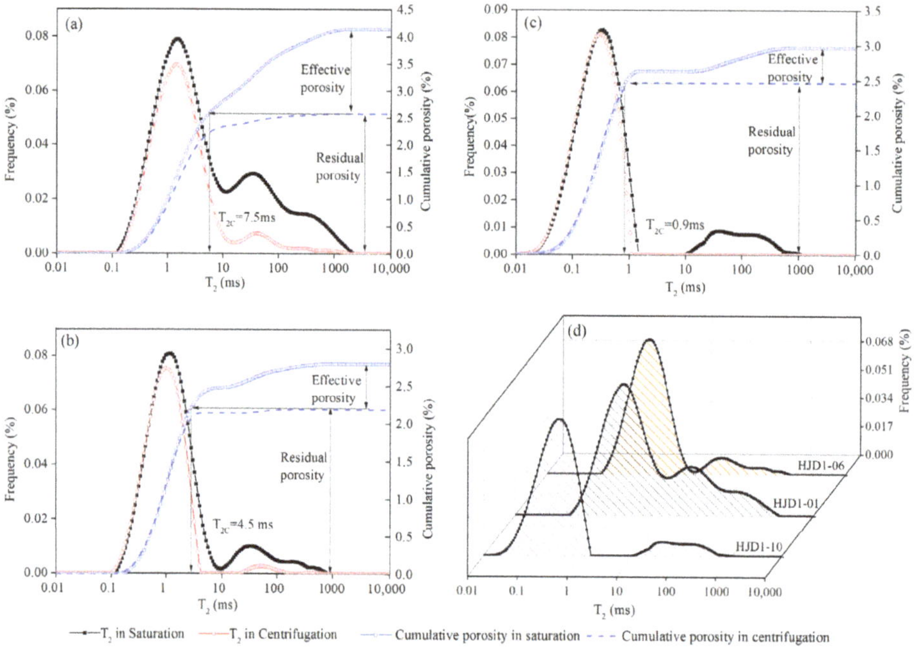

Figure 5. T_2 spectra showing the porosity of different sampling sandstones: (**a**) NMR measurements of HJD1-01 under the water-saturated condition and centrifugal condition; (**b**) NMR measurements of HJD1-06 under the water-saturated condition and centrifugal condition; (**c**) NMR measurements of HJD1-10 under the water-saturated condition and centrifugal condition; (**d**) contrast of T_2 spectra under the water-saturated condition between HJD1-01, HJD1-06, and HJD1-10.

T_2 distribution is closely related to the size of the pores. Generally, the longer T_2, the bigger the pore size. In contrast, the shorter T_2, the smaller the pore size. The characteristics of the pores in sandstone could be analyzed by the distribution position and the area of the T_2 spectrum [14]. The area of the peaks reflects the number of pores within a certain size; the larger the area, the greater the number of pores. The peak width suggests the distribution of a certain kind of pore, while the number of peaks reflects the continuity of pores at all levels. The two peaks of the T_2 spectra reflect the two main types of pores. Based on the T_2 spectra curve of HJD1-01, HJD1-06, and HJD1-10, micropores correspond to $T_2 < 11.09$ ms, and macropores correspond to 11.09 ms $< T_2 < 1889.65$ ms. The micropores peak of the T_2 spectra is the largest, indicating that the micropores are the most developed. The bigger the value of T_2, the bigger the pore size of the sandstone. The pore size of the micropores from big to small is HJD1-01, HJD1-06, and HJD1-10, The pore size of the macropores from big to small is HJD1-01, HJD1-06, and HJD1-10 as well.

The T_2 spectrum morphology of the peak in the saturated state is very close to that after centrifugation (Figure 5c). This means the residual water trapped in the pores cannot be removed by centrifugation, which indicates that the connectivity of the pores is poor. Otherwise, the T_2 spectrum morphology of the peak after centrifugation is very different from the peak in the saturated state; this indicates that the connectivity of pores is good, and the pores are very conducive to gas flow. According to Figure 5a–c, the macropore connectivity in the three testing samples is relatively good, which is significantly better than that of the micropores. Furthermore, there are obvious differences in the connectivity of micropores among the three sandstone samples; the micropores' connectivity is in the order from good to poor: HJD1-01, HJD1-06, HJD1-10.

The connection between the main peak and the subpeak could be used to identify the connectivity among pores [14]. For instance, there is a gap between the main peak and subpeak in the siltstone (HJD1-10) in Figure 5c, meaning the connection between micropores and macropores is poor. In contrast, the main peak is interconnected with the subpeak in HJD1-01 (Figure 5a) and HJD1-06 (Figure 5b), which suggests the pore connectivity of HJD1-01 and HJD1-06 is better than that of HJD1-10 between micropores and macropores. In general, the pores in the siltstone (HJD1-10) had poorer pore connectivity than that of the medium stone and the fine stone (HJD1-01, HJD1-06) [47].

The distribution position and area of the T_2 spectrum of medium sandstone, fine sandstone, and siltstone are different [47], which suggests that sandstone grains can affect pore characteristics. For example, the distribution area of T_2 spectra representing macropores in HJD1-1 (Figure 5a) is much larger than that of HJD1-6 and HJD1-10 (Figure 5b,c).

The porosity of sandstones can be analyzed based on LF-NMR T_2 [14,48]. Total porosity in the saturated state and centrifugal state can be obtained by separate NMR tests (Figure 5a–c). The maximum value of the cumulative porosity in the saturated water state is known as the total porosity [14,49]. There is residual water in the sample after centrifugation. The maximum value of the cumulative porosity after centrifugation could be seen as the residual porosity. The difference between the total porosity and the residual porosity is the porosity occupied by free water, which can be regarded as the effective porosity. The total porosity of HJD1-01, HJD1-06, and HJD1-10 was 4.14%, 2.80%, and 2.97%. The effective porosity of HJD1-01, HJD1-06, and HJD1-10 was 1.56%, 0.60%, and 0.51%. The proportion of the effective porosity to the total porosity for HJD1-01, HJD1-06, and HJD1-10 was 37.68%, 21.42%, and 17.18% (Table 3).

Table 3. Results of NMR analysis.

Sample	T_{2C}	Total Porosity	Residual Water Porosity	Effective Porosity	Pore Proportion (%)		
ID	(ms)	(%)	(%)	(%)	$T_2 < T_{2C}$	$T_2 > T_{2C}$	Effective Porosity/Total Porosity
HJD1-01	5.54	4.14	2.58	1.56	61.50	38.50	37.68
HJD1-06	2.77	2.80	2.20	0.60	79.07	20.93	21.42
HJD1-10	0.85	2.97	2.46	0.51	81.95	18.05	17.17

The T_{2C} value can be obtained by the following steps. First, draw a line parallel to the *X*-axis with the residual porosity (Figure 5a–c). This parallel line has an intersection point with the saturated cumulative porosity curve. The X value corresponding to this intersection point is T_{2C} [14,50]. T_{2C} can divide the T_2 spectrum into two segments: one segment corresponds to pores with water that could not be drained, and the other segment corresponds to pores with water that could be drained. Pores with water that could not be drained are known as immovable pores, pores with water that could be drained are called movable pores. Therefore, the pore size at the T_{2C} value is taken for the separation between the immovable pore volume and the movable pore volume [14]. The T_2 spectrum which is less than T_{2C} corresponds to micropores with poorer pore connectivity, which are the immovable pores. The T_2 spectrum which is bigger than T_{2C} corresponds to macropores with better pore connectivity which are the movable pores [15]. T_{2C} was 5.54 ms, 2.77 ms, and 0.85 ms for HJD1-01, HJD1-06, and HJD1-10, respectively. Almost all of the micropores were immovable pores in the three samples. The proportion of immovable pores to total pores for HJD1-01, HJD1-06, and HJD1-10 was 61.50%, 79.07%, and 81.95%, respectively. The proportion of movable pores to total pores for HJD1-01, HJD1-06, and HJD1-10 was 38.50%, 20.93%, and 18.05% (Table 3).

5. Discussion

5.1. Permeability Variation with Effective Stress and Grains

The permeability decreases with the increase of the effective stress; the relationship between the permeability and the effective stress is close to the linear relation when the effective stress is between 405 psi and 808 psi because the R^2 of all the samples is more than 0.9757 (Table 4, Figure 6).

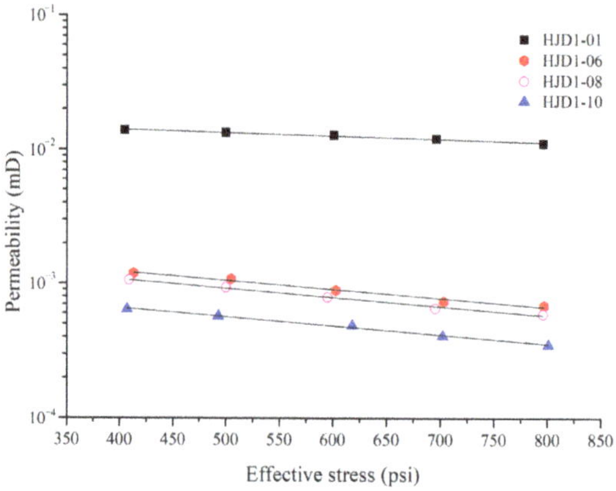

Figure 6. Relationship between the effective stress and permeability.

Table 4. Linear fitting results between permeability (y, mD) and effective stress (x, psi).

Sample	Fitting Liner Equations	R^2
HJD1-01	$y = -6.95 \times 10^{-6}x + 16.86 \times 10^{-3}$	0.9757
HJD1-06	$y = -1.39 \times 10^{-6}x + 1.76 \times 10^{-3}$	1
HJD1-08	$y = -1.24 \times 10^{-6}x + 1.55 \times 10^{-3}$	1
HJD1-10	$y = -0.75 \times 10^{-6}x + 0.94 \times 10^{-3}$	1

The grain size of sandstone has a significant impact on sandstone permeability [51]. The permeability decreases in the order of medium sandstone, fine sandstone, and siltstone under similar effective stress. For the medium sandstone (HJD1-01), the fine sandstone (HJD1-06), and the siltstone (HJD1-10), the slope in Figure 6 is -6.95×10^{-6}, -1.39×10^{-6}, and -0.75×10^{-6}, respectively, as the effective stress increases from nearly 400 psi to 800 psi.

The rock permeability damage by the stress sensitivity could be expressed by the permeability stress sensitivity coefficient. The greater the stress sensitivity coefficient, the better the stress sensitivity will be. The calculation formula is as follows [38]:

$$S_S = \frac{\left[1 - \sqrt[3]{\frac{K_i}{K_0}}\right]}{\log_{10}\frac{\delta_i}{\delta_0}} \tag{2}$$

where S_S is the permeability stress sensitivity coefficient, dimensionless; K_i is the permeability of the sample under δ_i, mD; K_0 is the permeability under δ_0, mD; δ_i is the effective stress at any time, psi; and δ_0 is the initial effective stress, psi.

Generally, S_s would increase for the same sample with the increase of the effective stress (Table 5). It seems that the sample with lower permeability and finer grains has a higher S_S. The permeability of the medium sandstone (HJD1-01) is much higher than that of the other samples (including HJD1-06 and HJD1-08) with similar initial effective stress, and the S_s of HJD1–01 is much lower than the other samples with similar effective stress. For the fine sandstone samples HJD1-06 and HJD1-08, the higher the permeability, the lower the stress sensitivity. This agrees with the reports that sandstones with very low permeability are affected by stress to a greater degree than those with higher permeability [52,53]. S_S could be affected by the sandstone grains. The S_s of the medium sandstone is a bit lower than that of the fine sandstone and the siltstone. However, the S_s of the fine sandstone and the siltstone is relatively close (Table 5).

Table 5. Stress sensitivity coefficient (S_S) under different effective stress between 405 and 801 psi.

Sample Number	Lithology	Effective Stress/psi	Permeability/ $\times 10^{-3}$ mD	S_S
HJD1-01	Medium sandstone	405	13.91	-
		500	13.41	0.1325
		601	12.86	0.1506
		696	12.11	0.1920
		796	11.16	0.2412
HJD1-06	Fine sandstone	413	1.188	-
		505	1.081	0.3546
		603	0.891	0.5563
		703	0.738	0.6352
		797	0.689	0.5816
HJD1-08	Fine sandstone	409	1.06	-
		500	0.932	0.4813
		595	0.788	0.5781
		695	0.656	0.6420
		796	0.594	0.6071
HJD1-10	Siltstone	407	0.643	-
		493	0.571	0.4662
		618	0.489	0.4808
		702	0.41	0.5884
		801	0.351	0.6215

5.2. Effects of NMR Pore Characteristics on Permeability Variation

Generally, the permeability of coal-measure sandstone is controlled by the number of movable pores, effective porosity, and total porosity. The total pore volume from big to small was HJD1-01, HJD1-06, and HJD1-10. The number of micropores and macropores was HJD1-01, HJD1-06, and HJD1-10 in descending order. Almost all of the micropores were the immovable pores in the three samples. The macropore connectivity affects the permeability of the tested samples. Nevertheless, the macropore connectivity in the three testing samples was relatively close. Therefore, the pore volume of the macropores or movable pores was the key factor to the permeability. Large pores are critical to the reservoir quality of the tight sandstone [50]. Movable pores, which are large scale pores with good pore connectivity, dominate the permeability of the reservoir [1,15]. The greater the number of movable pores and the larger the total porosity and the effective porosity, the bigger the permeability of the sandstone. The total porosity of HJD1-01, HJD1-06, and HJD1-10 was 4.14%, 2.80%, and 2.97%. The proportion of movable pores to total pores for HJD1-01, HJD1-06, and HJD1-10 was 38.50%, 20.93%, and 17.17%. The effective porosity of HJD1-01, HJD1-06, and HJD1-10 was 1.56%, 0.60%, and 0.51%. The permeability from big to small was HJD1-01, HJD1-06, and HJD1-10.

Grain size, compaction, and mineral composition could affect the pore structure characteristics [24,37,54]. Finer grain-size sandstones have experienced stronger compaction and cementation during diagenesis in comparison with sandstones with coarser grain size [55]. Stronger compaction and cementation tightly arranged the sandstone's particles and decrease the residual intergranular pores. Therefore, sandstone with a finer grain size develops smaller pore space and a narrower throat [55], which contributes to less total porosity and effective porosity. Mineral composition affects pore-throat structure parameters such as porosity, pore throat, and effective porosity as well [56]. Sandstones with a coarser grain size have larger rigid grains, such as quartz and feldspar [37]. In the contrary, finer grain-size sandstones have more ductile minerals such as mica and clays [54]. Rigid grains could sustain more pressure during compaction, which is beneficial for preserving intergranular pores [57]. Meanwhile, clay minerals occupy primary pores and cut pore throats, resulting in a decrease in effective porosity. Therefore, sandstone with a coarser grain size is more likely to develop bigger total porosity and effective porosity compared with finer-grained sandstone [26]. In the testing sandstone samples, the finer the grain sizes, the less the pore space. The coarser the grain size, the larger the pore space [55]. The coarser the grains, the larger the effective porosity and the larger the permeability. This suggests that the effective porosity of sandstone has a positive correlation with the permeability value. Therefore, the effective porosity of sandstone is a sensitive indicator for evaluating the permeability of the tight sandstone reservoirs.

The effective porosity has a dominant effect on the S_s, and S_s is negatively correlated with the effective porosity. S_s of HJD1-01 was only about 0.27–0.42 times that of the other samples when the effective porosity of HJD1-01 was 2.6–3 times that of other sandstone samples. It can be seen from HJD1-10 and HJD1-06 that S_s is not only affected by the effective porosity, but also by the total porosity. The effective porosity of HJD1-06 and HJD1-10 was 0.60 and 0.51, which appears closer. S_s increases as the effective stress gradually increases. However, in a certain stress scope, the S_s of the fine sandstone (HJD-06) with higher effective porosity may be bigger than that of the siltstone (HJD-10) with lower effective porosity. It is speculated that the total porosity of sandstone also has a certain influence on S_s due to the total porosity of HJD-10 being a bit bigger than that of HJD-06.

Another reason for the increase of S_s of fine sandstone and siltstone compared to medium sandstone may be the content of plastic minerals. Clay minerals tend to increase with the decrease of sandstone grain size. The content and type of plastic minerals are one factor that determines the difference in S_s; that is, the higher the content of plastic minerals such as mica and clay, the stronger the S_s of the tight rock reservoir.

6. Conclusions

An experimental study was conducted to investigate the pore characteristics and permeability in coal-measure sandstones based on the pulse attenuation gas permeability tester and LF-NMR. The permeability variation with the effective stress and grains and the effects of NMR pore characteristics on permeability variation were analyzed in detail. The following conclusions were obtained:

(1) Although the micropores are the most developed in sandstones from coal-measure, most of the micropores are the immovable pores. The pore characteristics are significantly affected by the grain size of sandstones in coal-measure. The movable pores and effective porosity increase with the grain size of testing sandstones. The pore connectivity ranks from good to poor with decreasing sandstone particle size.

(2) The relationship between the permeability and the effective stress is close to the linear relationship as the effective stress is between 405 psi and 808 psi. The greater the number of movable pores and the effective porosity, the bigger the permeability of the sandstone. The effective porosity of sandstone is a sensitive indicator for evaluating the permeability of tight sandstone reservoirs. S_s would increase for the same sample with the increase of the effective stress, whilst the effective porosity is negatively correlated with S_s. The particle size of sandstone from coal-measure has a great influence on permeability and S_s. Under similar stress conditions, the permeability of the sampling sandstones increases with the increase of the grain size. The sandstone with lower permeability and finer grains has a higher S_s.

Author Contributions: H.H. wrote the paper, conducted the experimental study, analyzed the results; Y.S. and X.C. wrote the paper, contributed to processing the raw data and completing the figures; Z.W., M.L. and S.Q. edited the original draft and made the tables. All authors have read and agreed to the published version of the manuscript.

Funding: This research was supported by the Fundamental Research Funds for the Central Universities (2019XKQYMS57), a project funded by the Priority Academic Program Development of Jiangsu Higher Education Institutions in China, Mining fissure field characteristics in high gas working panels in Jixi mining area and its influence on gas control (No. 2021060117), Influence of geological tectonic evolution on gas occurrence and migration in Xishan mining area, and the geological survey project of the China Geological Survey (No. DD20190101).

Informed Consent Statement: Not applicable.

Data Availability Statement: Not applicable.

Conflicts of Interest: The authors declare no conflict of interest.

References

1. Hou, X.W.; Zhu, Y.M.; Chen, S.B.; Wang, Y.; Liu, Y. Investigation on pore structure and multifractal of tight sandstone reservoirs in coal bearing strata using LF-NMR measurements. *J. Pet. Sci. Eng.* **2020**, *187*, 13. [CrossRef]
2. Su, X.B.; Li, F.; Su, L.A.; Wang, Q. The experimental study on integrated hydraulic fracturing of coal measures gas reservoirs. *Fuel* **2020**, *270*, 7. [CrossRef]
3. Law, B.E.; Curtis, J.B. Introduction to unconventional petroleum systems. *AAPG Bull.* **2002**, *86*, 1851–1852.
4. Sun, W.; Zuo, Y.; Wu, Z.; Liu, H.; Xi, S.; Shui, Y.; Wang, J.; Liu, R.; Lin, J. Fractal analysis of pores and the pore structure of the Lower Cambrian Niutitang shale in northern Guizhou province: Investigations using NMR, SEM and image analyses. *Mar. Pet. Geol.* **2019**, *99*, 416–428. [CrossRef]
5. Wan, L.M.; Hou, B.; Tan, P.; Chang, Z.; Muhadasi, Y. Observing the effects of transition zone properties on fracture vertical propagation behavior for coal measure strata. *J. Struct. Geol.* **2019**, *126*, 69–82. [CrossRef]
6. Hou, X.W.; Zhu, Y.M.; Yao, H.P. Coupled accumulation characteristics of Carboniferous-Permian coal measure gases in the Northern Ordos Basin, China. *Arab. J. Geosci.* **2018**, *11*, 13. [CrossRef]
7. Jiao, P.F.; Wang, P.W.; Zhou, S.W.; Wang, H.C.; Chen, X.Y. Study on the Microscopic Pore Structures of Coal Measure Reservoirs in the Shanxi Formation, Eastern Ordos Basin. *Front. Earth Sci.* **2022**, *10*, 903588. [CrossRef]
8. Jia, J.; Cao, L.; Sang, S.; Yi, T.; Zhou, X. A case study on the effective stimulation techniques practiced in the superposed gas reservoirs of coal-bearing series with multiple thin coal seams in Guizhou, China. *J. Pet. Sci. Eng.* **2016**, *146*, 489–504. [CrossRef]
9. Towler, B.; Firouzi, M.; Underschultz, J.; Rifkin, W.; Garnett, A.; Schultz, H.; Esterle, J.; Tyson, S.; Witt, K. An overview of the coal seam gas developments in Queensland. *J. Nat. Gas Sci. Eng.* **2016**, *31*, 249–271. [CrossRef]
10. Gao, H.; Li, H.A. Pore structure characterization, permeability evaluation and enhanced gas recovery techniques of tight gas sandstones. *J. Nat. Gas Sci. Eng.* **2016**, *28*, 536–547. [CrossRef]
11. Qin, Y. Research progress of symbiotic accumulation of coal measure gas in China. *Nat. Gas Ind. B* **2018**, *5*, 466–474. [CrossRef]
12. Mehmani, A.; Mehmani, Y.; Prodanovi, M.; Balhoff, M. A forward analysis on the applicability of tracer breakthrough profiles in revealing the pore structure of tight gas sandstone and carbonate rocks. *Water Resour. Res.* **2015**, *51*, 4751–4767. [CrossRef]
13. Liu, M.; Xie, R.; Guo, J.; Jin, G. Characterization of Pore Structures of Tight Sandstone Reservoirs by Multifractal Analysis of the NMR T 2 Distribution. *Energy Fuels* **2018**, *32*, 12218–12230. [CrossRef]
14. Yao, Y.B.; Liu, D.M.; Che, Y.; Tang, D.Z.; Tang, S.H.; Huang, W.H. Petrophysical characterization of coals by low-field nuclear magnetic resonance (NMR). *Fuel* **2010**, *89*, 1371–1380. [CrossRef]
15. Hou, X.; Zhu, Y.; Wang, Y.; Liu, Y. Experimental study of the interplay between pore system and permeability using pore compressibility for high rank coal reservoirs. *Fuel* **2019**, *254*, 115712. [CrossRef]
16. Kong, X.; Xiao, D.; Jiang, S.; Lu, S.; Sun, B.; Wang, J. Application of the combination of high-pressure mercury injection and nuclear magnetic resonance to the classification and evaluation of tight sandstone reservoirs: A case study of the Linxing Block in the Ordos Basin. *Nat. Gas Ind. B* **2020**, *7*, 433–442. [CrossRef]
17. Yao, Y.; Liu, D. Comparison of low-field NMR and mercury intrusion porosimetry in characterizing pore size distributions of coals. *Fuel* **2012**, *95*, 152–158. [CrossRef]
18. Xiao, D.; Lu, Z.; Jiang, S.; Lu, S. Comparison and integration of experimental methods to characterize the full-range pore features of tight gas sandstone-A case study in Songliao Basin of China. *J. Nat. Gas Sci. Eng.* **2016**, *34*, 1412–1421. [CrossRef]
19. Hinai, A.A.; Rezaee, R.; Esteban, L.; Labani, M. Comparisons of pore size distribution: A case from the Western Australian gas shale formations. *J. Unconv. Oil Gas Resour.* **2014**, *8*, 1–13. [CrossRef]
20. Li, M.; Wang, D.; Shao, Z. Experimental study on changes of pore structure and mechanical properties of sandstone after high-temperature treatment using nuclear magnetic resonance. *Eng. Geol.* **2020**, *275*, 105739. [CrossRef]
21. Yao, Y.; Liu, D.; Liu, J.; Xie, S. Assessing the Water Migration and Permeability of Large Intact Bituminous and Anthracite Coals Using NMR Relaxation Spectrometry. *Transp. Porous Media* **2015**, *107*, 527–542. [CrossRef]

22. Cai, Y.; Liu, D.; Pan, Z.; Yao, Y.; Li, J.; Qiu, Y. Petrophysical characterization of Chinese coal cores with heat treatment by nuclear magnetic resonance. *Fuel* **2013**, *108*, 292–302. [CrossRef]
23. Kleinberg, R.L.; Straley, C.; Kenyon, W.E.; Akkurt, R.; Farooqui, S.A. Nuclear Magnetic Resonance of Rocks: T1 vs. T2. In Proceedings of the SPE Annual Technical Conference and Exhibition, Houston, TX, USA, 3 October 1993.
24. Qu, Y.; Sun, W.; Tao, R.; Luo, B.; Chen, L.; Ren, D. Pore–throat structure and fractal characteristics of tight sandstones in Yanchang Formation, Ordos Basin. *Mar. Pet. Geol.* **2020**, *120*, 104573. [CrossRef]
25. Huang, W.; Lu, S.; Hersi, O.S.; Wang, M.; Deng, S.; Lu, R. Reservoir spaces in tight sandstones: Classification, fractal characters, and heterogeneity. *J. Nat. Gas Sci. Eng.* **2017**, *46*, 80–92. [CrossRef]
26. Zhang, N.; He, M.C.; Zhang, B.; Qiao, F.C.; Sheng, H.L.; Hu, Q.H. Pore structure characteristics and permeability of deep sedimentary rocks determined by mercury intrusion porosimetry. *J. Earth Sci.* **2016**, *27*, 670–676. [CrossRef]
27. Moosavi, S.A.; Goshtasbi, K.; Kazemzadeh, E.; Bakhtiari, H.A.; Esfahani, M.R.; Vali, J. Relationship between porosity and permeability with stress using pore volume compressibility characteristic of reservoir rocks. *Arab. J. Geosci.* **2014**, *7*, 231–239. [CrossRef]
28. Shao, X.H.; Pang, X.Q.; Jiang, F.J.; Li, L.L.; Huyan, Y.Y.; Zhene, D.Y. Reservoir Characterization of Tight Sandstones Using Nuclear Magnetic Resonance and Incremental Pressure Mercury Injection Experiments: Implication for Tight Sand Gas Reservoir Quality. *Energy Fuels* **2017**, *31*, 10420–10431. [CrossRef]
29. Li, Y.; Tang, D.Z.; Xu, H.; Meng, Y.J.; Li, J.Q. Experimental research on coal permeability: The roles of effective stress and gas slippage. *J. Nat. Gas Sci. Eng.* **2014**, *21*, 481–488. [CrossRef]
30. Ma, Q.; Harpalani, S.; Liu, S. A simplified permeability model for coalbed methane reservoirs based on matchstick strain and constant volume theory. *Int. J. Coal Geol.* **2011**, *85*, 43–48. [CrossRef]
31. Wu, S.; Tang, D.Z.; Li, S.; Wu, H.Y.; Hu, X.; Zhu, X.G. Effects of geological pressure and temperature on permeability behaviors of middle-low volatile bituminous coals in eastern Ordos Basin, China. *J. Pet. Sci. Eng.* **2017**, *153*, 372–384. [CrossRef]
32. Khormali, A.; Bahlakeh, G.; Struchkov, I.; Kazemzadeh, Y. Increasing inhibition performance of simultaneous precipitation of calcium and strontium sulfate scales using a new inhibitor—Laboratory and field application. *J. Pet. Sci. Eng.* **2021**, *202*, 108589. [CrossRef]
33. Li, X.; Fu, X.H.; Ranjith, P.G.; Xu, J. Stress sensitivity of medium- and high volatile bituminous coal: An experimental study based on nuclear magnetic resonance and permeability-porosity tests. *J. Pet. Sci. Eng.* **2019**, *172*, 889–910. [CrossRef]
34. Tao, S.; Wang, Y.; Tang, D.; Xu, H.; Lv, Y.; He, W.; Li, Y. Dynamic variation effects of coal permeability during the coalbed methane development process in the Qinshui Basin, China. *Int. J. Coal Geol.* **2012**, *93*, 16–22. [CrossRef]
35. Li, S.; Tang, D.Z.; Pan, Z.J.; Xu, H.; Huang, W.Q. Characterization of the stress sensitivity of pores for different rank coals by nuclear magnetic resonance. *Fuel* **2013**, *111*, 746–754. [CrossRef]
36. Shen, J.; Qin, Y.; Li, Y.P.; Yang, Y.H.; Ju, W.; Yang, C.L.; Wang, G. In situ stress field in the FZ Block of Qinshui Basin, China: Implications for the permeability and coalbed methane production. *J. Pet. Sci. Eng.* **2018**, *170*, 744–754. [CrossRef]
37. Meng, Y.F.; Luo, C.B.; Li, G.; Liu, H.B. An experimental study on stress sensitivity of tight sandstone gas reservoirs during nitrogen drilling. *Arab. J. Geosci.* **2019**, *12*, 11. [CrossRef]
38. Zhang, H.; Zhong, Y.; Kuru, E.G.; Kuang, J.C.; Sh, J.P. Impacts of permeability stress sensitivity and aqueous phase trapping on the tight sandstone gas well productivity—A case study of the Daniudi gas field. *J. Pet. Sci. Eng.* **2019**, *177*, 261–269. [CrossRef]
39. Zhang, K.; Sang, S.X.; Liu, C.J.; Ma, M.Y.; Zhou, X.Z. Experimental study the influences of geochemical reaction on coal structure during the CO_2 geological storage in deep coal seam. *J. Pet. Sci. Eng.* **2019**, *178*, 1006–1017. [CrossRef]
40. Li, S.J.; Wang, Z.H.; Sun, Y.X.; Xie, J.B. Stress Sensitivity of Low-permeability Sandstone Reservoir. In *Materials Processing and Manufacturing III, Pts 1–4*; Advanced Materials Research; Sang, X., Kim, Y.H., Eds.; Trans Tech Publications Ltd.: Durnten-Zurich, Switzerland, 2013; Volume 753–755, pp. 686–689.
41. Archer, R. Impact of Stress Sensitive Permeability on Production Data Analysis. In Proceedings of the SPE Unconventional Reservoirs Conference, Keystone, CO, USA, 10–12 February 2008. [CrossRef]
42. Fatt, I.; Davis, D.H. Reduction in Permeability with Overburden Pressure. *J. Pet. Technol.* **1952**, *4*, 16. [CrossRef]
43. Liu, J.-Q.; Zhang, C.-M.; Zhang, Z. Combine the capillary pressure curve data with the porosity to improve the prediction precision of permeability of sandstone reservoir. *J. Pet. Sci. Eng.* **2016**, *139*, 43–48. [CrossRef]
44. Davies, J.P.; Davies, D.K. Stress-Dependent Permeability: Characterization and Modeling. *Spe J.* **2001**, *6*, 224–235. [CrossRef]
45. Volokitin, Y.; Looyestijn, W.J.; Sli, J.; Kerman, W.; Hofman, J.P. A practical approach to obtain primary drainage capillary pressure curves from NMR core and log data. *Petrophysics* **2001**, *42*, 334–343.
46. Ghomeshi, S.; Kryuchkov, S.; Kantzas, A. An investigation into the effects of pore cconnectivity on T2 NMR relaxation. *J. Magn. Reson.* **2018**, *289*, 79–91. [CrossRef] [PubMed]
47. Zhang, Z.; Qin, Y.; Zhuang, X.; Li, G.; Wang, X. Poroperm characteristics of high-rank coals from Southern Qinshui Basin by mercury intrusion, SEM-EDS, nuclear magnetic resonance and relative permeability analysis. *J. Nat. Gas Sci. Eng.* **2018**, *51*, 116–128. [CrossRef]
48. Straley, C.; Rossini, D.; Vinegar, H.J.; Tutunjian, P.N.; Morriss, C.E. Core analysis by low-field NMR. *Log. Anal.* **1997**, *38*, 84–93.
49. Al-Mahrooqi, S.H.; Grattoni, C.A.; Moss, A.K.; Jing, X.D. An investigation of the effect of wettability on NMR characteristics of sandstone rock and fluid systems. *J. Pet. Sci. Eng.* **2003**, *39*, 389–398. [CrossRef]

50. Lai, J.; Wang, G.; Cao, J.; Xiao, C.; Wang, S.; Pang, X.; Dai, Q.; He, Z.; Fan, X.; Yang, L.; et al. Investigation of pore structure and petrophysical property in tight sandstones. *Mar. Pet. Geol.* **2018**, *91*, 179–189. [CrossRef]
51. Song, Z.Y.; Ji, H.G.; You, S.; Tan, J.; Wang, H. Experimental Study on Sensitivity to Temperature Stress of the Permeability of Weakly Cemented Sandstone. In Proceedings of the 3rd International Conference on Advances in Energy Resources and Environment Engineering, Harbin, China, 8–10 December 2017; IOP Conference Series-Earth and Environmental Science. IOP Publishing Ltd.: Bristol, UK, 2018; Volume 113.
52. Mclatchie, A.S.; Hemstock, R.A.; Young, J.W. The Effective Compressibility of Reservoir Rock and Its Effects on Permeability. *J. Pet. Technol.* **1958**, *10*, 49–51. [CrossRef]
53. Vairogs, J.; Hearn, C.L.; Dareing, D.W.; Rhoades, V.W. Effect of Rock Stress on Gas Production From Low-Permeability Reservoirs. *J. Pet. Technol.* **1971**, *23*, 1161–1167. [CrossRef]
54. Yin, S.; Dong, L.; Yang, X.; Wang, R. Experimental investigation of the petrophysical properties, minerals, elements and pore structures in tight sandstones. *J. Nat. Gas Sci. Eng.* **2020**, *76*, 103189. [CrossRef]
55. Li, Z.; Wu, S.; Xia, D.; He, S.; Zhang, X. An investigation into pore structure and petrophysical property in tight sandstones: A case of the Yanchang Formation in the southern Ordos Basin, China. *Mar. Pet. Geol.* **2018**, *97*, 390–406. [CrossRef]
56. Zhao, H.; Ning, Z.; Zhao, T.; Zhang, R.; Wang, Q. Effects of mineralogy on petrophysical properties and permeability estimation of the Upper Triassic Yanchang tight oil sandstones in Ordos Basin, Northern China. *Fuel* **2016**, *186*, 328–338. [CrossRef]
57. Zhong, D.; Zhou, L.; Sun, H. Influences of petrologic features on diagenesis and pore development: An example from the Triassic Yanchang Formation in Longdong Area, Ordos Basin. *Oil Gas Geol.* **2012**, *33*, 280–289.

MDPI
St. Alban-Anlage 66
4052 Basel
Switzerland
Tel. +41 61 683 77 34
Fax +41 61 302 89 18
www.mdpi.com

Energies Editorial Office
E-mail: energies@mdpi.com
www.mdpi.com/journal/energies